VHDL DESIGNER'S REFERENCE

VHDL
Designer's Reference

by

Jean-Michel Bergé
CNET, France

Alain Fonkoua
Institut Méditerranéen de Technologie, France

Serge Maginot
LEDA S.A., France

and

Jacques Rouillard
Institut Méditerranéen de Technologie, France

SPRINGER-SCIENCE+BUSINESS MEDIA, B.V.

Library of Congress Cataloging-in-Publication Data

VHDL designer's reference / by Jean-Michel Bergé ... [et al.].
p. cm.
Includes index.
ISBN 978-1-4613-6551-8 ISBN 978-1-4615-3498-3 (eBook)
DOI 10.1007/978-1-4615-3498-3
1. VHDL (Computer hardware description language) I. Bergé, Jean-Michel.
TK7885.7.V47 1992
621.39'2--dc20 92-10613

ISBN 978-1-4613-6551-8

Dacapo ® of DOSIS gmbh
Dazix ® of Intergraph
Ella ® of Computer General
Helix ™ of Silvar-Lisco
Ibm ® of IBM
Intergraph ® of Intergraph
Leda ™ of Leda SA
Lsim ™, of Mentor Graphics ®
Racal Redac ™ of Racal Redac
Synopsis ™ of Synopsys, Inc.
Texas Instruments ® of Texas Instruments, Inc.
Vantage ™ of Vantage Analysis Systems, Inc.
Verilog ®, Cadence ™ & Valid ™ of Cadence Design Systems, Inc.
Viewlogic ® of Viewlogic Systems, Inc.
Intermetrics ™ of Intermetrics Inc.

Printed on acid-free paper

à nos femmes et compagnes,
qui réussissent à être modèles sans être descriptibles ;
à nos enfants présents et futurs,
dont la conception ne doit rien à l'ordinateur ;
à nos parents,
à qui nous devons structure et comportement et pour qui conflit de génération ne rime pas avec erreur à l'élaboration ;

bref...

à tous ceux à qui nous n'arrivons pas à expliquer ce que nous faisons au bureau, et dont la lanterne ne sera pas éclairée par ce livre.

TABLE OF CONTENTS

LIST OF FIGURES

FOREWORD

... too vast, too complex, too grand... for description.
John Wesley Powell-1870 (discovering the Grand Canyon)

VHDL is a big world. A beginner can be easily disappointed by the generality of this language. This generality is explained by the large number of domains covered — from specifications to logical simulation or synthesis.

To the very beginner, VHDL appears as a "kit". He is quickly aware that his problem may be solved with VHDL, but does not know how. He does not even know how to start.

In this state of mind, all the constraints that can be set to his modeling job, by using a subset of the language or a given design methodology, may be seen as a life preserver.

The success of the introduction of VHDL in a company depends on solutions to many questions that should be answered months before the first line of code is written:

- Why choose VHDL?
- Which VHDL tools should be chosen?
- Which modeling methodology should be adopted?
- How should the VHDL environment be customized?
- What are the tricks? Where are the traps?
- What are the differences between VHDL and other competing HDLs?

Answers to these questions are organized according to different concerns: buying the tools, organizing the environment, and designing. Decisions taken in each of these areas may have many consequences on the way to the acceptance and efficiently use of VHDL in a company.

The purpose of this book is to bring some insight into all the above-stated problems.

Acknowledgments

Some chapters of this book are the result of work sponsored by the CEC's[1] ECIP2 ESPRIT project (parts of chapters 3, 8) and by the ELISE contract from the French Army department's DRET[2] (parts of chapters 9, 10). These projects were undertaken within the IMT.[3] Some of the work implying access to VHDL software was made possible thanks to the CNET,[4] IMT,[3] and LEDA.[5] The memo (chapter 11) is inspired by the one distributed by LEDA[5] in its VHDL modeling seminar.

Mike Newman (CEC) and *Philippe Sarazin* (DRET) were in charge of these projects and supported this publication. *Henri Felix* and *Yves Francillon* had the hard task of running these projects and of making things happen.

Mike Casey, publisher at Kluwer Academic Publishers, helped us bring this book to completion and provided useful comments, reviews, and deadlines.

We were lucky to find within the CNET, the IMT, and LEDA colleagues and friends to comment on, review, and discuss the contents of this book (and many other interesting subjects). Thanks to ECIP2 and to the CNET, we could also participate in the VHDL standardization process, where we were able to meet technical experts, and more important, make some good friends.

Therefore, thanks to *Roland Airiau, Michel Crastes de Paulet, Louis-Olivier Donzelle, Jean Lebrun, Wolfgang Mueller, Vincent Olive, Anne Robert, Alex Zamfirescu,* and *André Zénatti.*

We would also like to thank the anonymous reviewers provided by Kluwer, whose comments were very helpful, as well as the copy editor *Gerry Geer* for the quality of his corrections.

Nobody should underestimate the merits of *Jean-Louis Armand, Pierre Doucelance, Roland Gerber, Jean-Louis Lardy, Jean Mermet, Jean-Pierre Noblanc, Joël Rodriguez* and *Denis Rouquier* for supervising and supporting the authors.

Thanks are due to *Maureen Timmins* for the painful reading and correcting of the very first version of our English.

We want to address special thanks to *Alec Stanculescu* who was kind enough to host us, provide documents, survive tough discussions, and review drafts of the chapters concerning Verilog and UDL/I.

---

[1] CEC: Commission of the European Community - 200 rue de la Loi - Bruxelles, Belgium

[2] DRET: Département des Recherches et Etudes Techniques, 26 Bd Victor, 00460 Armées, France

[3] IMT: Institut Méditerranéen de Technologie BP 451, 13013 Marseille, France

[4] CNET: Centre National d'Etudes des Télécommunications (FRANCE TELECOM) - BP 42, 38240 Meylan, France

[5] LEDA: Languages for Design Automation - 35 av du Granier, 38240 Meylan, France

1. INTRODUCTION

1.1. VHDL STATUS

VHDL is today the most widespread hardware description language supported by CAD tool vendors, although some implementations do not yet fully comply with its standard definition.

It emerged in a market context of passive resistance by CAD vendors, who proposed at that time proprietary hardware description languages (HDLs). A standard and thus public HDL breaks the ties that keep customers captive of their CAD tool vendors through the use of owned and therefore private HDLs.

The customer sees many advantages of being independent from his tool supplier, but he must take into consideration the culture of his company and the technical performances of the language.

So far, some vendors have moved towards VHDL, and offer more or less complete solutions of simulation and synthesis. Others are less open and claim that VHDL is "just one more tool" they have to integrate onto their platform. However, one can hardly understand what could lead a designer to manage descriptions in two different languages, except for acquired experience and performance. But acquired experience is by nature perishable, and there is no reason why a VHDL tool should not reach state-of-the-art performance. As a language, VHDL's permanent evolution through the IEEE restandardization process allows it to match the capabilities of the best languages in most application fields.

A simple parallel can be made between HDL languages and programming languages: who could nowadays venture to write a large software program using a proprietary language, however powerful and fast it might be?

Of course, VHDL (and its tools) is not perfect: some other languages, in particular aspects, may provide quite superior performances. University studies end in much more appealing languages, and some commercial languages are more suited to fast simulation at given levels, or to logic synthesis and formal verification of hardware.

But a hardware description language is not only intended for managing man-CAD communication. It is also important for the transmission of design information between platforms from different vendors (tool-tool communication), between designers from different companies, or between designers from different time generations within the same company (man-man communication). All these communication requirements led the DoD to prescribe VHDL as the single hardware description language to replace the plethora of HDLs that were used by the many program contractors of the VHSIC (Very High Speed Integrated Circuit) project. The first draft of the language was released in August 1985, and was designed by Intermetrics, IBM, and Texas Instrument under exclusive contract with the DoD. From then on, the IEEE Computer Society decided to standardize VHDL and to promote its wide acceptance by the industrial world. The language was reviewed and improved by an IEEE Design Automation Technical Committee known as VASG (VHDL Analysis and Standardization Group). In December 1987, IEEE members, through their balloting procedure, approved the updated version of VHDL as an IEEE standard under the reference 1076-87.

1.1.1. The VHDL Standardization Process

IEEE bylaws dictate that standards be reviewed every five years. This deadline will be reached in 1992, and it happens that the DoD, at this time, no longer finances standardization[6]. Moreover, the standardization process has become international. Consequently, Europe is now participating, financed by public funds (European projects such as ESPRIT through ECIP2 are today interested in the standard redefinition).

Under control of the VASG, the ISAC (Issue Screening and Analysis Committee) is a small group of specialists identifying and fixing bugs and ambiguities of the language.

Following the internationalization of the VHDL standardization process, the VASG has organized the standardization activity into chapters — structures that coordinate activities at the continental level. From time to time plenary sessions of the VASG are held to set the clocks right.

The presence (just beginning) of an Asian chapter led by the Japanese should also be noted. The Japanese can be divided into two groups:

- The UDL/I group, led by NTT, tries to promote the use of UDL/I as a description standard in Japan. This language, as shown in chapter 10, is more adapted for synthesis applications and the description of integrated circuits than VHDL, since its semantics is largely defined in terms of

[6] IEEE working groups are made and led by "volunteers", which means of course that private companies are paying them to volunteer. Most of the administrative and technical support for VHDL' 87 was sponsored by the DoD through selected companies, directly or under related contracts.

equivalent hardware. But it presents weaknesses in functional specification of high-level systems.

- Other companies show an increasing interest in VHDL.

This dispersion results in an absence of offers in the VHDL tools market from Japanese CAD tool vendors.

1.1.2. VHDL CAD Tools

Since the VHSIC project is classified, only North American companies could take part in the VHDL 7.2 definition. As a consequence, the majority of skills in the languages were in the United States by the time the standard became civil and international. So offers in the VHDL market are dominated by U.S. tool vendors.

According to recent studies, VHDL tools will soon be proposed by all CAD vendors. VHDL competitors, namely Verilog and UDL/I (see chapters 9 and 10) could coexist with VHDL, due to the importance of the existing base of customers and to political pressures.

The offer in VHDL tools can be split into three main groups according to the kinds of companies involved: VHDL initiators, general EDA (Electronic Design Automation) CAD vendors, and VHDL-niched companies.

The VHDL initiators are the three original partners who designed VHDL 7.2: Texas Instruments, IBM, and Intermetrics. The general EDA CAD vendors are the main CAD companies currently dominating the electronic-design tools market (Cadence, Mentor-Graphics, Dazix-Intergraph, ViewLogic, Racal-Redac, and so on). The VHDL-niche companies are small or medium-size companies which make their business almost exclusively by selling VHDL tools they have successfully built from scratch (CLSI, Model-Technology, Vantage, and others). The CAD market is subject to frequent change. A complete review of VHDL products would be obsolete by the publication date of this book, as significative changes are occurring monthly.

Of all the initiators of the project, only Intermetrics has for some time proposed a VHDL simulator on the market. Since then, its VHDL department has been purchased by Valid, which after a while merged with Cadence. IBM is still using its own VHDL platform in-house.

A constant feature is that general EDA CAD vendors have great trouble interfacing their platforms with VHDL. They face many difficulties:

- they must cover the semantic gaps between the HDL that they own and VHDL;
- they must maintain compatibility with their own HDL, or at least provide means to recover the huge investment made in model libraries written in their proprietary HDL;

- they must face the pressure of their customers who have invested so much in model libraries and might move to competing platforms if not satisfied with the proposed solution;
- they are more used buying or merging with smaller or competing companies, rather than developing tools from scratch. And it happens that the VHDL tools for sale are not so many (the niche companies) are not so many, and the tools prove very, very difficult to develop.

1.1.3. The Model Market Faced by VHDL

With reference to the modeling business, VHDL raises some new problems. Up to now, model companies could protect their source text code by selling binary compiled version of their models. It happens that VHDL systems often offer "reverse analyzing" facilities to produce a text source code from its binary representation. Moreover, an intermediate format for VHDL is about to be proposed as a standard by the VIFASG (VHDL Intermediate Format Analysis and Standardization Group). Should this intermediate format be adopted by major companies, any customer could decompile its models.

Model companies are trying to keep their VHDL models secret by using tricks such as the use of the C interface proposed by some simulators.

1.2. THE VHDL SPECTRUM

Before choosing VHDL, one should be aware of the spectrum of the language, in terms of levels, styles of description, and application domains. One should also consider its use as a design medium and as a deliverable.

1.2.1. The Levels of Description

According to a now classical taxonomy, we can classify the levels of description by considering their timing models, from the system level, which deals with pure causality, through functional, dataflow, switch, and transistor levels, which deal with sequentiality, clock ticks, and discrete or continuous differential equations.

It is not clear whether or not VHDL is adapted to system-level description: users of specialized languages will find VHDL poor in certain aspects (global variables missing, hard-to-code random functions, and so forth). But most system-level formalisms can be mapped into VHDL by automatic translators, as shown in chapter 5, where the application of VHDL to system-level modeling is addressed.

It is also unclear whether VHDL is adapted to switch-level description: many papers have been published on this topic (see also [HAR91]). One approach to switch-level modeling in VHDL is to map switch devices into VHDL gates and functional models, sometimes at the expense of accuracy. For improved simulation performances, the basic switch models can be hard-coded in the simulator. This approach has already been taken in the field of hardware acceleration, allowing full-VHDL descriptions but accelerating only those expressed in a particular subset.

In any case, it is clear that VHDL is not really adapted to analog-level simulation, but fits well for the other logic functional and RTL levels.

1.2.2. The Description Styles

VHDL offers three styles of description: the behavioral, dataflow, and structural styles. In contrast to other systems, in which different styles cannot be mixed in the same description (and even often require different languages), the VHDL model may include any combination of the three aforementioned styles. A fourth style can even be found in the non-simulatable structural style[7].

1.2.2.1. The Behavioral Style

A behavioral description defines the functionalities of a device with a sequential algorithm (program) with no reference to any structural implementation. For example, an adder will be modeled with an addition operation.

If, for instance, OPERAND_1, OPERAND_2, CARRY_IN, CARRY_OUT, and OUTPUT are signals (defined elsewhere), and if the conversion functions (like BOOLEAN_TO_BIT) are defined, this can be written:

```
process
        variable RESULT : INTEGER := 0;
begin
        wait on OPERAND_1, OPERAND_2, CARRY_IN ;
        RESULT :=  BIT_TO_INTEGER(OPERAND_1)
                   + BIT_TO_INTEGER(OPERAND_2)
                   + BIT_TO_INTEGER(CARRY_IN) ;
        OUTPUT <= BOOLEAN_TO_BIT((RESULT mod 2)=1) ;
        CARRY_OUT <= BOOLEAN_TO_BIT(RESULT>1) ;
end process ;
```

This could be represented by the more-or-less black box of figure 1.1.

---

[7] A connecting mode of the objects of the hierarchy (linkage) is assigned to this description, which explicitly becomes a non-simulatable description.

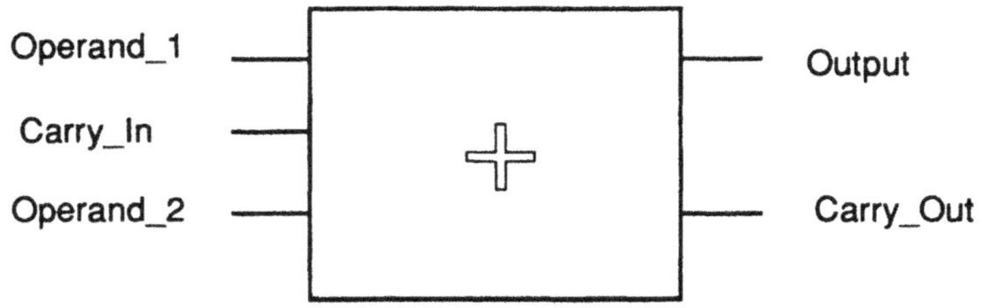

Fig 1.1 Functional Adder

1.2.2.2. The Dataflow Style

At the dataflow or RTL description, the system is represented by a concurrent set of equations involving user-defined functions and arithmetic and logic operators operating on signals of arbitrary complex types. These equations express the flow of information through RTL functional modules implied by the functions and operators. A direct hardware implementation can be derived by mapping signals into wires (or latches) and dataflow operators into RTL modules. Thus, it can be said that such descriptions are behavioral (the arithmetic and logical operators convey a meaning), and imply a structure in terms of interconnected RTL modules. The following dataflow model of an adder:

```
TMP <= OPERAND_1 xor OPERAND_2 after 10 ns ;
OUTPUT <= TMP xor CARRY_IN after 10 ns ;
CARRY_OUT <= (OPERAND_1 and OPERAND_2) or (TMP and CARRY_IN) after 20 ns ;
```

can be mapped into

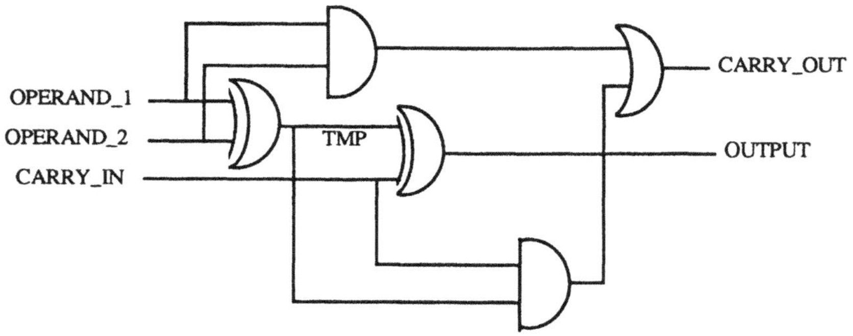

Fig 1.2 Dataflow Description

It must be noted here that **xor, and, not** are predefined language operators.

1.2.2.3. The Structural Style

In a structural style, the description lists the parts of the system and their interconnections. Strictly speaking, the functionalities of the components are not part of the description: the components are viewed as black boxes with regard to their interface. The resulting system is equivalent to an interconnected set of sockets. The functionalities of the system components are described elsewhere using distinct VHDL design entities (at least one design entity per component type occurring in the initial description). Our adder could be described as the interconnection of two half-adders and a gate.

```
INSTANCE1 : HALF_ADD    port map (OPERAND_1, OPERAND_2, TMP1, TMP2);
INSTANCE2 : HALF_ADD    port map (TMP2, CARRY_IN, TMP3, OUTPUT);
INSTANCE3 : OR_GATE     port map (TMP1, TMP3, CARRY_OUT) ;
```

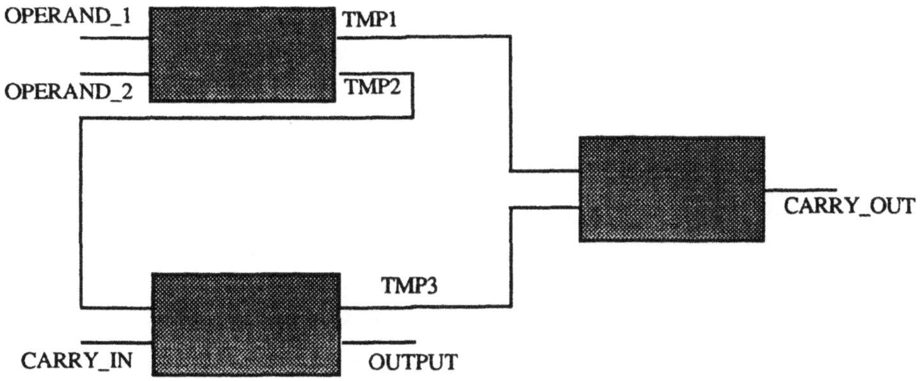

Fig 1.3 Structural Description

The corresponding diagram uses black boxes carrying no intrinsic semantics. The names "HALF_ADD" and "OR_GATE" are arbitrary identifiers without any particular meaning for the language. But they hold a meaning for designers and must be chosen with care to improve model readability, as explained in paragraph 3.9.1 of chapter 3.

VHDL provides a special mechanism called configuration to plug design entities into sockets in order to define their functionalities.

In VHDL, a description can mix different description styles, such as

```
INSTANCE1 : HALF_ADD port map (OPERAND_1, OPERAND_2, TMP1, TMP2);
INSTANCE2 : HALF_ADD port map (TMP2, CARRY_IN, TMP3, OUTPUT);
CARRY_OUT <= TMP1 or TMP3;
```

which is equivalent to the following schema, where the **or** gate is described in the "dataflow" style and bears a meaning.

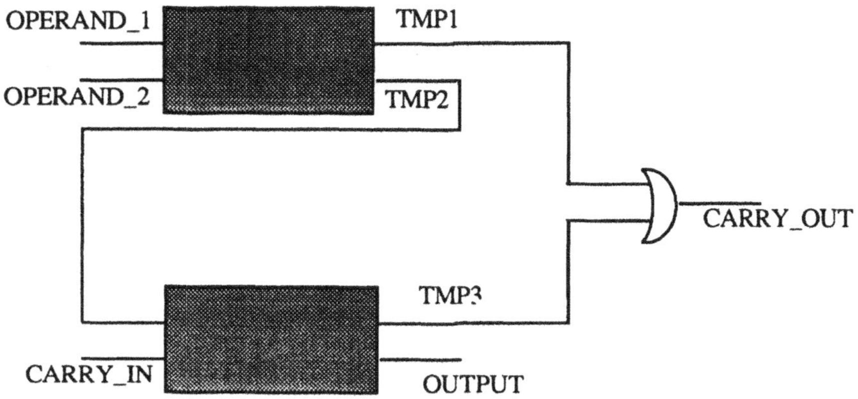

Fig 1.4 Mixed-Level Description

This way of describing hierarchical systems as an interconnection of boxes, which themselves contain interconnected set of boxes, and so forth, cannot be carried on endlessly: at the end, the functionalities of the leaf cells must be defined in either or both of the two ways: behavioral or dataflow.

1.2.3. Application Domains

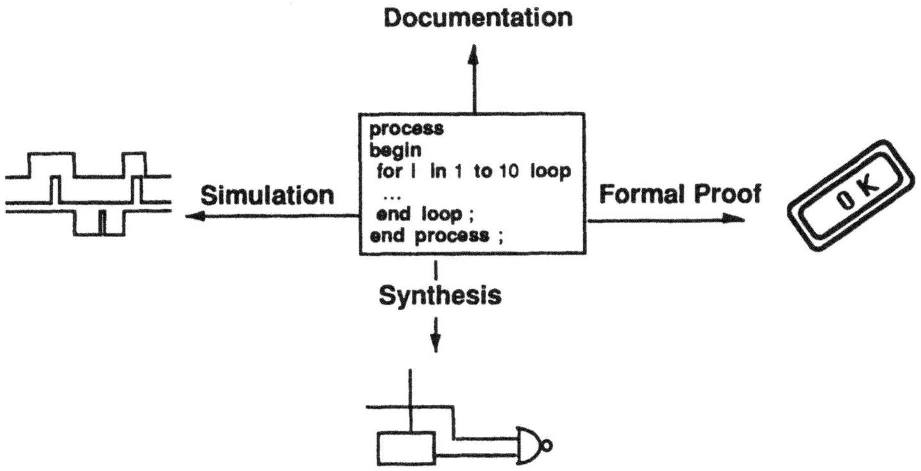

Fig 1.5 VHDL Application Domains

1.2.3.1. VHDL and Simulation

VHDL is a language whose semantics is biased towards simulation. That is, the power of expression of the language reaches its climax when one tries to describe the behavior of a system for simulation applications. But the complexity of the language raises a great concern as to the size of models that can be executed in a reasonable CPU time. The performances of a given simulator depend on its implementation technology and its intended use. Basically, a compromise has to be achieved between two parameters:

- the design time including time to code, compile, and debug a model, which tends to favor interpreted simulators;
- and the CPU time necessary to execute large models, which tends to favor compiled simulators.

Chapter 2 presents the two approaches and give some clues to an appropriate choice of VHDL platform.

1.2.3.2. VHDL and Documentation

The need for "machine readable" documentation is increasingly felt to be part of the important developments that involve numerous designers.

For some time now, the lifetime of integrated circuits has been longer than the lifetime of the designers within the company (whether they stop working due to a heart attack or to "turn-over"), and, often longer than the lifetime of the company itself or even of its customers. Communication between designers from different generations cannot take place by word of mouth, and it is not very wise to rely on proprietary languages whose existence depends on a vendor's health.

The volume of documentation is growing critically, and product documentation is now becoming unmanageable because of the complexity both of circuits and of their environment. One may compare (in a naive but meaningful way) the complexity of a large chip layout with a map of a large city such as Paris *intra muros,* with all the corridors and the buildings' rooms included. If the documentation is elaborated in a "natural language", no one can guarantee that it will be sufficiently safe for use to make the chip. If one uses a proprietary language, the whole work will be compromised if the proprietary language tool supplier goes out of business or abandons the language.

VHDL is the only answer to the problems raised by today's documentation needs. VHDL is the only (or at least the first!) HDL language supported by platforms of different CAD tool vendors, whose semantics is fully described by a public reference manual (the "Language Reference Manual", or LRM) that can be used to check a given implementation for compliance with its standard definition.

EDIF may be suitable for a purely descriptive documentation, but is not sufficient to initiate a re-design process, since this behavior is not yet part of the description.

1.2.3.3. VHDL and Synthesis

Synthesis is the automatic generation of hardware (or anything close to it, such as a netlist) from an initial description expressed in an appropriate language.

Some of these languages are designed so that each construct can be directly mapped into hardware structures. In this case, synthesis is equivalent to a simple logic optimization of the hardware described by the input. When the input language includes constructs to specify the desired functionalities of the chip with no regard to its hardware implementation, it must be interpreted in terms of equivalent hardware and optimized to yield the final chip.

Using VHDL to specify inputs to synthesis tools raises many issues. First, the semantics of VHDL is expressed in terms of a canonical simulator (described in the LRM) and not in terms of equivalent hardware constructions. Furthermore, some concepts hold meaning in simulation contexts, but are not acceptable in synthesis applications. Consider a signal assignment statement with delay: the delay is imperative (which makes sense in simulation: the signal must receive the given value exactly after the specified delay), whereas in a synthesis application only min-max delays can be used (a signal must receive its value no later than or no sooner than a specified delay). The problem of interpretation is even complicated by the great flexibility offered by VHDL to describe complex datatypes that must be mapped into bit string formats by synthesis tools.

However, the flexibility of complex types definition, attributes, and genericity qualify VHDL as a possible language for specifying abstract inputs to synthesis tools, as testified by the growing number of tools supporting VHDL as input. To be used in synthesis applications, a VHDL subset should be defined and methodologies must handle the distinction between two uses of VHDL:

- description, where VHDL is viewed as the textual description of a hardware portion to be produced as is;
- specification, where the text is interpreted as a partial description to be completed.

A growing number of synthesis tools currently available in the market support subsets of VHDL as their input language. In chapter 3, section 3.10 addresses these issues and proposes a VHDL subset and a possible interpretation in the field of logic synthesis.

1.2.3.4. VHDL and Formal Proof

Proving the correctness of a circuit with regard to functional design errors is equivalent to demonstrating that its implementation (for instance, described as a huge interconnection of gates) conforms to its specification. This can be achieved by exhaustive simulation but cannot be applied to circuits of complex size. Formal methods provide alternative solutions using mathematical means to prove *a priori* that the circuit specification is a logical consequence of its implementation.

All current approaches to formal verification of hardware design rely on a sound formal system with well-defined semantics and proof procedures in which the circuit specification and implementation must be reflected before the proof can be undertaken. The most commonly used formal systems include first-order logic, higher-order logic, and temporal logic. The use of VHDL as an HDL language for describing the specification and implementation of circuits raises many problems that are being addressed by current research on formal methods:

- The semantics of VHDL is biased towards simulation application and must therefore be adapted to support a formal system. Appropriate subsets must be derived and granted with well-founded formal semantics.
- The VHDL timing model allows asynchronous circuits, whereas most current HDL-based formal verification tools are limited to synchronous circuits with finite-state machine semantics [SAL91]. Formal systems must be extended to cover asynchronous-circuit semantics.

Commercial tools for formal verification of hardware have been announced [BUL92], but they are still limited to synchronous design. There is hope that the use of formal verification tools will be well developed in the near future.

1.2.3.5. Deliverable or Design Tool

VHDL can, of course, be used as a design medium in place of existing proprietary languages. Its wide range of styles and levels of use enable the designer to keep the same language throughout the design process, thereby permitting the management of designs with parts in different levels (i.e., system level, dataflow).

VHDL also can be a pure deliverable, built from an internal documentation once the design is finished. In fact, most of the VHDL code written today was produced as a deliverable.

1.3. MODELS, MODELING, AND MODELWARE

1.3.1. Specification and Description

Models can be split into two broad categories: those that specify and those that describe. These categories correspond to two overlapping sets of concerns: modeling what we want (specification) and modeling what we have (description, documentation).

Specification modeling finds its application in top-down design methodologies (one describes the main shapes of the desired chip, and then refines it). It is also used as input to automated synthesis and formal verification tools.

The initial specification model will evolve to a more descriptive model as details are added to reflect the choices dictated by the design process. In this use, the specification model is viewed as a partial description that can be completed in one of many ways to yield the full description of a chip satisfying the specification. Each way determines a different solution. The design process performs a selection guided by additional design constraints.

Descriptive models are produced to document the design or the use of a chip. In the first case, they give precise structural information concerning the architectural implementation of the chip in its smaller details. Such models encapsulate strategic information that could be used to produce the chip.

To document the use of a chip, a model must, in addition to performing the same functionalities as the chip with respect to input-output traces and timing characteristics, provide all the information and indication necessary for a proper use of the chip. Such models are destined to be integrated as components of more complex models that are executed to assess the architectural trade-offs made in ASIC (Application Specific Integrated Circuit) and PCB (Printed Circuit Board) design.

Because they are used in larger models, such models need to be efficient in terms of simulation. Furthermore, they are often distributed along with the chip they document and thus must not unveil the sensible technological and architectural information concerning the chip implementation.

That is why the structural properties of the chip are often abstracted away to yield a more functional model annotated with timing characteristics. This proves to be more efficient in terms of simulation performances than the corresponding fully structural model. Thus the documenting model can be regarded as a model — of a model.

Models are commonly used to exchange information between designers of the same team or between two or more companies. But what is the meaning of a model? The true model dwells in the designer's mind. It is the task of the project leader to ensure the integrity of the information conveyed by the models exchanged within his team. This is achieved by adopting a set of conventions and formalisms. Like the cartographer who must use agreed-upon signs and symbols to indicate the elements of the relief, the designer must conform (or must be compelled to conform!) to the adopted set of conventions in order to produce models that are more than plain certified exact programs.

These conventions are hierarchical; although they are numerous and possibly informal within a given team (names, types, style conventions), they are fewer but often contractual between different companies.

The quality of a model is difficult to assess. The criteria to be considered depends on the finality of the model:

- when documenting the design of a chip, a good model will reflect the system structure;
- when documenting the use of a chip, a good model will essentially be rapid and will reflect the functions and timing properties of the chip;
- when used to specify the desired functionalities of a chip as well as the architectural choices made in a design process, a good model will be commented upon and explained abundantly, and will lend itself to a division into subsystems;
- when used to specify inputs to a synthesis or formal verification tool, a good model will be adapted to the tool.

1.3.2. Modeling in VHDL

The first thing to know about VHDL is that it is not really a description language, but rather a kit for defining one. For instance, VHDL allows nearly arbitrary datatypes for signals, of which only a limited set of scalar types is provided in the predefined package STANDARD. The other, more complex types must (can) be defined by the user.

Furthermore, technology-dependent effects such as the effective value of signals driven by many distinct sources (open collector, charge storage) are modeled in VHDL with resolution functions that must be defined by the user. In the state/strength approach, signal values can be modeled by multi-valued logic: two values ('0' and '1'), three ('0', '1', 'X'), four (with 'Z') or more ('U', like uninitialized), the introduction of weak and strong '0' and '1', or a type with 46 values representing intervals, etc. (see chapter 4).

This flexibility of type definition entails the obligation of managing conflicts: for example, what is the effective value of a signal driven simultaneously by a weak '0' and a strong '1'? A complete multi-value logic system must be defined including a value set and common logic operations (what is the logic value of a logic **or** between a weak '0' and a strong '1'?) as well as logic tests: if 'X' stands for indeterminate state, does 'X' equal 'X'? In addition, conversion functions must be defined to interface with models relying on different multi-value logic systems. Chapter 4 covers these various issues in depth.

Designers must be relieved from the burden of defining the logic system. In fact all the implied programming effort fall out of their concern. Other HDLs like the three HDLs presented in this book (M, Verilog, and UDL/I, in chapters 8, 9, and 10, respectively) usually offer predefined logic systems.

That is why the introduction of VHDL into a company must be realized in two steps:

- first, a team including software inclined engineers will define a set of libraries and VHDL packages to build an environment more fitted to the designers' needs and conforming to the company's policy (portability, performances, connections with other descriptions, information hiding, etc.);
- then, the designer can use this environment without bothering about the problems of implementing the logic system in VHDL. Still, he must be trained to understand the underlying rationale of the logic system he uses in order to properly interface with models coming from external sources.

Thus, VHDL introduces what we may call “modelware”, meaning software engineering practices applied to the modeling of electronic devices. Practices that proved useful in the software engineering world, such as project management and consultations, and the methods developed for the Ada programming language from which VHDL is derived, will find a valuable application in today's ill-structured world of model crafting. These issues of VHDL modeling are further discussed in chapter 3.

1.4. THE OTHER LANGUAGES AND FORMATS

This book contains a set of chapters (8, 9 and 10) comparing competing hardware description languages with VHDL (Verilog and UDL/I as public languages, M as a paradigm of proprietary language). Let us try to classify the world of existing description languages.

1.4.1. Non-Proprietary Languages

EDIF is often mistaken as a hardware description language. In fact, EDIF is not a language, but a format, as its name suggests: Electronic Design Interchange Format. This slight difference is considerable, and means that EDIF is not supposed to be written or read by a human operator. It was created at the instigation of a set of CAD tool vendors as a means to allow many tools operating on data with different formats to exchange design data. For that, each platform should provide software utilities to translate data from their internal format to the standard EDIF format and vice versa. On the surface, EDIF syntax looks much like that of LISP expressions. Its power is to propose several description layers, and especially to be actually accepted as a format for interchanging design data at the most used layers (netlist and graphics).

The main drawback of EDIF is the huge size of the data required to describe a circuit with normal size at the structural or physical level (netlist, layout): according to SPECTRUM magazine (11/90), a modern circuit might require 100 M disk space for its description at the schematic level. The description at the layout level, at this ell, must need ten large 6500 bpi tapes. That is why the EDIF compression techniques are used to reduce the volume of transmitted data.

UDL/I and ELLA have the common feature of being two national languages (ELLA is backed in Great Britain, UDL/I in Japan). Each one is or was promoted by one company (Praxis for ELLA, NTT for UDL/I). In both cases, the arguments against VHDL are the same and can be summarized as follows:

- VHDL is too complicated;
- the semantics of the features of both languages can easily be interpreted in terms of equivalent hardware, which is more useful to synthesis tools;
- a national standard is a good shield against American hegemony;
- the time is not ripe to define a world-wide standard;
- one can live with several description languages.

This book tries to ascertain these claims, and in chapter 10, UDL/I is analyzed and compared to VHDL in more detail.

Verilog HDL has now been made public by Cadence. The apparent strategy is simple and may usefully be compared with the assertions of the previous paragraph: there will finally be a sole standard (or very few), and by definition this standard will not be a proprietary language. The only chance to survive is to raise second-source CAD tools by opening the language to the public, which is equivalent to competing with other tool vendors or even with previously captive customers!

Several companies took up that challenge and are now developing Verilog compatibles.

Verilog is analyzed in detail in comparison to VHDL in chapter 9.

1.4.2. Proprietary Languages

Proprietary HDL languages similar to VHDL are not legion; none of those on the market is posterior to VHDL.

Among them are the SDL/HHDL couple, from a Pascal background; HELIX input languages, which served as a reflexion basis to the designers of VHDL; DACAPO which is also built on Pascal; and the M language of Lsim from a C background (see chapter 8 regarding M).

1.4.3. The Outsiders

Structural programming languages (Pascal, Ada, C) are less often used to describe hardware. The result is a "simulating program".

At a very abstract description stage, designers with computer science backgrounds can use an environment such as SMALLTALK or C++ to describe their systems. The object-oriented programming methods are inestimably helpful in cases where the design problem is poorly defined or structured at the beginning of the project. Declarative languages such as PROLOG may prove useful and natural to express the functionalities of certain kind of circuits. Some proprietary languages are specialized to particular domains such as network protocol description, finite-state machines, etc.

1.4.4. Summary of the Comparisons

VHDL will be compared to three other hardware description languages: M, Verilog, and UDL/I. Following are some tables that list the main points of these comparisons.

Warning: Each point discussed below will be treated in greater detail in the corresponding chapters (8, 9, 10), and the reader should refer to them in order to avoid misunderstandings due to presentation shortcuts.

	VHDL	M	Verilog	UDL/I
Origin	IEEE Standard	SCS-Mentor Graphics	Cadence Design Systems	Japanese Standard
Domain	Public	Proprietary	Proprietary now Public	Public
Designers Acceptance ('92)	Real use is starting	Widely used	Widely used	Unknown

The histories of these languages are very different: VHDL and UDL/I were designed by standardization committee or association, whereas M and Verilog were designed as a language for a proprietary simulation product (Verilog was later placed in the public domain). Only M and Verilog have already been widely used by IC designers.

Reference Manual	VHDL	M	Verilog	UDL/I
Tool independent	Yes	No	Yes	Yes
Easy to read	No	-	Yes	Yes
Designer friendly	No	-	Yes	Yes
Tool builder friendly	Yes	-	Average	Average
Accuracy of definitions	Good	-	Average	Average

The M language is a proprietary language: a very good documentation is available, but this reference manual is not tool independent. This documentation has thus not been included in the comparison.

The other three HDLs are defined by a Language Reference Manual. The VHDL reference manual is hard to read and almost designer hostile, containing too much "language", not enough "hardware", and very few examples. On the other hand, it contains much information for the tool builder (a canonical event-driven VHDL simulator is described: driver mechanism, elaboration, initialization, and simulation cycles). Definitions, although difficult to find in the manual, are also very accurate.

The Verilog reference manual has quite opposite characteristics. It is very easy to read and provides many examples that have straightforward meaning for the designer. But some definitions are not very accurate, and the tool builder's point of view is not really handled by the manual.

It is more difficult to evaluate the UDL/I manual, because the release we used was not the definitive one (some sections have not yet been translated from the original Japanese manual).

	VHDL	M	Verilog	UDL/I
Looks like	Ada	C	C, Pascal	Nothing
Level of Complexity	Difficult	Easy	Easy	Average
Straightforward Hardware Meanings	Not really	Partially	Partially	Yes
Requires Software Background	Yes	More or Less	More or Less	No

VHDL is certainly not easy to learn, because it requires the designer to have or to get some software habits (separate compilation, use of a strongly typed language, overloading, or TEXTIO). The other languages, especially UDL/I, are closer to the hardware notions (in UDL/I: synchronous and asynchronous assignments, latches, registers and RAMs, and clock expressions).

Semantics of	VHDL	M	Verilog	UDL/I
Simulation	Yes	Yes	Yes	Yes
Synthesis				Yes

Only UDL/I has clear and unambiguous synthesis semantics. Among the three other languages, Verilog can be used more easily for logic synthesis than VHDL.

Level of Description	VHDL	M	Verilog	UDL/I
Functional Level	++	+	+	
Behavioral Level	++	++	++	+
Register Transfer Level	++	++	++	++
Logic Level	++	++	++	++
Switch Level		++	++	++
Electrical Level		++		

Some papers and books illustrate the use of VHDL at the switch level, but its efficiency is questionable. Electrical or analog modeling is not implemented at all in VHDL. A VHDL extension for mixed-mode simulation is under study and should appear by 1994. All other languages provide efficient switch-level modeling, and M even includes analog modeling.

What can be described	VHDL	M	Verilog	UDL/I
IC	++	++	++	++
PCB	++	+	+	+
System	+			

VHDL has the most general purpose and can be used in a wide range of applications (integrated circuits, PCBs, and even system modeling). The other HDLs are more focused on IC design — especially UDL/I, which is completely dedicated to integrated circuits ("I" in "UDL/I" stands for Integrated circuits).

Implementation of	VHDL	M	Verilog	UDL/I
Concurrent Domain	++	+	++	++
Sequential Domain	++	++	++	

The semantics of the concurrent and sequential worlds are very clearly defined in the VHDL reference manual. The sequential domain is not implemented in UDL/I, and the behavioral concurrent domain of M is restricted to the use of UNIX coprocesses. Verilog is very close to VHDL from this point of view.

Predefined Features	VHDL	M	Verilog	UDL/I
Basic Gates		++	++	
Logic Operators	+	++	++	+
Arithmetic Operators	+	+	+	++
Resolved Logic Type		++	++	++

The VHDL language has a very poor set of predefined subprograms or types, and does not provide any predefined gates: this is the counterpart of its general modeling purpose. The other languages have many predefined gates or operators, which are in return very oriented toward IC modeling.

Actually, VHDL may be seen as a *kit to build its own HDL.* The general philosophy of this language is to let the user define and build (or let build) all the specific constructs he or she needs: multi-value logic type, abstract data types, subprograms and operators, generic and/or technology dependent components and entities, and so forth.

Therefore, an efficient and necessary methodology for using VHDL in a company is to determine all the constructs that are specific to the target application or domain and to develop (or let develop) the corresponding VHDL libraries. Once such an *environment* is set, then most of the designers will only need to reuse these "predefined" constructs (not predefined in the language, but in the built environment). Due to the language capabilities of VHDL, such a specific environment can be even more powerful than the implicit environment provided by other HDLs.

2. VHDL TOOLS

2.1. INTRODUCTION

The rush to the VHDL standard raised many problems for both HDL users and tool vendors.

The HDL users applauded to the emergence of a standard HDL that provided them with more freedom from their CAD tools suppliers. But they had to pay the price of migrating to VHDL. The problems faced are different depending on the situation. Which kind of simulator or synthesis tool should be purchased? The answer to this question depends on the intended use, and many aspects should be considered, such as the degree of compliance to full-VHDL semantics, the model development time, the performances on large models, the graphical environment, and debugging aids. When model libraries in non-standard HDL exist, their migration to VHDL must be considered: should the HDL users build a specific translator? Should they wait for the VHDL release announced by their tool supplier? Some of these questions can also be addressed using benchmarks to be developed: How should these benchmarks be built? The first part of this chapter will give some answers to these questions.

The tool vendors had to upgrade their platform with VHDL capabilities or develop brand-new ones from scratch. Developing brand-new tools implied further choices concerning the simulator: Should compiled or interpreted approaches be used? How could HDL users recover the investment made in model libraries written in their private HDL? Should a source-to-source translator be used? Or should the capabilities of private HDL simulation kernel be augmented to also accept VHDL models? In this latter case, the simulator can execute models written in either or both of the two HDLs, but what are the advantages to such an approach?

All these problems arose mainly because of the complexity of VHDL (which is derived from Ada), and some vendors encountered many difficulties because they overlooked this real complexity. In the second part of this chapter, we will first give some examples to highlight the complexities of VHDL as a language. After that, the different choices that face tool vendors when deciding on their VHDL tool strategy will be discussed: What are the particularities of the interpreted and compiled approaches to simulation? What does source translation or cosimulation imply? This last part will give some extra hints to the many questions that must be addressed in order to choose a VHDL platform. However, this section requires of the reader some knowledge of programming languages and their implementation.

2.2. EVALUATING VHDL TOOLS

2.2.1. VHDL Subset

One of the first consequences of VHDL compilation complexity (which will be detailed later in paragraph 2.3.1) is the notion of the VHDL subset supported by the compiler. This is the first criterion to evaluate on a given platform. Tools as close as possible to full-VHDL are required most of the time.

It may happen, however, that a subset is enough — for example, if the long-term strategy of the company is not to use VHDL as a design tool, but as a deliverable. This situation occurs when the company is happy with a proprietary language and uses one of the add-on techniques described later.

Having a full-VHDL is more important for system design teams. The need to use complex abstract types is one of the reasons.

The subset syndrome is typical of a not yet stabilized VHDL market, and we can assume its obsolescence in future years as most of the products reach the full-VHDL.

The VHDL features most commonly not implemented are user-defined types, access types, aliases, disconnection clauses, and subelement associations.

2.2.2. Simulation Environment Change

Many people choose to wait for the VHDL tools that are supported by their own CAD vendor. This attitude implies an economical strategy; if learning another environment is avoided, VHDL can be more easily accepted by designers.

Nevertheless, the full integration of VHDL brings some dangers that should be mentioned. VHDL is supposed to be portable and has all its semantics in the LRM. CAD environments often offer many utilities such as primitives to control the simulation kernel, or to get information from other tools in a nice and coherent way (e.g., delays depending on layout information). These primitives are available in the proprietary languages where the separation between the language and the functionalities of the simulator is fuzzy. CAD vendors often deliver with their VHDL tools "packages" to provide for the same functionalities, and this is the problem: such packages could not be written in VHDL, and therefore the whole VHDL description relying on them is not portable.

Such a package usually has only its specification (package declaration) written in VHDL. The body does not exist in the VHDL world; exported subprograms are systems calls to classical programming languages (often C). The consequences are disastrous: the management has chosen VHDL because it is a standard allowing interoperability, but in fact, the use of this language by the designer makes him dependent on the CAD environment. Even if he thinks he is using VHDL, he is in fact writing his descriptions in a kind of super-set of VHDL, which is a new proprietary language.

Waiting for VHDL tools integrated in the familiar environment is a good strategy if the following precaution is observed: never use a subprogram if it is not possible to write it in VHDL.[8] Even if timing performances are worse or the functionalities are a bit poorer than in the VHDL form, this precaution will provide interoperability. One can run the embedded version of these subprograms on the CAD environment (and benefit from the good timing performances for debugging) but export one's description to other platforms by merging a VHDL package emulating the functionalities.

Why is it impossible to write such subprograms in VHDL? The power of VHDL is not at stake, but VHDL is only an HDL. Many proprietary implementations of HDL confuse language and simulator (or other tools). The border between language and tool is vague. Information specific to a tool is easily available from the language, and the semantics of this information does not have to be precisely defined: a proprietary language only works with a given tool by nature. Various information, such as capacity values, backannotation timing values, or even critical paths can be obtained using such a tool. We can take two examples to illustrate this idea:

• My CAD environment proposes a "VHDL" package for menu management. Using this package, it is very easy for me to create a menu, use icons, and manage a dialogue with a user. Of course, the body of such a package is not VHDL code: there is no standard possibility for making graphical interfaces in VHDL. Nevertheless, it is possible for me to use this package for a given design (and for all its debugging phases). Once the design is completed, I will have the possibility of using it (with this menu management) on my CAD system or of transferring it to other systems. This second solution means that I can write the package body of the menu management in VHDL. In fact, this package body code will only emulate menus by textual dialogues, but will keep all the functionality of the previous package (to acquire and present information).

---

[8] If it is possible, the CAD vendor has done it already. Look at your documentation.

I achieve a compromise: a better environment on my CAD environment, but interoperability with other environments with roughly the same functionalities.

• My CAD environment gives me the possibility of using a specific "package" with some functions able to provide fan-out during simulation. Of course, the body of this package is not VHDL code. It is made up of kernel functions acceding to the simulator mechanism. No such mechanism is provided in VHDL'87, and the possibility of emulating such calls does not exist. No compromise is possible: give up or use this super-set of VHDL as a "proprietary language", i.e. one that prohibits all interoperability notions.

This problem often appears when translating old proprietary language models into VHDL. Depending on the interaction between this proprietary language and the simulator, the translation into "portable" VHDL can be impossible.

2.2.3. Design Cycle Efficiency

The organization of parsing, analyzing, compiling, and simulating phases is very different from one platform to another, mainly due to the kind of approach, compiled or interpreted, that is used (see paragraph 2.3.4). Some products have a slow (but general-purpose) front-end analyzer; others spend a lot of time in translating VHDL into C code and compiling it before simulation; and others are not optimized at all for datatypes other than a particular given one.

These differences in product implementation make design cycle, time measurement the only significant criterion in comparing timing performances.

Debugging can also be considered as a part of this design cycle and the debugging facilities are various. The possibility of going down into the structure of a block, the possibility of controlling certain parameters, or the impossibility of displaying an element of array are examples of how debugging facilities may be different.

The kind of design targeted, ranging from a behavioral design with many processes to a structural one with numerous interconnected components, can greatly influence the simulation phase timing. Benchmarks can be used to point out differences on this point.

We suggest three short benchmarks. Of course, it is impossible to test every feature of a given platform using only some tens of code lines, but a rough overview of its main good and bad points can be given.

The first benchmark shows the ability of the simulator to deal with a structural description using many interconnected components. The following example is the description of a chain of inverters instantiating 1000 components. The modeling consists of five design units. An entity declaration named BENCH associated with an architecture named STRUCT uses a component INV. This component is bound (by a configuration unit CONF) to the entity declaration INVERTER associated with the architecture BEHAVIORAL. All the CPU time is consumed during the initialization phase (NOW=0) to compute the initial values of all the nodes of the chain.

```
entity INVERTER is
        port (INPUT : in BIT; OUTPUT : out BIT);
end INVERTER;
architecture BEHAVIORAL of INVERTER is
begin
        OUTPUT <= not INPUT;
end BEHAVIORAL;
entity BENCH is
        generic (N : POSITIVE :=1000);
        port (INPUT : in BIT :='1'; OUTPUT : out BIT);
end BENCH;
architecture STRUCT of BENCH is
        signal NODE : BIT_VECTOR(1 to N-1);
        component INV
            port(INPUT : in BIT; OUTPUT : out BIT);
        end component;
begin
S1 : for I in 1 to N-2 generate
        INST : INV port map (NODE(I), NODE(I+1));
end generate;
S2 : if N>2 generate
        INST_1 : INV port map (INPUT, NODE(1));
        INST_N : INV port map (NODE(N-1),OUTPUT);
end generate;
S3 : if N=1 generate
        INSTANCE_1 : INV port map (INPUT, OUTPUT);
end generate;
end STRUCT;
configuration CONF of WORK.BENCH is
for STRUCT
        for S1
            for all : INV use entity WORK.INVERTER(BEHAVIORAL);
            end for;
        end for;
        for S2
            for all : INV use entity WORK.INVERTER(BEHAVIORAL);
            end for;
        end for;
        for S3
            for all : INV use entity WORK.INVERTER(BEHAVIORAL);
            end for;
        end for;
end for;
end CONF;
```

Some compilers do not accept ports in the upper-level entity declaration. In this case, the following update can be made to the benchmark.

```
entity BENCH is
        generic (N : POSITIVE :=1000);
--      port (INPUT : in BIT :='1'; OUTPUT : out BIT);
--      first update
end BENCH;

architecture STRUCT of BENCH is
signal INPUT : BIT :='1'; signal OUTPUT : BIT;
--      second (and last) update
signal NODE : BIT_VECTOR(1 to N-1);
component INV
        port(INPUT : in BIT; OUTPUT : out BIT);
end component;
--      Other possible update (to spare configuration unit CONF)
--      for all : INV use entity WORK.INVERTER(BEHAVIORAL);
begin
...
```

The second benchmark must point out the performances of the simulator with multi-process descriptions. It always consists of modeling a chain of inverters, but each one is described as a process. This must be conceptually the same problem for the simulator as for the first benchmark, but many simulators implement optimization in the case of instantiation of components. The timing performances of this benchmark should be worse in this case. The code source is divided into two design units: the same entity declaration named BENCH associated with an architecture named BEHAV. Once again, the CPU time is consumed during the initialization phase.

```
entity BENCH is
        generic (N : POSITIVE :=1000);
        port (INPUT : in BIT :='1'; OUTPUT : out BIT);
end BENCH;
architecture BEHAV of BENCH is
        signal NODE : BIT_VECTOR(1 to N-1);
begin
S1 : for I in 1 to N-2 generate
        NODE(I+1) <= not NODE(I);
end generate;
S2 : if N>2 generate
        NODE(1) <= not INPUT;
        OUTPUT <= not NODE(N-1);
end generate;
S3 : if N=1 generate
        OUTPUT <= not INPUT;
end generate;
end BEHAV;
```

The last benchmark is wellknown in classical programming languages. Only one process is involved in the recursive computation of the Fibonacci series: FIBO(0)=FIBO(1)=0 and FIBO(n)=FIBO(n-1)+FIBO(n-2). Computation and function calls are tested. In fact, this benchmark tests the classical programming features of the VHDL implementation. In some simulators where resolution functions are treated exactly like classical subprograms, the result of this benchmark is particularly important. It consists of two design units: an entity declaration named FIBO associated with an architecture named RECURSIVE. The process, spawned at the initialization phase, computes the value of FIBO(32) and becomes inactive.

```
entity FIBO is
        generic (N :NATURAL :=32);
end FIBO;
architecture RECURSIVE of FIBO is
begin
        process
          variable RESULT : POSITIVE;
          function FIBO(N : NATURAL) return POSITIVE is
          begin
            if N=0 or N=1 then
              return 1;
            else
              return FIBO(N-1)+FIBO(N-2);
            end if;
          end FIBO;
        begin
        RESULT := FIBO(N);
        wait;
        end process;
end RECURSIVE;
```

Note: The computation of FIBO(32) is suggested. Other computations are possible, but it is important to be conscious of the explosive increase in CPU time consumption when incrementing the parameter N.

These three benchmarks give some idea of the performance of the simulator. However, the main evaluation of platform performances must be a design cycle value of time. One idea is to use a bugged text of VHDL as a benchmark. The feature to be retained on a given platform will be the time taken as a whole to:

- write the VHDL code;
- compile it;
- read and understand the error message;
- correct the bug;
- recompile;
- set up the monitors to control the run;

- run the simulation;
- visualize the problem;
- use debug facilities to point out the error;
- correct the error;
- recompile;
- rerun the simulation;
- display correct results.

The following architecture contains two bugs. The first is a static one: the key word **begin** is missing. This bug will be discovered (hopefully!) during compilation. The second concerns the semantics of this inverter where the output is not the inverse of the input. The correct assignment (according to the conventional semantics of an inverter) is

```
OUTPUT <= not INPUT;
```

This line has to be corrected.

To run this bugged architecture, use the first benchmark given earlier with the previous design units CONF, BENCH, STRUCT, INVERTER.

```
architecture BEHAVIORAL of INVERTER is
        OUTPUT <= INPUT;
end BEHAVIORAL;
```

We have to be very clear: the time to be measured here is not the CPU time consumed, but rather the real time spent completing the sequence — writing the code and compiling it, finding the first static bug and correcting it, recompiling and running the simulation, visualizing the nodes and correcting the second bug, recompiling the architecture, rerunning the simulation, and visualizing the correct node values.

2.2.4. Librarians

The complexity of a VHDL design must not only be measured by the number of source code lines or the degree of their concurrency, but also by the number of design units used. All design units (entities, architectures, package declarations, package bodies, and configuration units) are strongly dependent on each other. This dependency implies an order (usually, more than one exists) of compilation that must be respected for two main reasons:

- Optimizing the number of design unit recompilations: only updated design units and related units are recompiled.
- Avoiding obsolete design units: if the order of compilation is incorrect, consistency of the whole set of design units will not be achieved, and some of them will be said to be obsolete.

Of course, a smart designer has a very accurate knowledge of the dependencies of his design units; but when he must use design units he has not written or when the number of design units is very high, it can be very helpful for him to rely on an automatic mechanism to perform the necessary recompilation in an optimized order. This recompilation command is usually provided by the librarian of theVHDL system, but this has to be checked.

In the same way, the exact relationship between the librarian and the text editor must be known: the possibility of optionally storing the last correct source code of your design unit in the library when compiling can be very useful in providing an answer to the question: "What is the exact code I am running now?"

2.2.5. Text Editors

Text editors are the most basic tool for writing VHDL code. Actually, any classical text editor can be used for this purpose, but due to the verbose nature of VHDL, a full-page editor with efficient cut/paste commands is recommended. Multi-window editors are more effective at dealing with modularity, which is another characteristic of VHDL.

Language-oriented editors are sometimes proposed. Automatically closing a structure you have opened or refusing a declaration in a bad declarative part can appear very attractive to the beginner. Once skilled, however, the designer usually comes back to a more manual mode.

For documentation purposes, source code pretty printers are very useful and should be considered as part of the widely used VHDL tools.

2.2.6. Debuggers

A very important phase in the design process depends on debuggers. With reference to programming language production, three new problems regarding debuggers have to be faced:

- VHDL handles sophisticated objects. Constants and variables are well known in programming languages, but this is not the case with signals. A good debugger must be able to visualize and modify all the VHDL objects. Visualizing a driver can be an attractive feature of a VHDL debugger.
- VHDL design is concurrent. A good VHDL debugger must provide the possibility of placing multiple breakpoints and visualizing the current position in multiple source codes (one per process). This currently implies multi-window debugging.
- The VHDL simulation begins before TIME equals zero. There is a big-bang phase called elaboration during which a certain code is running.

Nothing is more frustrating for the designer than to see an error occurring during this phase and to be unable to debug it. The elaboration phase must be handled by the debugger.

As we will see in paragraph 2.3.3, debugging functions are often better implemented on the compiler using the interpreted approach, because this implementation is more straightforward.

2.2.7. Intermediate Format

The intermediate format is very important for people who develop tools. The way to access (and even the possibility of accessing) is very different from one platform to another.

One good way to appreciate this functionality is by the quality (and importance) of the documentation of this intermediate format.

The amount of information available in the intermediate format depends on the generality of the front-end analyzer. For optimization reasons, some analyzers only store the information necessary for simulation, i.e., executable code and some symbol-tables for debugging and separate compilation. This can be a problem when implementing other tools, e.g., for architecture synthesis or formal proof. The availability of the information needed for a given tool has to be checked.

2.2.8. Extra Tools: Translators

Some VHDL platforms currently propose source code translators from VHDL into a proprietary language or vice versa. The usefulness of such a tool is very dependent on the context, but the designer must be aware of two points: the target source code will certainly be less efficient than the origin one, and the resulting source code will be humanly unreadable.

Paragraph 2.3.5 will give more information on the technology of such translators.

Some vendors also propose a translator that converts an EDIF netlist to structural VHDL (and vice versa). This kind of tool (to convert one standard to another) must be part of the VHDL tool box.

2.2.9. Graphical User Interfaces

VHDL is a complex language, and some companies face training problems in introducing it. For certain designers, using such a general language, with so many ways to model the same thing, is very disconcerting. In some domains

(PCB design, for example), where simulation is often not commonly used, the acceptance of such a modern language is also a problem.

Graphical user interfaces can be an answer. By hiding the language itself from the designer, this interface captures the architecture (structural description) as a collection of boxes interconnected with wires. Graphical facilities make it possible to go down into the hierarchy of boxes and to describe internal views as interconnected boxes.

Concepts of components (and libraries) are also currently implemented as graphical items. With the same userfriendliness concept, interactive facilities are provided to submit patterns of stimuli or to visualize results. In fact, at a low level of description, a good graphical interface can completely hide the language. If this VHDL tool is available on the VHDL platform previously used, the complexity of VHDL can be invisible to a designer. At the end of the design, the system is able to provide the VHDL source code corresponding to the schema. This source code can be compiled on every VHDL platform with or without a graphical environment.

But it is necessary to moderate this assertion. Even if the previous schema seems attractive, it is not quite realistic.

First, the kind of design covered is often restricted to low-level structural descriptions using primary data structures (BIT, BIT_VECTOR, etc.). When using behavioral descriptions or abstract data structures, it is necessary to explicitly write VHDL code.

Second, the quality of VHDL code produced by graphical inputs is debatable. The non-readability of this VHDL code can heavily compromise a good debugging process.

Finally, the use of graphical interfaces reduces the number of people working concurrently on the same platform and considerably increases the cost of such a workstation.

In addition to these possible graphical advantages, a good graphical input can ease system-level modeling: VHDL has drawbacks in this domain, especially because it is not possible to express causality in one single place in a VHDL text. A Petri-net can be expressed graphically, then translated into an equivalent VHDL text.

2.2.10. Synthesis Tools

In spite of persuasive statements from vendors, only logic synthesis tools are really available. The existence of such a tool on a given platform can be a significant advantage. The choice of such a tool can even lead to the choice of platform.

All VHDL features cannot be synthesized, and this will never be the case; access types or files are abstractions that are too unconnected with hardware. Consequently, each synthesis tool builder defines his own subset of the VHDL language as the input of his product. The first criterion to respect when choosing a synthesis tool is the size of this subset. (Does it accept more than one wait statement per process? Which predefined attributes are allowed?). A large subset gives more freedom when refining the system description and so defines the modeling style.

Note that a large subset does not imply optimum efficiency of the result. This criterion is also very important but is often difficult to evaluate.

Another strategic point is the existence and number of possible target libraries. Being linked to a given technology can be a very severe drawback of a synthesis tool.

Synthesis tools also offer many facilities that can be taken into account. A non-exhaustive list includes the following:

- The possibility exists to control the hierarchy of the resulting design. A flat description is not always desirable.
- The concept of black boxes that are not synthesized by the tool but are imported from silicon compilation libraries can be differently implemented.
- The efficiency when synthesizing some critical (because expensive) operators such as adders, multipliers, or comparators varies widely from one tool to another.
- Some products propose an integrated backannotation mechanism to allow more accurate simulation.
- Integrated test features (JTAG, BIST, and so forth) also appear as options of synthesis tools.
- Automatic generation of test patterns for a given coverage rate can be proposed.

2.3. TECHNOLOGY OF PLATFORMS

2.3.1. Specific VHDL Compilation Difficulties

Before execution, a VHDL model must be analyzed and elaborated. During analysis the VHDL code is checked for syntactic and semantic errors. VHDL is a strongly typed language supporting modularity and separate compilation (see chapter 3). This increases the complexity of analyzing a VHDL text because the many checks involved require information resulting from previously analyzed units. Elaboration is a process through which all objects declared by the VHDL model come into existence in the main memory prior to execution. The semantic rules of VHDL are organized into three groups depending on when they should be checked: Locally static checks are performed during analysis;

globally static checks are performed during elaboration; and dynamic checks are performed during execution of the model. But the boundaries between the three sets are not clear-cut: some rules may be checked earlier than during the phase at which they are due to apply. For example, in many cases, dynamic or globally static constraints can be checked at analysis or elaboration time when sufficient information is available.

VHDL rules such as the following are due to be performed at elaboration:

- a function cannot perform a wait statement, even indirectly;[9]
- a process with a sensitivity list cannot explicitly wait, even indirectly;
- a function cannot have access to external objects, even indirectly.

No assumption can be made prior to elaboration about the body of the called subprograms (if declared in a package). But if the package bodies are available, or if the called subprograms are defined locally, the checks *can* be performed by a smart compiler. Thus, some checks that resort to a given phase might be performed during an earlier or later phase.

The result of the analysis of a VHDL text is to produce code that will be interpreted or executed during elaboration and simulation of the model. At the end of the elaboration phase, a VHDL model is equivalent to a concurrent set of processes communicating through signals. The semantics of executing a VHDL model is expressed in terms of a logical agent called kernel process which manages the advance of time, propagates the values of signals, and executes the processes in turn.

The effective value of signals is computed by the kernel process and may imply invocation of user-defined resolution functions. Each process is executed in turn until it suspends in a wait statement. When it resumes, execution proceeds after the point at which it suspended earlier. So the kernel process must keep track of each process context, including the set of variables and the current position in the control flow. Now, consider a process calling a procedure with a wait statement, as illustrated below:

[9] For example, a function cannot call a procedure that includes a wait statement.

```
procedure TheProcedure (...) is
begin
...
wait for 10 ns;
...
end;
```

```
process
begin
 TheProcedure (...);
end process;
```

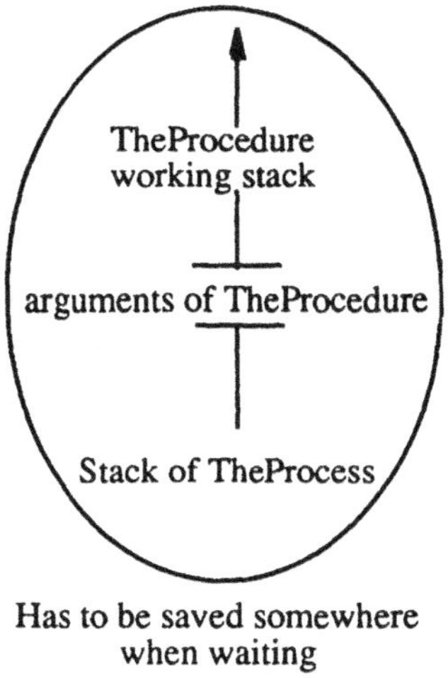

Fig 2.1 Waiting Procedure

Here, the process calls a procedure, which in turn waits for a certain period of time. Of course, during this lapse of time other processes may execute and perform some actions. The whole context of the current process (variables, program pointer) together with the context of the called procedure (variables, program pointers) has to be saved somewhere, for reuse at resumption time.

Things can be even worse when recursive calls and unconstrained array objects are involved, as shown by the following example.

```
procedure  TheProcedure (...)   is
 variable  S : String (1  to AnArgument);
begin
 ...
 wait for  10 ns;
 TheProcedure(...) ; -- recursion
 ...
end ;
```

```
process
begin
 TheProcedure(...);
end process  ;
```

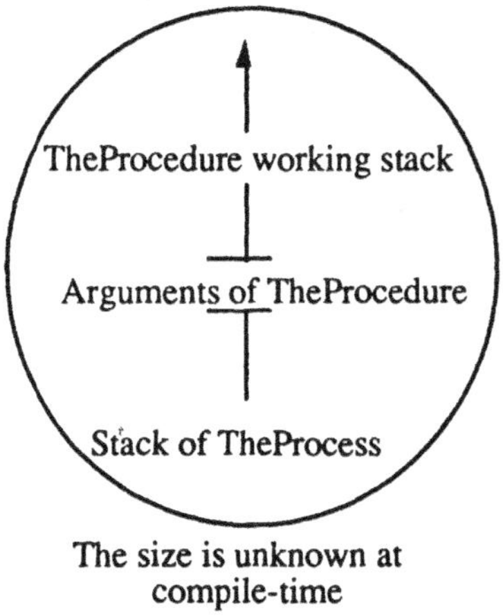

Fig 2.2 Worst Case of Waiting Procedure

Now, obviously, the size of the context cannot be known at compile time: the recursive calls as well as the size of the unconstrained object could be dependent on any value.[10] Since all the processes in a model are concurrent, many threads of control and contexts must be managed.

The implementation of stand-alone VHDL simulators uses one of two techniques: compilation or interpretation [SIN91].

The first approach compiles the models in machine code modules that must be linked with the object code implementing the kernel process to obtain an executable machine program that must be run to simulate the model.

The second approach implements the kernel process as a single executable program. The models are then analyzed to produce data that are fed to the simulator to execute the model.

After an overview of the two techniques with the encountered problems, a comparison is made to determine their merits with regard to complexity, portability, design time, and model execution speed.

2.3.2. The Compiled Approach

The rationale behind adopting a compile approach is to obtain maximum efficiency during execution of the models, as is the case for compiled programs in high-level languages (HLLs). In an ideal case, a VHDL text is translated into machine code modules that are linked with compiled modules representing the kernel process and then executed.

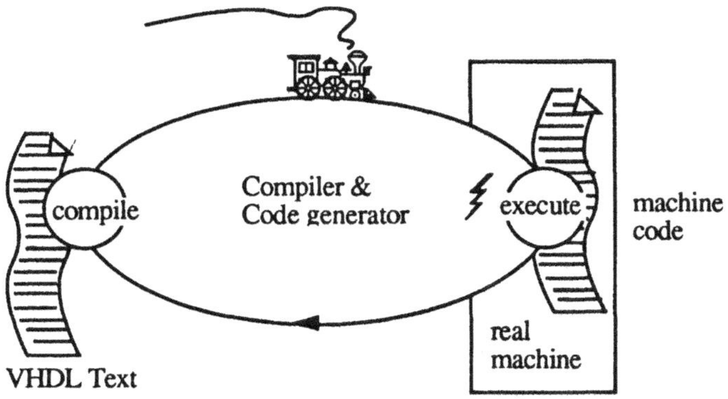

Fig 2.3 Compiled Implementation

But it happens that no existing implementation uses such a direct strategy: the translation to a particular processor is by nature very machine dependent,

---

[10] And local variables are efficiently handled when put on the stack, because their space is automatically reclaimed at the end of the subprogram.

and CAD companies try to keep machine independence for their software. Instead of directly translating into assembly code, actual implementation translates into high-level programming languages such as C (an old version of a VHDL system even translated into Ada!) with possibly a few parts in assembly language. The code produced in high-level programming language then requires a compiler on the host machine. Thus, machine independence is achieved at the expense of an HLL compiler.

The following picture shows the full compilation cycle of a typical "compiled" VHDL, with all intermediate steps and disk accesses.

The main difficulty of this approach is that the high-level language used (i.e., C) implements a program as a set of sequential instructions executing over a single stack, whereas a single VHDL model may encompass many processes executing concurrently. Implementing such processes using the stack execution scheme of HLLs would imply managing one stack per process (there may be tens of thousands of processes). Furthermore, these stacks should be allowed to grow independently and should cope with situations in which sizes could not be known statically (i.e., a process with call to recursive routine).

A mechanism must be found to map the many concurrent stacks of the processes into the single stack over which the model considered as a single program will be executed. Since the size of the stacks may not always be known statically, the stacks and contexts of processes must be managed with pointers to statically and dynamically allocated data structures.

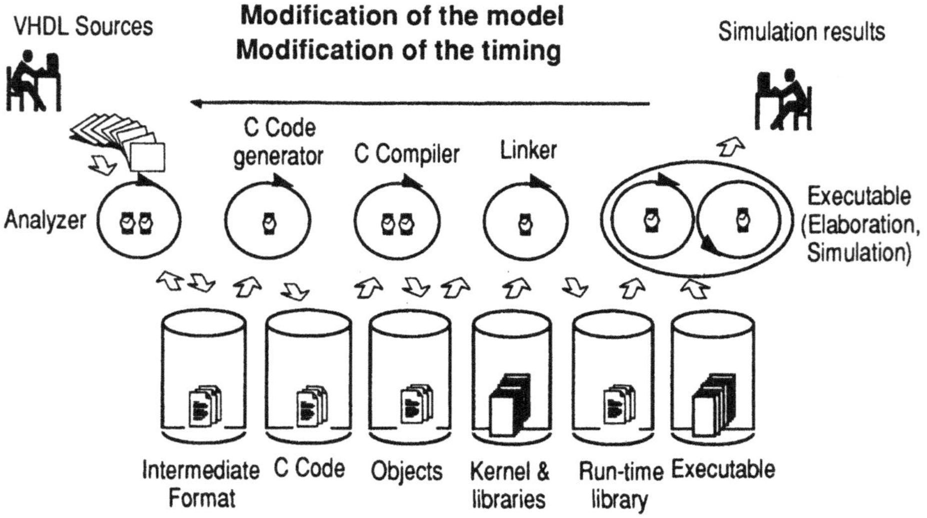

Fig 2.4 Development Cycle

Since the model is translated into a program executing over a single stack, all the processes are mapped into HLL subprograms and wait statements transformed into calls to the kernel module. The different process entry points are implemented with branching GOTO instructions at different places in the corresponding subprogram [MAR90].

The kernel module includes the necessary code to perform swapping between different process contexts. This swapping includes saving the current status of the execution stack of the old process in dynamic memory and relocating the new process context in the execution stack.

This swapping is in many ways similar to the way computers handle interruption routines. Execution of models containing concurrent processes with many wait statements may therefore be very slow due to context-switching overheads.

2.3.3. The Interpreted Approach

An interpreter, in the classical sense, is supposed to take immediate actions from the input text. This, of course, would be inefficient for VHDL because of the abundance and complexities of the many checks to be performed. So the VHDL text is actually translated into an intermediate format (pseudo-code) that is interpreted by the simulator, which implements the virtual machine that accepts the pseudo-code as its instructions. This pseudo-code is produced by an analyzer. The simulator is the same for all VHDL models and consumes the pseudo-code produced by the analyzer to simulate a model. Its functionalities are very close to the LRM description of the kernel process, but also include diagnostics and debugging functions.

The general view of an interpreted system is shown in the following figure: the VHDL text is translated into a specific pseudo-code, which is fetched, decoded, and executed by the interpreter.

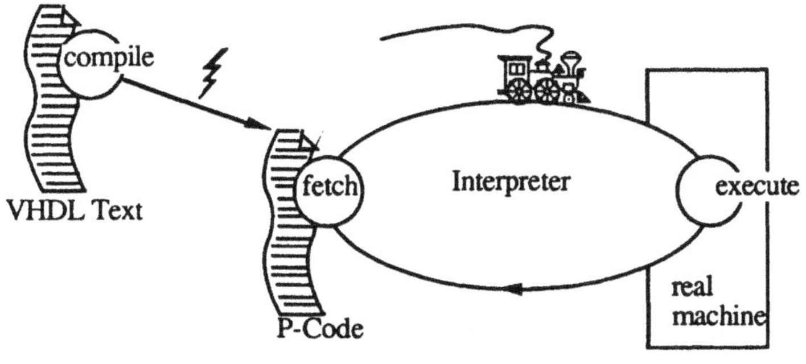

Fig 2.5 Interpreted Implementation

The translation step is faster in this case because no HLL compilation is required. Furthermore, designers of the pseudo-code master its abstraction level and can make it very close to VHDL (i.e., one-to-one correspondence between VHDL concepts and intermediate code statements). The problems of context swapping between concurrent processes can be handled by the interpreted simulator in a more flexible and effective way, since it is the only active program using the execution stack: the contexts and the code generated for the processes are handled as simple data exploited by the simulator on behalf of the model processes.

The semantic level at which the models are interpreted has a great impact on flexibility and performance. If the intermediate is on a very high level, there is a one-to-one correspondence between VHDL concepts and virtual machine statements. Semantic errors can be determined only during execution. But a great flexibility is obtained because the intermediate code depends only on VHDL and not on the host machine language [KLI81]. Furthermore, it is easier to implement debugging functions. On the other extreme, if the intermediate code is very close to the level of assembly language or machine language, the translation phase is more complex and may perform more static checks. There is a one-to-many correspondence between VHDL statements and virtual machine instructions. The corresponding virtual machine is simpler to implement, since more basic features are required. The lost in flexibility is compensated by the gain in efficiency of model simulation.

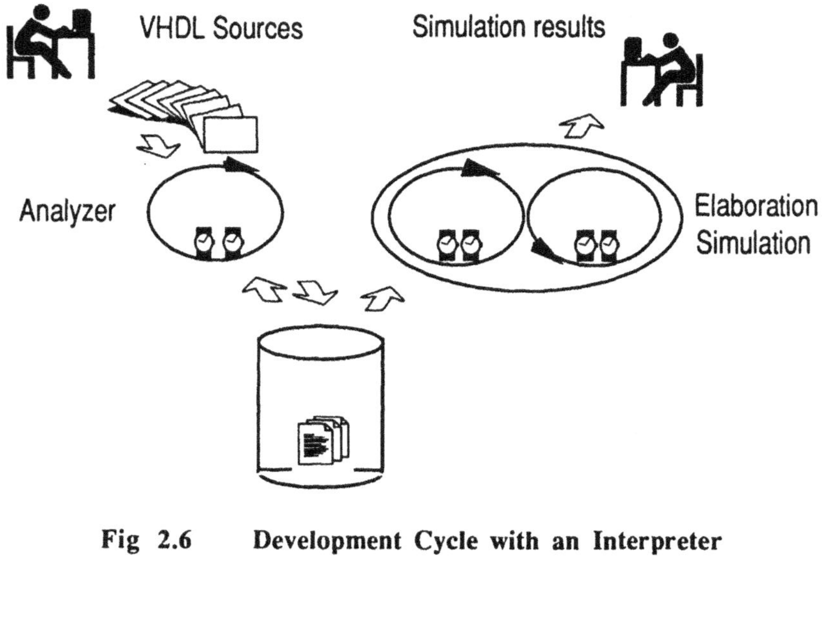

Fig 2.6 Development Cycle with an Interpreter

2.3.4. Comparing the Two Approaches

2.3.4.1. Model Design Time

The goal of the compiled approach is to achieve maximum efficiency by mapping models into executable HLL programs. When the models are very functional with few wait statements and large sections of sequential statements, the simulation speed is maximal because the VHDL models look more like Ada written programs.

But things are more difficult to appreciate when the models encompass large sets of massively communicating processes due to the overhead of context-switching. Because execution time is of primary concern in the compiled approaches, the analysis phase is longer, since maximum checks are performed and some elaboration actions are anticipated, such as unrolling the loops of iterate statements. This results in a longer analysis time (see figure 2.4). But once the models succeed in the analysis-compilation and link phases, they are usually ready for conceptual errors detection. So the debugging phase tends to be shorter, and most of the errors are flagged during the analysis-compilation and link phases.

The interpreted approach shortens the design time but at the expense of model execution time. Most of the current interpreted implementations tend to defer many static checks to the simulation phase. This results in a very short time to analyze, but extends the debugging time because a model that succeeds in the analysis phase is not guaranteed to be free of static errors. At any rate, for the model development phase, the interpreted approach seems more pleasant for the designer, who has the impression of rapidly testing the code he has written and who does not really care about the distinction between static and dynamic errors.

Also, the interpreted approach is more adapted to debugging functions. The simulator need only be extended a little to offer debugging capabilities: the interpreter is the only active program and is in charge of managing all the data on behalf of the model processes.

2.3.4.2. Model Simulation Speed

Considering the issue of model execution speed, a recent benchmarking of interpreted and compiled simulators from the market revealed that interpreted simulators could be even faster on certain kinds of models [HUE91]! This is not really surprising, since interpreted simulators handle context-switching in a more effective way than the compiled simulators. But in general, compiled simulators are considered faster for highly functional models because the models are directly executed by the hardware facilities.

2.3.4.3. Memory Requirements

The interpreted approach has the advantage that any code can be loaded and thrown away when necessary. In order to reduce its memory requirements, it can take advantage of:

- the sequentiality of the three phases: elaboration, initialization, and simulation. In particular, the elaboration code can be forgotten once the simulation phase has begun.
- the locality of VHDL models, which requires each process to be executed in turn.

The compiled approach has to rely on the host operating system paging and swapping mechanisms, if any. Moreover, all this code must be linked together, and this takes time! Of course, smart operating systems have incremental linkers and can guess segment of code to be swapped out in the compiled approach, but not all operating systems are smart.

In addition, the segmentation of the code is decided at compile-time on a per-unit basis. At run-time, this segmentation often leads to a bad locality of references [SIN91].

2.3.4.4. Complexity

As is the case for the implementation of HLL [KLI81], the compilation approach performs the translation of a VHDL model defined by a high-level set of semantic rules into an equivalent executable program defined by another, low-level set of semantic rules. The interpreter, on the other hand, deals with only one set of semantic rules, which is more tractable than two. So compiled simulators are by nature more difficult to implement than interpreted ones. Furthermore, VHDL semantics is biased towards the interpreted approach: the VHDL kernel process is a rough specification of an interpreted simulator.

2.3.4.5. Portability

When portability is an issue, the two approaches solve this problem in different ways. Compiled approaches defer the portability problem to the portability of the HLL modules generated. So they are compelled to use only portable subsets of the HLL in which the models are generated. The code generated by the interpreted approaches is independent of any particular hardware. But the interpreter must be implemented in the target computer before models can be simulated, with the same dilemma: either use a high-level portable language, or use an efficient, low-level assembly code.

2.3.5. Adding VHDL Capabilities to Front-Ends

Many CAD vendors have their own proprietary language, with a corresponding simulator. Their customers have sometimes invested huge amounts of money in libraries in these languages. Adding VHDL capabilities on such systems makes sense only if the original investment is not lost (at least in the beginning). Various implementation strategies exist, as described below.

2.3.5.1. Source-to-Source Translation

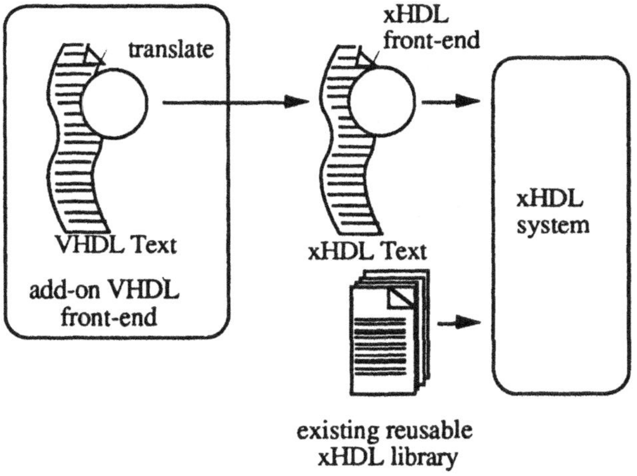

Fig 2.7 Source-to-Source Translation Approach

The basic idea in this architecture is quite simple: let us translate the original VHDL source into an equivalent xHDL source, and let us use this second model as an input for the simulator. Thus the existing models will be able to communicate with the VHDL models. Another benefit is that this kind of translation does not need any implementation-dependent information: no system-dependent binary files are read or written.

Alas, things are not so simple! The translatable VHDL subset may be quite small:

- VHDL timing semantics is quite complex and difficult to map in a previously defined HDL.
- The rich set of user-defined types is found in very few proprietary HDLs, especially when they define the value set of signals. Many non-HDL

simulators propose a limited and exclusive set of predefined types for signals.

- The mechanism of user-defined resolution function is not common in other HDLs. Most of the time signals are resolved by functions hard-wired in the simulator.
- The debugging may be difficult, because what is actually simulated is the translated version of the VHDL model.

The development cycle is augmented by one translation phase.

2.3.5.2. Source-to-Intermediate Format

An interesting alternative is to translate the VHDL input into the intermediate format, if any, of the target simulator. The subset of VHDL may be much larger, and we keep the interoperability of VHDL and xHDL.

The price paid is that only the original manufacturer of the target language has the information to perform this mapping, and some modifications may be necessary in this intermediate format to accommodate for new semantic features of VHDL, or at least for the debugging environment. Some commercial tools are available to retarget a given kernel to VHDL.

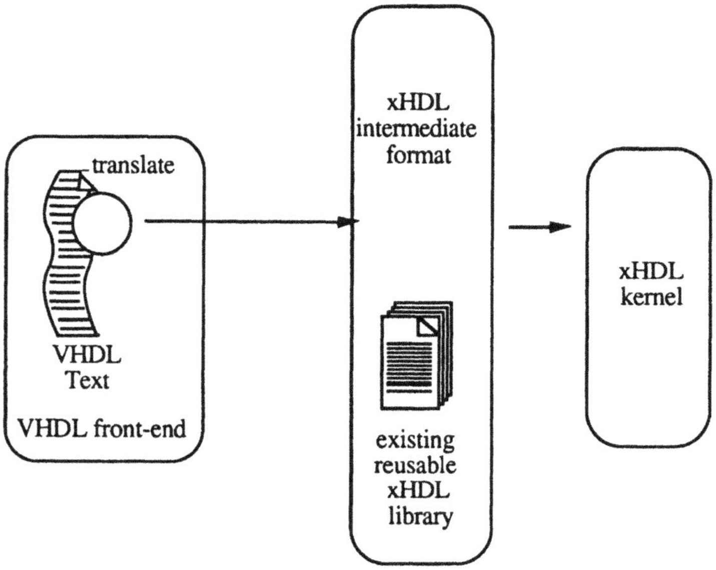

Fig 2.8 Source-to-Intermediate-Format Architecture

2.3.5.3. Cosimulation

As seen before, it is nearly impossible to map all the semantics of VHDL onto an existing system. Should a vendor want to mix its own descriptions and full-VHDL ones, and avoid entirely rewriting the existing simulator, the only practical solution is to use cosimulation techniques: a brand-new VHDL simulator is written, and then interfaced (at the event level) with the old one. This solution has drawbacks in terms of speed, but keeps the two simulators nearly separate. This technique is, of course, usable for interconnection of third-party simulators. Some companies propose such tools on the market.

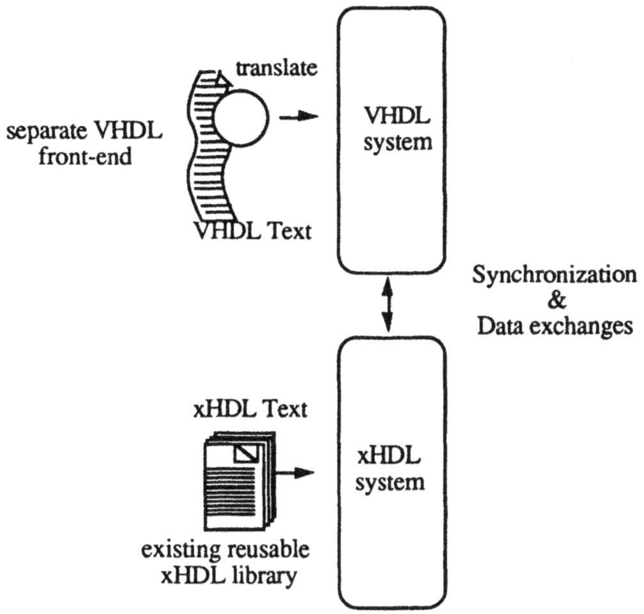

Fig 2.9 Cosimulation Architecture

3. VHDL AND MODELING ISSUES

3.1. INTRODUCTION

The designer's view of an electronic device varies according to the design phases and the intended use. Three views may be distinguished: specification, conception, and documentation.

Specification defines the interface and functional characteristics of a desired system. This includes three main aspects. First, the interface defines the communication with its environment. Next, the behavior of the chip is expressed with no regard to its implementation. Last, the constraints are chosen that will govern the choice of an appropriate architecture for the chip. When the chip is designed with automatic synthesis tools, care must be taken to ensure that the specification is accepted by the tool. This may imply restricting the use of certain language constructs.

Conception implements a behavioral description in terms of interconnected components. The components may be taken from an existing library or specified for further conception.

For later use, a designed chip must be documented with a functional model reflecting its timing characteristics. This abstract model may be obtained after backannotation of its initial specification.

VHDL as a hardware description language can be used to describe each of the previously stated views of the system.

VHDL was strongly influenced by the Ada language, from which it borrows many software concepts such as types, subtypes, subprograms, and packages.

These features make available in VHDL many of the software engineering practices known to Ada programmers, such as abstract types, modularity, and reusability.

After an overview of the basic VHDL concepts for modeling hardware devices, this chapter will discuss some issues that must be understood for a better use of VHDL to model electronic devices in simulation and/or synthesis applications.

3.2. CORE VHDL CONCEPTS

VHDL is a hardware description language aimed at describing digital electronic devices at levels ranging from logic level to functional level through structural level. Despite its complexity, the semantics of VHDL can be understood in terms of a core VHDL subset.

In fact, a VHDL model is equivalent to a set of processes executing asynchronously with respect to each other and communicating through a network of signals. The semantics of all the concurrent VHDL statements is given in terms of block statements and equivalent process statements. Block statements model portions designs by grouping a set of concurrent statements in the same declarative region. The semantics of all VHDL hierarchical and structural descriptions is defined with blocks or nested blocks. All the other concurrent statements are equivalent to one or more processes. For example, the following VHDL model:

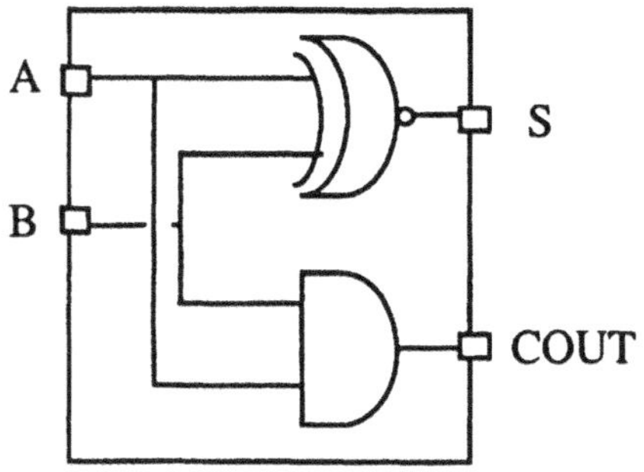


```
entity HALF_ADDER is
   port ( A,B : in BIT ;
          S, COUT : out BIT) ;
end HALF_ADDER ;
architecture DATAFLOW
    of HALF_ADDER is
begin
   S <= A xor B;
   COUT <= A and B;
end DATAFLOW ;
```

Fig 3.1 Half Adder

is equivalent to:

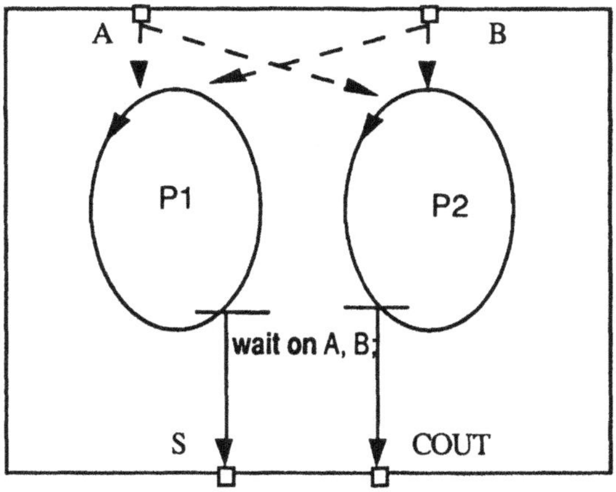


```
HALF_ADDER : block
  port (A, B : in BIT ;
    S,COUT : out BIT ) ;
begin
  P1 : process
  begin
      S<=A xor B;
      wait on A, B;
  end process ;
  P2 : process
  begin
     COUT <=A and B;
     wait on A, B ;
  end process ;
end block ;
```

Fig 3.2 Semantics of VHDL Half Adder

Each concurrent signal assignment statement is replaced by a process statement.

3.2.1. Process

```
IDENT: process
        <declarative region>
begin
        <sequential statements>
end process IDENT;
```

A process is an independent section of sequential code following the syntax given above that executes continually throughout the simulation. It is used to describe the behavior of a portion of hardware. For that purpose it has a declarative region delimited by the reserved words **is** and **begin** where a set of datatypes, subprograms, and variables can be declared. The variables defined in this region identify simple memory locations that can be read and updated only within the process. They are provided during the elaboration phase and persist throughout the simulation period. All the descriptive power of high-level programming languages is made available within the process statement. A process can be viewed as an infinite loop: during simulation, execution resumes with the first statement when the last statement is executed. The time does not advance during the execution of the process sequential statements unless a wait statement is encountered.

3.2.1.1. Synchronization

Synchronization is handled by the execution of a wait statement, which causes the process to suspend. In the example given above, the two processes contain the statement

```
wait on A, B;
```

which causes them to suspend until either of the signals A or B changes value. Other forms of wait statement include waiting for a delay,

```
wait for DELAY;
```

where the process is suspended for the duration of the specified delay. The statement

```
wait until CONDITION;
```

suspends the process until the specified condition is satisfied. The condition is evaluated any time a signal occurring in the expression of the condition is modified. The more general case is

```
wait on SIGNAL_LIST until CONDITION for DELAY ;
```

which suspends the process until the condition is verified or the delay is over. The condition is reevaluated each time a signal of the on list is modified.

3.2.1.2. Interprocess Communication

The process statements of a VHDL model can be viewed as concurrently executing, performing wait statements for synchronization, and communicating with each other through signals. In fact, signals defined within a block can be read and updated within any process of the block. Signal assignment statements are used within a process to modify the projected future values of signals.

3.2.2. Signals

Physical wires can be modeled in VHDL by the concept of a signal. A VHDL signal is an object with a past history of values acting as a medium for waveforms propagation. It can be created by a declaration of the following kind:

```
signal S : BIT:='1';
```

The declaration specifies the name (here S) of the signal and the set of values that it may take (here BIT). The expression following the assignment symbol ":=" is optional and defines the expression associated with the signal. When present, this expression is used to initialize the drivers of the signal as explained below.

3.2.2.1. Discrete Representation of Waveforms

At the electric level, a waveform can be represented as follows:

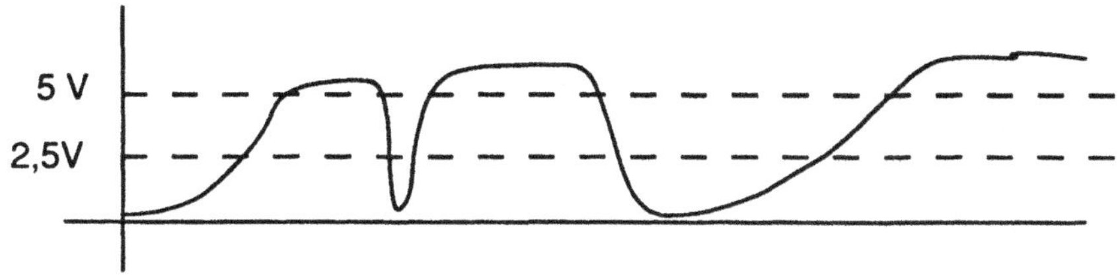

Fig 3.3 Physical Waveform S

In VHDL, it would be described with a discrete set of values. For example, at the logic level, using two values, it would read as

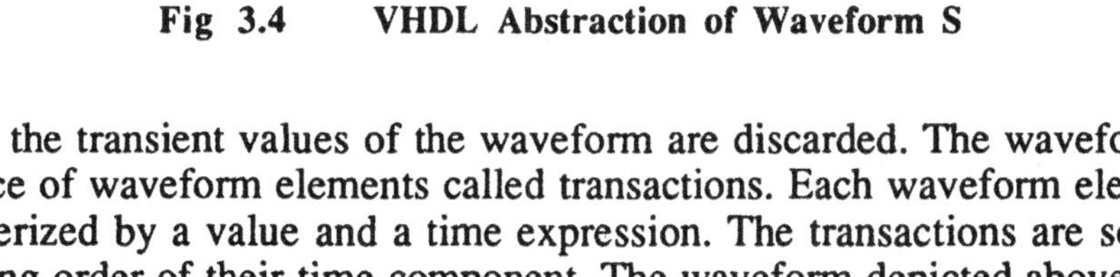

Fig 3.4 VHDL Abstraction of Waveform S

All the transient values of the waveform are discarded. The waveform is a sequence of waveform elements called transactions. Each waveform element is characterized by a value and a time expression. The transactions are sorted in ascending order of their time component. The waveform depicted above can be described by the following sequence:

('0',0 ns), ('1', 15 ns), ('0', 33 ns), ('1', 38 ns), ('0', 63 ns), ('0', 108 ns)

An **event** occurs on a given signal each time its value changes. Here, five events occur on signal S. This discrete representation of waveforms and signals in VHDL greatly simplifies the simulation of a model: there is no need to deal with the complex computations of transient values determined by differential equations, as is the case with analog simulation. Events on signals are used to trigger the simulation.

3.2.2.2. Propagation of Signal Values

In VHDL, the propagation of waveforms is achieved by signal assignment statements. The current value of a signal represents the current value of the waveform running through the signal. It is the **effective value** of the signal obtainable by a simple read of that signal in an expression. The past portion of the waveform is the past history of the signal. The yet-to-come part of the waveform defines its projected future value. A signal assignment statement never modifies the past or the current value of a signal; it only affects its projected future values.

The above waveform can be produced by the following statement:

```
S <= '0', '1' after 15 ns, '0' after 33 ns, '1' after 38 ns,
     '0' after 63 ns, '1' after 108 ns;
```

3.2.2.3. Driver

In fact, associated with each signal assignment is a **driver**, which is used to store the projected future waveform. Many signal assignments will share the same driver if they affect the same target signal and occur in the same process statement. During elaboration, one driver is created per signal appearing as a target of a signal assignment statement.

A driver is characterized by two parameters:

- A queue to store the sequence of transactions defining a waveform. As simulation time advances, transactions are consumed from the queue. This ensures that at any given simulation time, the first transaction of the queue is the one whose time component is closest to but not greater than the current simulation time. A driver is said to be active every time a transaction is removed from the queue.
- A current value that is the value field of the first transaction. During the initialization phase, this value is set to the expression associated with the signal. This expression may be explicitly defined by the user in the signal declaration. If it is not specified, for scalar signals, the leftmost value of their type is taken. Each time the driver acquires a new value as a consequence of time advancing, it is said to be **active**.

As time advances during simulation, the values of signals are updated according to the values of their sources.

Signals that are not resolved (see paragraph 3.2.5) may only have one source: their effective value is equal to the driving value of their single source.

At simulation time 34 ns, the driver of S will appear as shown in figure 3.5.

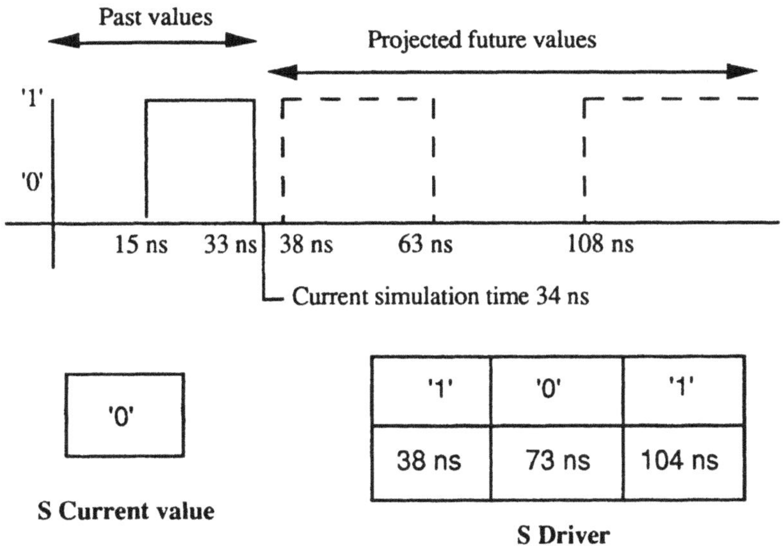

Fig 3.5 Status of S at Simulation Time 34 ns

3.2.2.4. Signal Attributes

VHDL allows references to past values of signals by the use of attributes. The expression

```
S'DELAYED(10 ns)
```

yields another signal derived from s by a delay of 10 ns. Such an attribute is an implicit signal, as no assignment is allowed on it. It is directly monitored by the simulator. Other attributes are used to detect whether the signal acquires a new value (S'ACTIVE, S'TRANSACTION); whether a change occurred on the signal value (S'STABLE, S'EVENT); the elapsed time since the last change of the signal value (S'LAST_EVENT); and so forth.

Some of these attributes are implicit signals ('QUIET, 'TRANSACTION, 'STABLE, 'DELAYED), and others are functions ('EVENT, 'ACTIVE, 'LAST_EVENT, 'LAST_VALUE, 'LAST_ACTIVE).

3.2.3. VHDL Timing Model

VHDL simulation is event driven, which means that events on signals trigger the simulation. That is, the simulation of a model takes into account only the instant at which events occur on signals or at which processes resume execution after a timeout period of a wait statement. For example, the following process

```
process
begin
S4<= not S;
wait on S;
end process;
```

which is equivalent to the concurrent signal assignment in a block statement or architecture body

```
BL : block
begin
S4<=not S;
end block;
```

will feed the driver associated with S4 with the value of not S. Every time a change (event) occurs to S, the S4 driver will be updated consistently.

3.2.3.1. Delta Delay

Although no time is specified for the signal (default value is 0 fs) assignment statement, its execution does not affect the current value of S4: its effects occur after an infinitesimally small delay called **delta delay** .

3.2.3.2. Preemptive Timing Model

The driver of a given signal is updated whenever an event occurs on one of the signals involved in the right-hand side of the corresponding signal assignment statement. After such an event, the new waveform will cancel the relevant portion of projected future values stored in the driver. A projected future waveform may be cancelled by another execution of the signal assignment, as illustrated by the following example:

```
BL : block
        signal A, S : BIT;
begin
        process
        begin
            A<='1', '0' after 10 ns ;
            if A='1' then
                S<= '1', '0' after 6 ns, '1' after 40 ns ;
                wait for 5 ns;
            else
                S<= '0' after 15 ns ;
                wait on  A ;
            end if;
        end process;
end block;
```

Fig 3.6 Signal S at Simulation Time 5 ns

At simulation time 5 ns, the projected future waveform (in dotted lines) of S is defined by the first branch of the conditional assignment statement, since signal A has value '1'. At time 10 ns, an event on signal A triggers a new execution of the second signal assignment statement, which modifies the projected waveform as shown below. So the projected waveform contained in a driver can be superseded by the execution of a signal assignment affecting the same driver.

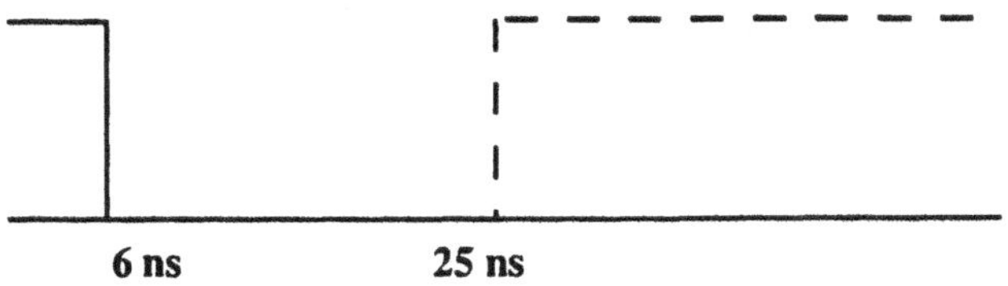

Fig 3.7 Signal S at Simulation Time 10 ns

Two timing models — namely, inertial and transport delays — are supported by VHDL.

3.2.3.3. Inertial Delay

The inertial delay is appropriate for modeling switching circuits. In this mode, a pulse shorter than the switching time of the circuit is not transmitted. This is well illustrated by the following example: signal S2 is derived from S using a delay circuit.

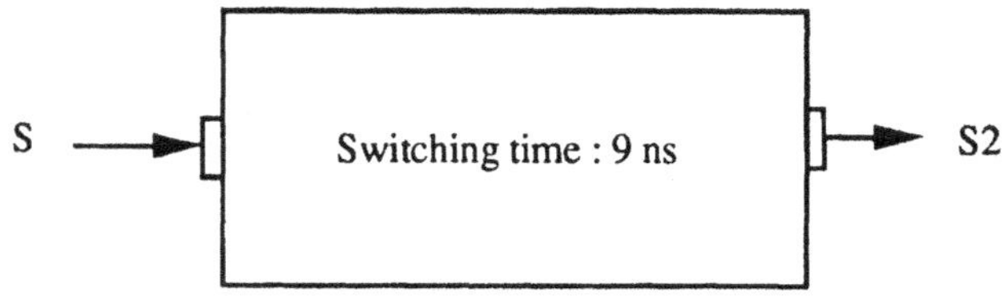

Fig 3.8 Inertial Delay

This is achieved by the statement

```
process
begin
S2<= S after 9 ns;
wait on S;
end process;
```

equivalent to the concurrent statement

```
S2<= S after 9 ns;
```

By default, a signal assignment in VHDL follows the inertial delay mode. The inertial delay is the time specified after the reserved word **after**. Here it is 9 ns. The waveform of S2 does not contain the 5-ns-width pulse occurring at time 33 ns (cf. figure 3.9) because it is shorter than the inertial delay.

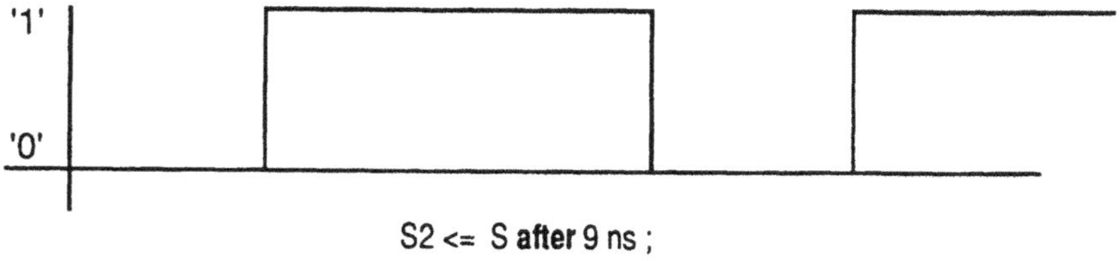

Fig 3.9 VHDL Inertial Delay

3.2.4. Transport Delay

In some cases, it might be necessary to model a transmission line where every pulse, whatever its duration, should be transmitted. VHDL provides the transport delay mode for this case. It is achieved merely by adding the reserved word **transport** after the signal assignment symbol as follows:

```
S3<= transport S after 9 ns ;
```

Simulation results are shown in figure 3.10:

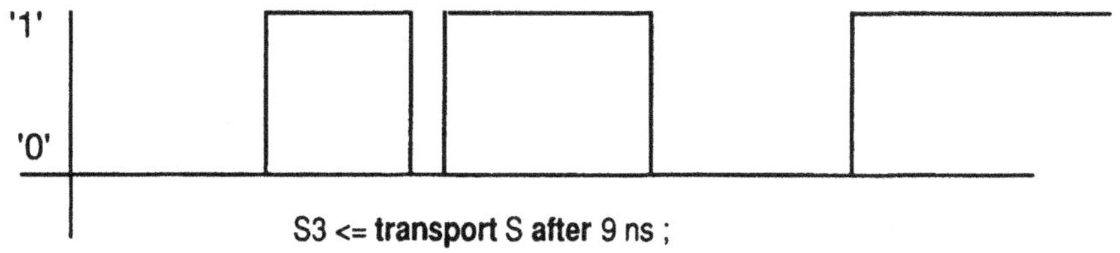

Fig 3.10 VHDL Transport Delay

3.2.5. Resolved Signals

Because a signal is accessible to all the process statements, it may be driven by many different sources. VHDL requires each such signal to be **resolved** — that is, to be associated with a **resolution function** that will be used to compute the value of the signal from the value proposed by the various sources. A resolution function for a given type is a function returning a value of the given type and accepting as its single parameter an unconstrained array of elements of that type. For type BIT, the statement

```
function WIREDOR(ARG : BIT_VECTOR) return BIT ;
```

declares a function that can be used as a resolution function.

Association of a signal with a resolution function can be done in one of two ways:

- By using a resolved type — a type associated with a resolution function as follows:

```
subtype RESOLVED_BIT is WIREDOR BIT;
signal RESOLVED_SIGNAL : RESOLVED_BIT;
```

• By directly associating the signal with the resolution function:

```
signal RESOLVED_SIGNAL : WIREDOR BIT;
```

The simulator computes the effective value of a resolved signal by calling the resolution with an array containing the values of its active sources as an actual parameter.

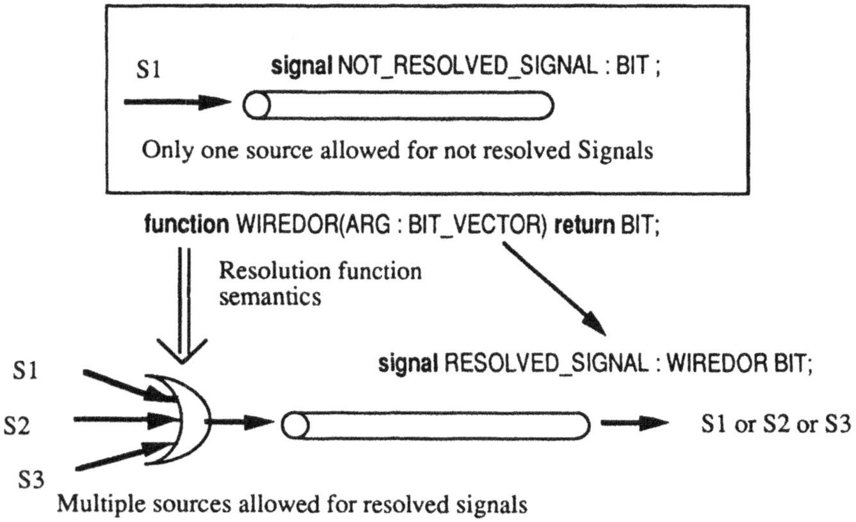

Fig 3.11 VHDL Resolved Signals

3.2.6. Guarded Signals

Guarded signals were introduced in VHDL to support the hardware bus and register model. These features also support high-level state machine models as explained in chapter 5.

The resolved signals that are described earlier model standard physical-device outputs where the sources always drive the associated resolved signal. This is not very convenient for the Bus model, where Tristate output devices are used to set a node in a high-impedance state. A source in a high-impedance state does not drive the bus and thus does not affect its effective value. It is even insufficient to support the register model, which requires the possibility of disconnecting the sources from the memory device. VHDL proposes the concept of guarded signals to overcome these limitations.

There are two kinds of guarded signals: the **bus** and the **register**. They are declared as follows:

```
signal B : RESOLVED_BIT bus;
signal R : RESOLVED_BIT register;
```

VHDL requires such signals to be modified only within guarded blocks. Guarded blocks are blocks associated with a boolean condition, also called the GUARD condition, that is stored in an implicit signal named GUARD. Within a guarded block, guarded signals may only be modified by guarded-signal assignment statements (reserved word guarded after the arrow).

```
BL : block (TRIGGER)
begin
B<=guarded '1';
end block;
```

In the example above, the guarded assignment is equivalent to the process:

```
process
begin
  if GUARD then
       B<= '1';
  else
       B<=null;
  end if;
  wait on GUARD;
end process;
```

The signal B is driven with '1' when the GUARD condition (equal to TRIGGER) is true. When this condition is false, B is driven with a **null** transaction. The semantics of this transaction is to disconnect the driver. Notice that the else part is relevant only for guarded signal; if the signal is not guarded, the else part is discarded. The instant of disconnection can be controlled by a disconnect statement. For example, the following specification

```
disconnect B : RESOLVED_BIT after 5 ns;
```

causes the driver of B to be disconnected 5 ns after the guard condition switches to false. A driver that is disconnected is no more active.

The effective value of guarded signals is computed as in the case of resolved signals; but only the active-driver values are considered. A question arises: what happens when all the drivers of a guarded signal are disconnected?

For signals of the **bus** kind, the resolution function is called with a null argument and must return a value that defines the effective value of the signal. For signals of the **register** kind, the resolution function is not called: the signal keeps its last value and thus exhibits a memory functionality.

3.2.7. Composite Signals

In VHDL, a composite signal (record or array) is viewed as the juxtaposition of its scalar subelements acting as individual signals. So the driving value and effective value of the composite signal is defined in terms of the driving value and effective values of its scalar subelements.

```
type COMPLEX is record
        R,I : REAL;
        end record;
signal A: COMPLEX ;
signal B : BIT_VECTOR(1 to 5);
type COMPLEX_VECTOR is array (POSITIVE range <>) of COMPLEX;
signal C : COMPLEX_VECTOR(1 to 6);
```

From the definition above, A is a composite signal formed by the juxtaposition of the two scalar signals A.R and A.I. Signal B is an array of five elements. Each of these elements are scalar signals of type BIT. Signal C is an array of six elements. Each of its element is also a composite signal with two scalar subelements. These scalar subelements are signals.

The driver of a composite signal is formed by the set of drivers of its scalar subelements. A composite signal may be modified in one of two ways: as a whole or by parts. In the first case, scalar components of the aggregate waveform are extracted and assigned to the relevant signal scalar subelement. The concurrent statement

```
A <= (3.4, 7.8) ;
```

assigns 3.4 to A.R and 7.8 to A.I. In the second case, a signal subelement is assigned a value, as in

```
A.I <= 9.0 ;
```

The effects of this instruction depend on whether or not A is a resolved signal.

3.2.7.1. Unresolved Composite Signals

Here, unresolved means that the signal taken as a whole is not resolved, but any of its scalar subelement may be resolved. Unresolved composite signals act as individual signals. A resolved scalar subelement may have multiple sources. But an unresolved subelement cannot have more than one source. The question then is how the driving and effective values of the aggregate signal are computed.[11]

The driving value of an unresolved composite signal is the aggregate value of its subelements driving values. Its effective value is equal to its driving value.

3.2.7.2. Resolved Composite Signals

If a composite signal is resolved, then the driving and effective values of its subelements are derived from the driving and effective values of the composite signal. Subelements no more behave as individual signals for two reasons:

- A concurrent statement modifying a subelement of a composite signal also affects the other subelements because a driver is created for all the other subelements. The current values of the drivers of all the unmodified subelements is equal to the default values associated with these subelements. So the collection of all the drivers constitutes a single source of the resolved signal.
- If a subelement is itself resolved, its associated resolution function is not used to compute its driving and effective values. The resolution function of the composite resolved signal hides the resolution function of its subelements. In fact, once the driving value of the composite signal is computed with the resolution function, the subelement values of the returned aggregate driving value represent the driving values of the subelement signals.

Consider the following VHDL block statement:

```
type TWO_BIT is record
   ONE : BIT;
   TWO: WIREDOR BIT;
end record ;
```

```
B : block
begin
   AGG.ONE<= '1' ;
   AGG.TWO <='1' ;
   AGG.TWO <='0' ;
end block ;
```

The following figures illustrate the distinction between unresolved and resolved signals. Notice that subelement TWO of signal AGG is resolved with a WIREDOR resolution function.

If AGG is declared as an unresolved signal, the three concurrent statements yield one source for signal AGG.

---

[11] Here we overlook the problems raised by the type conversion function. These will be tackled in paragraph 3.2.8.4.

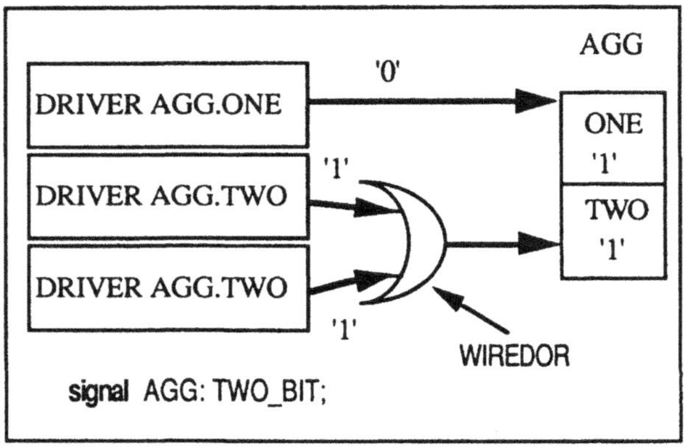

Fig 3.12 Unresolved Aggregate Signal

If AGG is resolved using another resolution function WIREAND operating on type TWO_BIT, three distinct sources are produced for signal AGG, although the statements only modify its subelements.

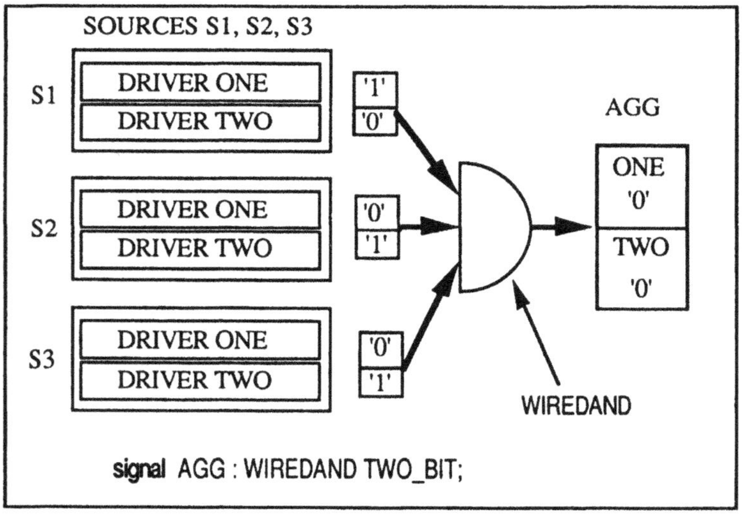

Fig 3.13 Driving Value of Resolved Composite Signal

If a composite signal is guarded, so are its subelements, and they are of the same kind (**bus** or **register**).

3.2.8. Ports and Port Association Lists

Another category of signals, called ports, is used to define the interface and to control the communication of a given portion of design with its environment. Ports are created by port declarations in entity declarations, component declarations, and block statements. For example, the following VHDL text

```
PART: block    port(A : in BIT; B: inout WIREDOR BIT; C : out BIT) ;
               port map (A=>S1,B=>S2, C=>S3);
        signal INTERNAL_S : BIT;
     begin
        INTERNAL_S <= B ;
        ...
     end block;
```

defines a portion of hardware communicating with the overall design through ports connected to signals from the environment. Connection is accomplished by a **port map** clause comprising a port association list. The port association list associate ports with signals from the outside environment. These signals may be ports defined in an upper level.

The mode of the port determines the direction of information flow: ports of mode **in** may receive information from the outside world; ports of mode **out** may transmit information to the outside world; ports of mode **inout** or **buffer** are used for bidirectional communication with the environment.

Here, it makes sense to distinguish between the driving value and effective value of a signal. The effective value of a signal is the value obtained by a simple read on that signal. Within the block above, it makes sense to speak of the effective value of ports A or B, since they can be read within the block. Port S has no effective value within the block because it cannot be read. The driving value of a signal is the value proposed by that signal as a source to another signal.

Ports of mode **out**, **inout**, and **buffer** have driving values for signals with which they are associated in the port map clause. Within the block, the effective value of **in** or **inout** ports is also the driving value for the signals to which they are assigned.

3.2.8.1. Effective Value of Ports of Mode IN

Consider the following block statement:

```
B1 : block
signal N1, N2 : INTEGER;
function VECTOR(ARG : INTEGER)
return BIT_VECTOR is ...end ;
begin
 N1<= 4 ;
 N2<= 7;
 INTERNAL_BLOCK: block
    port (A : in BIT:='0';
  B: in BIT_VECTOR(1 to 8);
  C: in INTEGER;...);
    port map (   B=>VECTOR(N1),
   C=>N2;... );
     begin
       ...
   end block;
end block ;
```

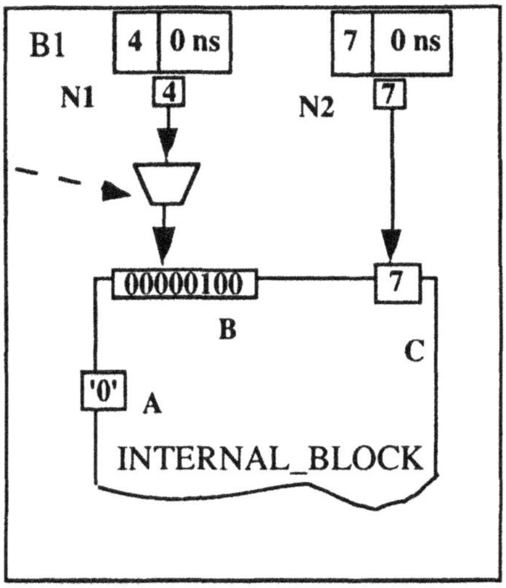

Fig 3.14 Effective Value of Ports of Mode IN

The internal block (INTERNAL_BLOCK) contains three ports of mode **in** that are associated with signals defined in the enclosing block. The port map clause does not only associate ports with signal of compatible or identical type. Port B, which is of type BIT_VECTOR(1 to 8), is associated with signal N1 of type N1. The association is achieved with the type conversion function VECTOR that will be used by the simulator to convert signal N1 into a BIT_VECTOR(1 to 8) value. The declaration of formal port A includes an assignment symbol followed by an expression. This expression is used as the default associated with port A. Since port A is not connected, this value defines its effective value. In VHDL, unconnected ports of mode **in** should include a default expression in their declaration.

For all the connected ports of mode **in**, their effective value is obtained by evaluating the effective value of the actual part with which they are associated in the port map clause. This actual part may include a type conversion function (VECTOR(N1) in the case of port B). For port C, its effective value is obtained by applying function VECTOR to the effective value of N1. For B, its effective value is the same as that of N2.

3.2.8.2. Driving Values of Ports of Mode OUT

The following block illustrates the computation of driving values of ports of mode **out**. These ports are used to transmit information to the outside of the block in which they are declared. They are associated with outside signals or ports in a port map clause. This association may include a type conversion function to convert the value of the **out** port into the type of the target signal or port with which it is associated. Output ports serve as a possible source of the signals with which they are connected. The driving value of that source is the driving value of the formal part of the relevant association element. This formal part may include a conversion function for converting the driving value of the port into a value of the type of the target signal. Here, VALUE is used to convert the driving value of port E into an integer. The driving values of **out** ports act as simple sources to other signals. Signal S2, for example, has two sources, one of which is the port F. If an expression is associated with the port in its declaration, that expression defines the initial value of its driver. If such an expression is absent, the value considered is the leftmost value of the type of the port if the port is of scalar type or an aggregation of the leftmost value of the type of the scalar subelements. If an **out** port is not assigned within the block in which it is declared, the initial content of its driver defines its driving value.

```
B2 : block
signal S1,S3: INTEGER;
function VALUE(ARG:BIT)
   return INTEGER is ...end ;
signal S2 : WIREDAND BIT;
begin
 ...
 S2<=...;
  INTERNAL_BLOCK: block
 port ( D : out INTEGER:=0;
  E: out WIREDOR BIT;
  F: out BIT);
 port map ( VALUE(E)=>S1),
  F=>S2 );
   begin
        E<=...;...;E<=...;...;F<=...;
   end block;
end block ;
```

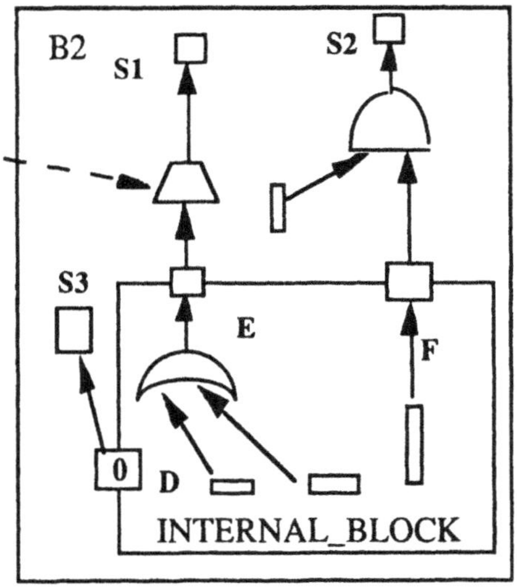

Fig 3.15 Driving Values on Output Ports

3.2.8.3. Ports of Mode INOUT or BUFFER

Ports of mode **inout** or **buffer** have an effective value representing the value of the information flowing from the outside, and a driving value representing the value of the information transmitted to the outside. The driving value is computed first, since it is one of the sources that determines the effective value of the associated signal. The driving value is obtained after evaluating the driving value of the actual part of the relevant association element. This may imply a call to a type conversion function. This value is proposed by the **inout** port to the signal with which it is associated as a source value. The effective value of the **inout** port is the effective value of the formal part of the relevant association element. Once more, a type conversion function may be called to convert the value of the actual parameter to a value of the port type. The figure 3.16 illustrates the computation of the driving value and effective value of an **inout** port. The driving value flows out and the effective value flows in. Here port A is a WIREDOR resolved port associated with the WIREDAND resolved signal N1. Port A has two internal sources that resolved to '1' as a driving value. Signal N1 has another source beside A; the two resolved to a '0' value. Although port A drives N1 with a '1' value, its effective value is '0'. The main distinction between **buffer** and **inout** ports is that the former can have at most one source and the latter may have any number of sources.

```
B1 : block
signal N1 : WIREDAND BIT;

begin
        N1<='0';
        INTERNAL_BLOCK: block
        port (   A : inout WIREDOR BIT;
                 ... );
        port map(   A=> N1, ... );
        signal S1 : BIT;
        begin
          S1<= A;
          A <= '1';
          A <= '0';
          ...
        end block;
end block;
```

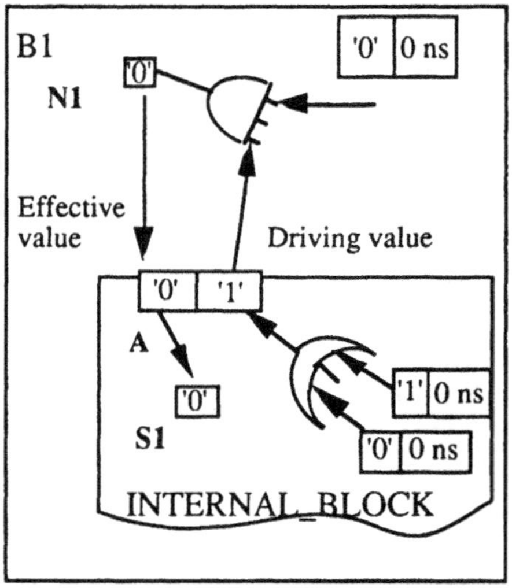

Fig 3.16 Driving and Effective Value of INOUT Port

3.2.8.4. Ports and Composite Signals

Association of an unresolved formal port of composite type with an unresolved signal of the same type is equivalent to the association of subelements when no type conversion function is involved. Subelements behave like individual signals. A signal subelement is active if and only if the associated matching signal subelement is active.

If a type conversion function is used in the association, the port is associated with the signal as a whole. If the associated signal is a source of the port (mode **buffer**, **inout**, **out**), then any time the signal is active the port subelements will become active. The effective value of the port subelements will be derived from the port effective value that is obtained by calling the type conversion on the effective value of the associated signal. Similarly, if the port is a source to the signal (ports of mode **inout**, **out**, **buffer**), the effective value of the signal subelements will be obtained after calling the type conversion function on the driving value of the port.

For resolved composite signals, association with a port is also equivalent to subelements associations. But the signal subelements do not behave as independent signals. If the port is of mode **out**, **inout**, or **buffer**, the composite drivers of the port subelements constitute a single source to the resolved signal. That is why if a subelement of a resolved signal is associated with a port of mode **out**, **buffer**, or **inout**, then all the other subelements of the resolved signals must be similarly associated exactly once in the same port map clause. The collection of all the associated ports constitutes a single source to the resolved signal.

3.2.9. Simulation Cycle

A VHDL model can ultimately be viewed as a set of concurrent processes communicating through signals. The execution of the model is described in terms of a conceptual agent called a kernel process, which after initializing signals performs a sequence of steps called simulation cycles. At the beginning of the simulation, time is assumed to be 0 ns. Each simulation cycle runs in three main steps:

- The time advances to the next point at which a driver becomes active (i.e., proposes a new value to a signal) or a process waiting for a delay resumes. Simulation stops when time reaches the value TIME'HIGH.
- Signals value are updated. An event occurs on any signal that changes value. As a consequence, all the processes that were suspended in a wait statement on that signal resume.
- Each of the resumed process is executed until it suspends.

3.2.10. Conclusion

This section presented basic VHDL concepts that are used to model hardware behavior. Other concepts, such as packages, will be introduced later when necessary. The remaining part of the chapter presents some critical modeling issues faced by designers, as well as some solutions offered by VHDL.

3.3. ABSTRACTION

The complexity of designed ICs and other electronic devices favored the development of new design methodologies based on abstraction to cope with the huge size of information required to describe a circuit. Abstraction denotes the ability to retain the main features of a design, discarding the less relevant or minor details. It reduces the amount of information that must be addressed by the designer. Many levels of abstraction have been defined to deal with the huge size of detailed information describing electronic devices. Each level is characterized by a set of primitive components and by the size and nature of information processed at that level. Despite their differences, all of the methodologies rely on some or all of the following levels of abstraction:

- The *system level* encompasses block-processing elements such as processors, memories, IO devices, and buses. At this level, the overall system is split in a set of loosely coupled nodes with well-defined functionalities. The main focus is to work out practical specifications and to achieve a good compromise between parameters such as processing time, well-balanced functional decomposition into subsystems, required storage space, and resource utilization.
- The *RTL level* deals with words of bits using combinational and sequential devices (word gates, multiplexers, decoders, arithmetic operators, registers, etc.). The objective is to design a datapath and controller implementing a given functional or behavioral description.
- The *logic level* is well defined by switching theory based on boolean or multi-valued algebra. It involves all of the logic gates processing boolean or multiple-value bit-information. The logic networks must be optimized.
- The *electric level* falls in the analog world, which is not yet covered by VHDL. At this level, current and voltages are processed by transistors, resistors, and capacitances.
- The layout level is mostly relevant for integrated circuit design and represents the final physical description of the geometrical masks that must be used to produce the final chip.

Three strategies may be distinguished in design methodologies:

- Top-Down proceeds hierarchically from an abstract level to a more detailed one by successive decomposition into subsystems;
- Bottom-Up achieves the design of more complex systems by assembling less complex ones;
- Meet-in-the-middle decomposes a system into subsystems until the parts of the resulting decomposition can be designed using a library of

components and standard cells. This last strategy is the most commonly used.

VHDL support for abstraction is highlighted by four main features. VHDL entity and component declarations abstract the functions of a hardware device.

3.3.1. Entity Declaration

VHDL entity declarations give an external view of a system. They only describe the interface of the system with its environment; the internal structure or behavior of the device is completely hidden. However, an entity declaration may include passive concurrent statements (i.e., assertions or processes with no signal assignment statements) to enforce checks on the operating condition (setup checks). A VHDL entity declaration such as

```
entity HALFADDER is
        generic(DELAY:TIME:=5 ns);
        port(A: in BIT; B: in BIT; S: out BIT);
end HALFADDER;
```

is equivalent to a block statement with the same **generic** and **port** clauses. It defines a system as a black box.

The function computed by the entity is specified in a separate unit called architecture body. For a single entity declaration, many different architectures may exist. This feature is particularly useful to describe a design at its various stages, as explained in a top-down methodology.

3.3.2. VHDL Datatypes

To describe the nature of information exchanged with the outside world, VHDL use the concept of datatype found in all high-level programming languages. VHDL proposes a wide range of types, from scalar types such as enumeration and numeric types to structured types such as records, arrays, and access types. Types specify the nature of information exchanged or processed with no regard to structural choices, such as the size of the ports or the primary hardware module used to perform the operations. Abstract datatypes are available in VHDL through the possibility of overloading predefined operators (such as "=", "+"), thereby extending their use on user-defined types. This topic is covered in depth in paragraph 7.1.6 of chapter 7 ("Tricks and Traps").

3.3.3. Behavioral Statements

VHDL behavioral statements encompass most of the high-level sequential statements (e.g., loops, if-statements, and subprograms). In fact, VHDL borrows many features from the Ada language. Behavioral statements define the simulation semantics of models. Any VHDL model representing a design hierarchy cannot be simulated unless its leaf cells are functionally described with behavioral or dataflow statements.

The high-level flavor of VHDL sequential statements is useful for specification in order to express the functionality of electronic systems with no regard to their internal structure.

3.3.4. Structural and Dataflow Statements

VHDL structural statements unveil the internal structure of a VHDL model in the form of an interconnected set of components. Component declarations serve as templates for creating instances of components. Interconnection of the instances is achieved by generic map and port map associations where actual signals are associated with formal ports. The structural components are sockets into which a design entity (entity/architecture pair) can be plugged through configuration statements. Configuration statements also include a port map aspect that associates local parameters (parameter of a component declaration) with the parameters of the design entity.

3.4. HIERARCHY

Hierarchy is useful for splitting an initial, complex problem into simpler sub-problems that can be worked out separately to achieve a solution to the initial problem. The structure of electronic devices is often described in terms of interconnected subsystems. VHDL reflects hierarchy with the presence of structural statements that define a design as a set of interconnected components.

3.5. MODULARITY

For very large models, it is useful to split the whole code into many files that can be compiled separately. Furthermore, when many people are working on the same model, they must be able to work concurrently on different parts of the design and to have access to each other's work. VHDL offers five kinds of design units that can be compiled separately; that is, each design unit represents a compilation unit. A good modeling practice is to use one file to store a single design unit to avoid recompilation overheads, as will be explained later.

3.5.1. Design Units

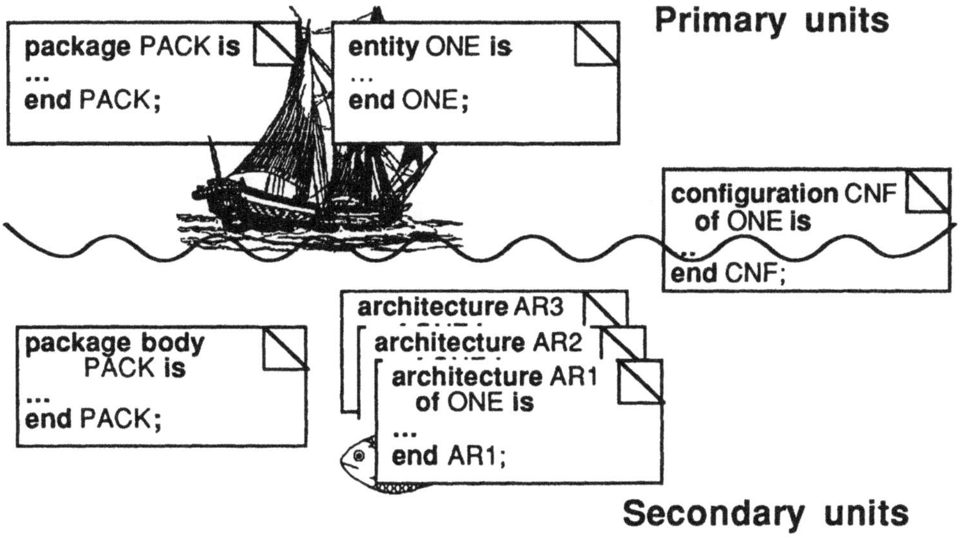

Fig 3.17 VHDL Library Units

Two design units are primary, namely, package declaration and entity declaration. Their purpose is to define the interface with the outside world.

The entity declaration defines the type and direction of information exchanged by a hardware design entity with its environment.

The package declaration specifies a set of items such as types, component declarations, and subprogram declarations that can be used in the description of other design units.

Two secondary units are available to express the implementation of the primary units.

Many different architectures can be designed for a given entity declaration.

One package body contains the code performed by the subprograms declared in the corresponding package declaration. If the package declaration does not contain any subprogram declaration or deferred constant, there is no need for a package body.

The last kind of design unit is a configuration declaration, which gathers all the entity/architecture mappings to the components of a given design entity.

An important advantage of modularity lies in the separate compilation. The design units are related by dependence links. A design unit A depends on design unit B in two possible cases:

- if A is a secondary unit corresponding to B (i.e., A is a package body implementing the package declaration A);
- if A makes use of items declared in B or in another design unit depending on B.

3.5.2. Compilation

Design units can be compiled in any order with respect to the dependence links. That is, a design unit A can be compiled only after all the design units on which it depends are compiled. An entity declaration will be compiled before the relevant architecture bodies. Similarly, the package declaration will be compiled before the package body.

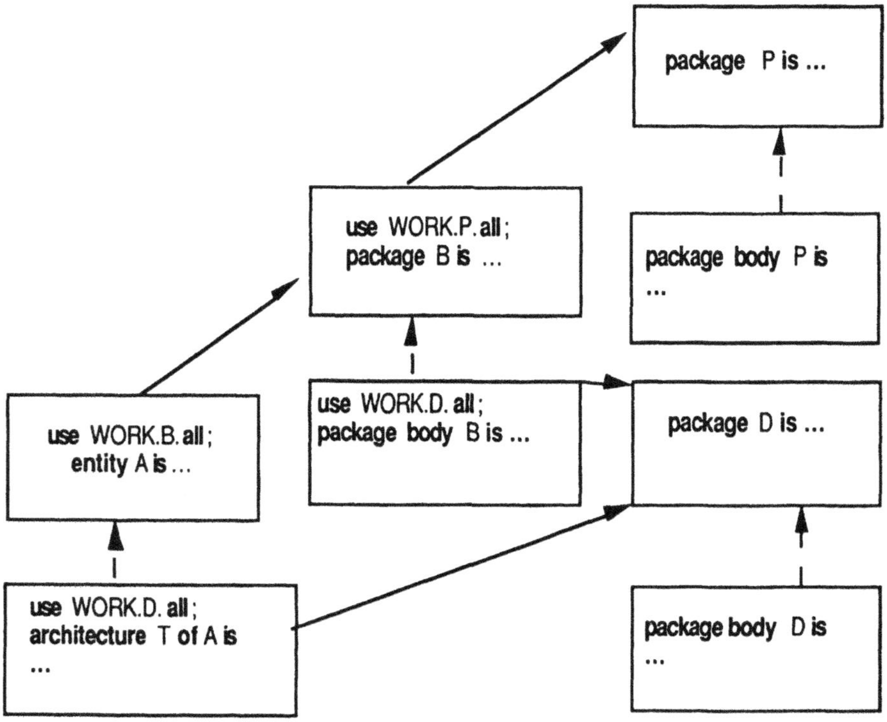

Fig 3.18 Dependence Links between Design Units of a Project

Consider the application described by figure 3.18. The arrows represent the dependence links between the library units. Thus, architecture body T depends on package declaration D and entity A. To compile this application, according to the dependence links, package declaration P and package declaration D must be compiled first, in any order, because they do not depend on any other design unit. Then, package declaration B, package body P, and package body D can be compiled in any order. Next, entity A and package body B can be compiled in any order. Last, architecture T of A can be compiled.

3.5.3. Application to Project Development

Notice that although package P is used within B, package declaration B does not depend on package body P. Once package declaration P has been compiled, package declaration B can be compiled and is not affected by the compilation of package body P. This enforces the complete separation between specification and implementation: the implementation of a given primary unit (here package P) does not affect the way the outside world uses it. This is very useful for a team working on the same project.

Once the design units are specified, the primary units can be compiled first, and then each member of the team can work on his or her portion without disturbing the work done by the other members. For example, our application may be developed by a team of four people, one for each package and the last for entity A and architecture T. The team must first define the package declarations and entity declaration that determine the dependence links between them. These dependence links are represented by arrows with continuous lines. Next, all the primary units are compiled: package declaration and entity declaration. Then each member of the team can independently develop his or her secondary unit. The specification phase is very important because of the recompilation overheads, as explained below.

When a design unit is recompiled, all the units depending on it become obsolete and thus must be recompiled. Recompilation of any secondary unit involves no recompilation of any other unit, because the dependence links are always oriented towards primary units; secondary units merely represent implementation of primary units. But recompiling a primary unit can cause many other units to be recompiled. In our example, recompiling package declaration P will cause package declaration B and package body P to become obsolete. Since entity declaration A and package body B depend on package declaration B, they will become obsolete. Last, architecture body T, which depends on entity declaration A, will become obsolete. This highlights the importance of the initial definition of design units and the trick of using one file per design unit to avoid such recompilation overheads. Once the primary units are determined and compiled, the secondary units can be developed concurrently with no recompilation effects on the other units being developed.

3.6. REUSABILITY

The complexity of designed electronic devices underlines the need to adopt effective methods to cut the design time. This need led to the development of design methodologies based on libraries of standard cells and macro-generators. The idea is to save design efforts by the reuse of components taken from libraries. For that purpose, library cells are designed with care to cover a wide range of needs to increase their usability. Similarly, VHDL includes many features that favor the reuse of previous code to model new designs. The concept of a library gathers a set of VHDL models and packages that can be used as resources in other VHDL descriptions.

3.6.1. Libraries

The VHDL LRM defines a library as "an implementation dependent storage facility for previously analyzed design units". A design unit may have access to any number of libraries.

Two libraries are implicitly declared for any design unit: one of them is the library with the logical name "WORK" that is supposed to receive the analyzed (and correct) design units; the other is the library STD, which contains the packages STANDARD and TEXTIO. The other libraries are resource libraries and must be explicitly declared by a library clause at the beginning of the design unit. The library clause follows the syntax

```
library LOGICAL_NAME_LIST ;
```

Here, LOGICAL_NAME_LIST is a list of identifiers separated by commas and representing the logical names of the resources libraries. At the beginning of all design units, the following clause is implicitly assumed:

```
library WORK, STD;
```

Library clauses make available all the resources contained in the libraries named in the clauses.

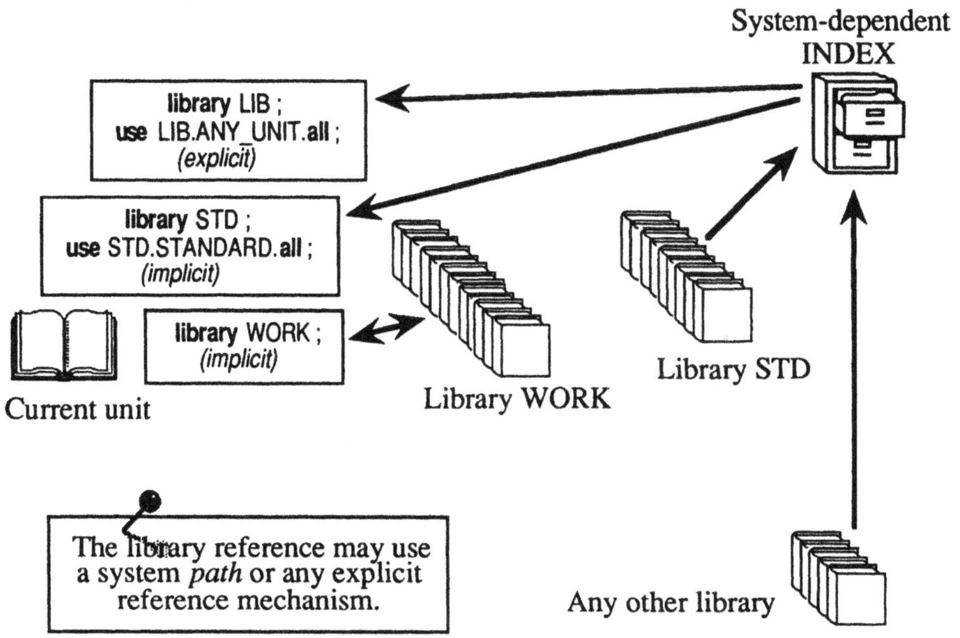

Fig 3.19 VHDL Libraries

Access is ensured by selection using the full name of the desired item. For example, if a type T is defined in package declaration P compiled in library L, this type can be referenced to declare a signal as follows:

```
library L;
entity E is
  signal S : L.P.T;
end E;
```

Use clauses can be used to shorten the references to the items, as illustrated by the following example:

```
library L ;
use L.all ;
entity E is
  signal S : P.T;
end E ;
```

The clause **use** L.**all** makes directly accessible all library units contained in library L. Furthermore, items declared in a package can be made directly accessible by another use clause:

```
library L;
use L.all ;
use P.all ;
entity E is
  signal S : T;
end E ;
```

Another form of the use clause is restricted to specific items of a library unit:

```
library L ;
use L.P.T;
entity E is
  signal S : T;
end E ;
```

The advantage of this last form is to control the items that are made directly accessible and thus to limit the possibility of conflict in case the same identifier denotes items defined in many distinct library units. In effect, if many items with the same identifier are made directly visible as the consequence of a use clause, they are hidden and can only be accessed by selection.

The design units that can be found in a library are of one of the following kinds:

- Entity declaration
- Architecture body
- Configuration declaration
- Package declaration
- Package body

The first three may be reused as hardware components of a given design hierarchy. The package declaration defines a set of software and hardware items that can be used to design any other model.

3.6.2. Design Entities and Configuration

An electronic device can be modeled as a set of interconnected components. In VHDL, this is achieved by the use of component instantiations. The characteristics of such an instance are specified by a component declaration, which is not at all a model of a hardware component. It is rather a socket that merely constrains the interface properties of chips that can be plugged into it.

```
entity ADDER is
        port (A,B,CIN : in BIT; COUT, SOUT : out BIT);
end ADDER;
```

```
architecture STRUCT of ADDER is
        component HALF_ADDER
            port(A,B : in BIT; S,COUT : out BIT);
        end component;
        for HALFADD1 : HALF_ADDER use entity WORK.HALF_ADDER(DATAFLOW) ;
        signal C,S : BIT;
begin
        HALFADD1 : HALF_ADDER port map (A,CIN,C,S) ;
        HALFADD2 : HALF_ADDER port map (B,C,COUT, SOUT) ;
end STRUCT;
```

This plugging in is done in VHDL by the configuration specification. Its effect is to associate a design entity that is an entity/architecture pair into a given component instance. Thus, a VHDL code describing a design entity is used to specify the functional characteristics of a new model.

```
for HALFADD1 : HALF_ADDER use entity WORK.HALF_ADDER(DATAFLOW) ;
```

For more complex architectures, VHDL proposes the concept of configuration declaration to collect all the binding information concerning a given design hierarchy in a library unit. This library unit can then be used to configure components of other models.

3.6.3. Packages

VHDL is a strongly typed language, which means that every object must be declared before it can be used. The declarations must state the set of values the object can hold and the operations that can be performed on it by specifying its type. To factor out the declaration requirement deriving from its strongly typed feature, VHDL borrows the concept of package from Ada. A package is a library unit (design unit) holding a collection of items that can be used in other design units. All the items specified in the package declaration are accessible to other design units either by selection or directly after an appropriate use clause. A package can include component declarations, signal declarations, disconnection specification, attribute declarations, type declarations, and subprograms. The package is thus a means of sharing portions of code that are used frequently. An example of this is the package STANDARD, which is predefined and available to all design units by the implicit clause

```
use STD.STANDARD.all ;
```

that is assumed at their beginning. Type BIT, BIT_VECTOR, and other constructs are defined in this package. Another example of a package is the package TEXTIO contained in library STD, which encapsulates all the necessary subprograms and declarations to process text files. Unlike the package STANDARD, it is not directly accessible to a design unit; a use clause must be inserted in a design unit to make it directly accessible. Other examples of packages will be given in the next chapter.

The most commonly used items of a package include types and subtypes, subprograms and component declarations. To achieve a greater usability, these items should be designed with care.

3.6.4. Types and Subtypes

3.6.4.1. Types

A type defines a set of values and a set of operations that can be performed on those values. For example, type integer represents the set of integers on which common arithmetic operations can be performed. Types are used in VHDL to constraint the values that signals, variables or constants may take. Every time an assignment is performed on a signal or variable, a check is done to ensure that the proposed new value satisfies the type requirements. Also all the expressions involving values and operations are checked for type compatibility. The rules underlying these checks can be summarized as follows:

- a value assigned to a signal or variable must belong to the type of that signal or variable;
- the operations allowed on values are determined by their types.

For example, the following two types are distinct:

```
type A is array (INTEGER range 10 to 20) of INTEGER ;
type B is array (INTEGER range 10 to 20) of INTEGER ;
```

Although the values they denote share the same structural properties, elements of type A cannot be assigned to objects of type B. Thus, given the declarations:

```
variable T: A ;
variable K : B ;
```

the assignment

```
T := K ;
```

is illegal because T and K have different types. Furthermore, operations like

```
T & K
```

are forbidden for the same reason.

3.6.4.2. Subtypes

The notion of subtype can help overcome this limitation. A subtype is derived from a type by some additional constraints. In the standard package, type BIT_VECTOR is defined as

```
type BIT_VECTOR is array(NATURAL range <>) of BIT ;
```

The subtype BIT10

```
subtype BIT10 is BIT_VECTOR(0 to 9) ;
```

defines a set of elements of type BIT_VECTOR but with subscript ranging from 0 to 10. All operations defined for BIT_VECTOR are allowed for BIT10 elements. Type BIT_VECTOR is an unconstrained type that can be used to derive many subtypes.

The advantage of subtypes over multiple type definitions is that they share the same basic set of operations. Given the declarations

```
subtype BIT10BIS is BIT_VECTOR (9 downto 0) ;
subtype BIT20 is BIT_VECTOR (0 to 19) ;
variable A10 : BIT10 ;
variable A10B : BIT10BIS ;
variable A20 : BIT20 ;
```

the following statements are legal in VHDL:

```
A20 := A10 & A10B ;
A10 := A20(10 to 19) ;
A10B := A10 ;
```

So to increase reusability, it is better to declare subtypes rather than types whenever possible.

3.6.5. Subprograms

Two features of VHDL can be exploited to increase the reusability of subprograms: unconstrained parameters and default formal parameter values. Also, a process can be replaced by a concurrent subprogram call.

3.6.5.1. Unconstrained Parameters

Packages may be used to hold subprograms that can be shared by many different design units. To increase their usability, such programs must try to cover a wide range of situations. Using unconstrained parameters and relying on predefined attributes to operate on these parameters is a clean way of achieving this, as shown in paragraph 7.2.4 of chapter 7. This is especially useful to express resolution functions which are required to have one unconstrained array parameter.

```
function WIRED_OR(ARG : BIT_VECTOR) return BIT is
begin
   for i in ARG'RANGE loop
      if ARG(I)='1' then
         return '1';
      end if ;
   end loop ;
   return '0' ;
end WIRED_OR ;
```

This example displays a function that may be used as a resolution function. Notice the use of the attribute 'RANGE, which returns the range of values of the index of the parameter subscript. The formal parameter is left unconstrained.

3.6.5.2. Default Parameters

Whenever a VHDL subprogram declaration includes a default value for an **in** mode parameter, this subprogram can be called without providing a value for the corresponding parameter. For example, consider the declaration

```
procedure READ ( I : out ITEM ; SIZE : NATURAL :=8 ) ;
```

A call such as

```
READ (TOTO) ;
```

is equivalent to

```
READ(TOTO,8);
```

So the default parameters in the subprogram add flexibility to subprogram calls.

3.6.5.3. Processes and Subprograms

VHDL process statements that are used very often can be replaced by concurrent procedure calls with signal parameters. This topic is addressed in paragraph 6.2.5 of chapter 6.

3.6.6. Generic Parameters

The generic parameters of entities and component declarations represent another way of writing models that can apply to a wide range of situations.

3.6.7. Association Lists

Association lists are used in two possible contexts: hardware and software.

In the hardware context, association lists specify port map clauses of component instantiation statements or configuration specification. In the software context, they are used to associate formal parameters of subprograms with actuals.

The rules of association elements add flexibility in the use of design entities and subprograms.

3.7. PORTABILITY

Portability is favored in VHDL by the clean definition of its semantics contained in the VHDL language reference manual [IEE87]. Any simulator fully conforming to this semantics should produce the same effects for a given valid VHDL model.

Due to the complexity of the VHDL type system, appropriate methodologies and datatypes must be agreed upon to allow different models emanating from many different environments to communicate without any loss of information.

Furthermore, VHDL allows the use of packages written in foreign languages, provided that the subprograms conform to a VHDL parameter passing mechanism. These packages represent a limit to the portability of VHDL descriptions.

Some marginal portability problems are pointed out in chapter 7 ("Tricks and Traps").

3.8. EFFICIENCY

The complexity of the models developed results in an increase of CPU time consumed to simulate them. This raises the issue of the efficiency of VHDL concepts used for modeling.

3.8.1. Signals and Variables

Another factor that may increase efficiency is to use variable rather than signals whenever possible. Each concurrent assignment to a given signal creates a driver that must be monitored by the simulator.

3.8.2. Using Signal Attributes

A set of attributes is defined for signals in VHDL. Some of these attributes are equivalent to functions and others are equivalent to signals. A close look at the meaning of these attributes reveals that some signal attributes have a function counterpart. Whenever possible, it is better to use attributes of function kind rather than attributes of signal kind.

3.8.3. Writing Efficient Process Statements

Within a given process statement, it is useful to insert as many wait statements as possible to make sure any activation of the process produces a useful action.

3.8.4. Process versus Concurrent Signal Assignment

In certain conditions, it may be better to replace a concurrent signal assignment statement with a process statement. The concurrent statement

```
S <= A and B when (not CLK'STABLE and CLK='1' and ENABLE='1') else S ;
```

is equivalent to the process statement

```
process
begin
  if (not CLK'STABLE and CLK='1' and ENABLE='1') then
      S<=A and B;
  else
      S <= S;
  end if ;
  wait on A, B, CLK, CLK'STABLE, ENABLE, S ;
end process;
```

The equivalent process is sensitive to any change on all the signals. This results in multiple and useless evaluations of the process. In fact, the statement could be replaced with a more efficient code, such as

```
process
begin
  wait until (not CLK'STABLE and CLK='1' and ENABLE='1');
  S<=A and B ;
end process ;
```

3.9. DOCUMENTATION

VHDL may be used to document a design. In this prospect, it is a man-to-tool as well as a man-to-man and tool-to-tool communication medium. For tool-to-tool communication, VHDL is used as a format. But VHDL is also a language with a semantics that must be shared by all the people in order to communicate. The distinction between format and language must be made clear.

3.9.1. Language or Format: Where is the Semantics?

Let us consider this clean, obvious description of an RS gate:

```
entity LATCH is
 port (R, S : in BIT; Q, QB : inout BIT);
end LATCH;

architecture STRUCT of LATCH is
 component NAND_CMP port (INPUT1,INPUT2 : in BIT; OUTPUT : out BIT);end component;
begin
 -- Q and QB are inout ports for the sake of simplicity (no intermediate signals)
 NAND1 : NAND_CMP port map (R, QB, Q);
 NAND2 : NAND_CMP port map (S, Q, QB);
end STRUCT;
```

If we simply change some identifiers, the whole understanding of the behavior is altered. However, the semantics is the same for the compiler: identifiers are user-defined, and no behavior is attached to them. We could as well configure the NOR_CMP with an entity holding the behavior of a NAND. This would be quite confusing for the reader, but very explicit for the simulator (neither erroneous nor ambiguous):

```
entity LATCH is
      port (R, S : in BIT; Q, QB : inout BIT);
end LATCH;
```

```
architecture STRUCT of LATCH is
        component NOR_CMP
            port (INPUT1,INPUT2 : in BIT; OUTPUT : out BIT);
        end component;
begin
        -- Q and QB are inout
        NOR1 : NOR_CMP port map (R, QB, Q);
        NOR2 : NOR_CMP port map (S, Q, QB);
end STRUCT;
```

Let us emphasize the fact that the semantics of this description is poor: the next two descriptions are the same (again), and it becomes obvious that the semantics is not tied to the strict LRM interpretation of the VHDL text. Note that we try to keep comments consistent with the style of the VHDL.

```
entity X001 is
 port (X002, X003 : in BIT; X004, X005 : inout BIT);
end X001;

architecture X006 of X001 is
 -- X004 and X005 are inout otherwise compilation fails
 component X007 port (X008, X009 : in BIT; X010 : out BIT);end component;
begin
 X011 : X007 port map (X002, X005, X004);
 X012 : X007 port map (X003, X004, X005);
end X006;

-- RS id # 01001000 - rev 3.2 -- code GJKbis.
entity O0O1 is port(O0l0,O0l1:in BIT;O1O0,O1O1:inout BIT);end O0O1;
architecture O1l0 of O0O1 is component O1l1 port(l0O0,l0O1:in BIT;l0l0:
out BIT);end component;begin l0l1:O1l1 port map(O0l0,O1O1,O1O0);l1O0:
O1l1 port map(O0l1,O1O0,O1O1);end O1l0;
```

Next we provide the same VHDL text, and identifiers have been turned back to comprehensive plain Bamileke. Oh, you don't read Bamileke?

```
entity KAMTE is
 port (KOYEM, KOYEPOUE : in BIT; TEM, QUETEM : inout BIT);
end KAMTE;

architecture NE of KAMTE is
 component SOMONEPOUBO port (KOYEM, KOYEPOUE: in BIT; TEM: out BIT);end component;
begin
 -- TEM pou QUETEM be inout ne gue a yin poupoun.
 SOMONEPOUYEM       : SOMONEPOUBO port map (KOYEM, QUETEM, TEM) ;
 SOMONEPOUYEPOUE  : SOMONEPOUBO port map (KOYEPOUE, TEM, QUETEM) ;
end NE;
```

Perhaps you think that the trick here is that we use a structural description, whose actual behavior comes from the configuration. Let us see this new architecture:

```
use WORK.SOME_BIT_PKG.all ;
entity LATCH is
        port (R, S : in SOME_BIT; Q, QB : inout SOME_BIT);
end LATCH;
architecture STRUCT of LATCH is
begin
        Q <= R nand QB ;
        QB <= S nand Q ;
end STRUCT;
```

"No way", you think. Well, let us look at this package:

```
use WORK.SOME_BIT_PKG.all ;
package PACK is
        function "nand" (I1, I2 : in SOME_BIT) return SOME_BIT ;
end PACK;
package body PACK is
        function "nand" (I1, I2 : in SOME_BIT) return SOME_BIT is
        begin
            if I1 = '1' or I2 = '1' then    -- notice the "or" function!
                return '0' ;
            else
                return '1';
            end if ;
        end "and";
end PACK;
```

If this package can be seen by some use clause, we have turned out nand-based latch into a nor-based one.

This shows that not only the correctness but also the actual semantics of a VHDL text depends strongly on its environment. In VHDL, it is particularly obvious at configuration-time (the meaning of a description is given afterwards), and with the use clauses (the meaning of a description is given elsewhere). It could be specified elsewhere and afterwards. It could be left unspecified (with or without default behaviors).

We can see in these examples the real difference between a language and a format: the format is (supposed to be) self-consistent, and communicates data between machines; the language communicates concepts between men or programs, who must share a common background.

As a consequence, it is a good idea to insert comments in VHDL texts, and to choose identifiers that will make sense to the people who will need to read the VHDL text.

3.9.2. Backannotation

Backannotation is a process through which timing information is gathered from physical designs of chips or hardware components to document their functional model. Software is often used to achieve this. The functional model can be described using user-defined attributes and a deferred constant package. Deferred constants are constants declared in package declaration but without a value. The package body defines their values. This facilitates backannotation, since recompiling a package body has no compilation overhead on other VHDL units using that package.

3.10. SYNTHESIS

A synthesis application (figure 3.20) is a process through which an abstract specification is interpreted in terms of an equivalent optimized hardware implementation satisfying the constraints accompanying the specification. The inputs to a synthesis system include the following:

- A functional specification of the desired chip. This may be viewed as a partial description of how the hardware is to be built and as such may fit a wide range of possible hardware solutions.
- The constraints (area, speed, power consumption, etc.) that will be used to guide the synthesis process. These provide the criteria that will govern the choice of an optimal solution.
- The library of components that will provide the basic hardware modules of the final design. This library must include a model of each hardware module describing its physical characteristics (timing, area, power, etc.). These characteristics are accessed during the synthesis process to choose the appropriate components.

The result of a synthesis process is often given as a netlist of components to be processed by back-end tools.

The emergence of VHDL as a worldwide standard hardware description language makes it appealing as a formalism to specify the inputs and outputs of synthesis applications. It would thus provide a mean to check the synthesis results for consistency with the specification through simulation.

When a VHDL text shows how a piece of hardware is built, it is called a VHDL description [SIG90]. A synthesis tool must implement it without any interpretation, and the description must provide all the necessary information details.

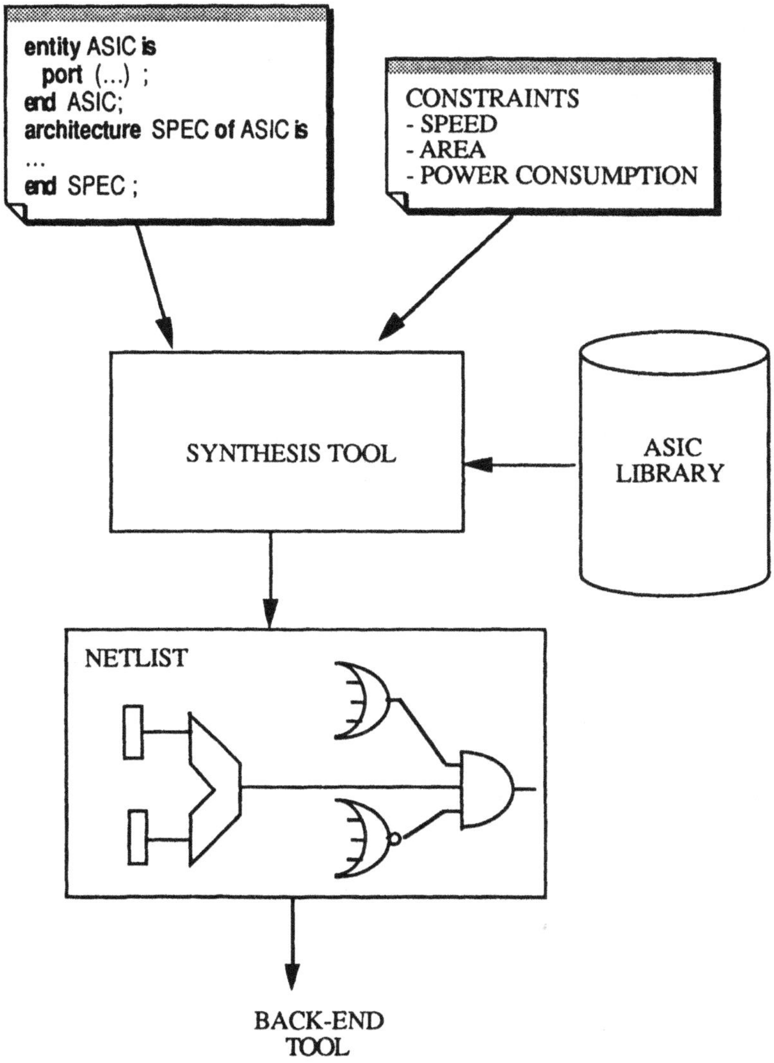

Fig 3.20 VHDL Synthesis Application

When VHDL is used to state the functionalities of a piece of hardware, it is called a VHDL specification. This is a partial description that must be interpreted to generate a hardware solution satisfying the constraints.

An input to a synthesis system can be viewed as a mix of description and specification. A synthesis methodology must propose a way to delimit the two subsets and to propose an interpretation to the features identified as elements of specification. But all the VHDL features do not fall into one of the two categories above, and some can be used in both of them. A methodology must define the conventions used to distinguish between the two uses of VHDL in a given VHDL representation. Furthermore, it must define the VHDL subset used for synthesis application.

This section introduces a VHDL subset that can be used as input for logic synthesis tools. The mappings between VHDL constructs and hardware that are given in this section are only suggestions; other solutions are possible and depend on the synthesis tool that is used, but the limitations defining the subset itself should be valid for all tools.

3.10.1. VHDL Datatypes for Logic Synthesis

As a high-level hardware description language, VHDL offers abstract datatypes similar to those encountered in high-level programming languages to model information data flowing in the electronic device being modeled. In a synthesis perspective, all these datatypes should be translated into a bit, word, or RAM format.

Thus, two-value enumeration types can be mapped to a bit format; other enumeration types as well as scalar types, can be mapped to words of appropriate size. A record can be viewed as a juxtaposition of data formats corresponding to subtype indications applying to the fields. Arrays can be mapped to RAM or registers depending on the types of elements.

Access types and file types can be discarded from the accepted VHDL datatype subset, as they are very difficult to handle. In fact, access types were added to VHDL to support system-level modeling, specifically to support abstract datatypes such as stacks, queues, and trees. File types were added in part to support communication with the environment, especially during simulation and debugging.

3.10.2. VHDL Expression Subset for Logic Synthesis

As far as logic synthesis is concerned, most of the VHDL expressions can be accepted. But two main restrictions should be observed:

All expressions involving datatypes excluded from the subset must not be supported: operations on files (read, write) or allocators ('new') on variables of access types are concerned.

Other expressions are closely related to simulation and therefore cannot be supported for synthesis, e.g., signal attributes like 'Transaction, 'Active, 'quiet.

VHDL expressions include a set of predefined operators that might be mapped to a set of logic operators provided by the target system. These logic operators might be implemented as hardware components or as "macros" to be expanded into a logic network. As a result of this mapping, some further restrictions may be set, depending upon the implementation issues.

Function calls may be mapped to a logic network.

3.10.3. Mapping of VHDL Sequential Statements into Logic Network

3.10.3.1. Introduction

A VHDL process statement is characterized by a set of signals read and/or driven by an algorithm. For logic synthesis purpose, the VHDL subset must be defined to restrict the algorithms of the VHDL process to descriptions equivalent to combinational or sequential logic. In this view, a naive synthesis program may transform a process statement into an equivalent logic network. Some processes will yield a combinational logic network and others a sequential logic network. Some signals will be latched, and others not. In the following, the sequential constructs of VHDL proposed for logic synthesis are listed and illustrated by some examples. The restrictions to full VHDL are also mentioned. The rationale behind these restrictions is to keep the synthesized logic network consistent with the simulation semantics of the corresponding VHDL process statement. Note also that the translations in terms of hardware proposed for the VHDL constructs are only examples: various alternatives may be considered according to the synthesis system. If possible, several equivalent implementations for the same construct will be proposed.

3.10.3.2. Signals

In the transformation of a process statement into a logic network, signals can be represented either by a wire or by a latch. As a general rule, a signal will be transformed into a latch in the following situations:

- The signal is driven by a process containing a "**wait until**" statement.
- The signal is driven by a process not containing a "**wait until**" statement, but the lifetime of values assigned to the signal might expand to many simulation cycles. Such situations arise when the signal is not updated at every execution of the process — for example if a signal is updated by some but not all of the alternative paths of a conditional branch (see example below). The signal should be updated within a conditional statement, where the relevant conditions will be used to control the latch.

Some of the attributes defined by VHDL for signals are not likely to be supported by synthesis systems because they reflect VHDL simulation concepts: 'QUIET, 'TRANSACTION, 'ACTIVE, 'LAST_ACTIVE.

```
entity ADD is
        port(A : in BOOLEAN ; B,C : in INTEGER ; S : out INTEGER) ;
end ADD;
architecture BEHAVIOR of ADD is
begin
        process(A,B,C)
        begin
            if A then
                S <= B+C;
            end if;
        end process;
end BEHAVIOR;
```

A possible implementation of the above VHDL description could be

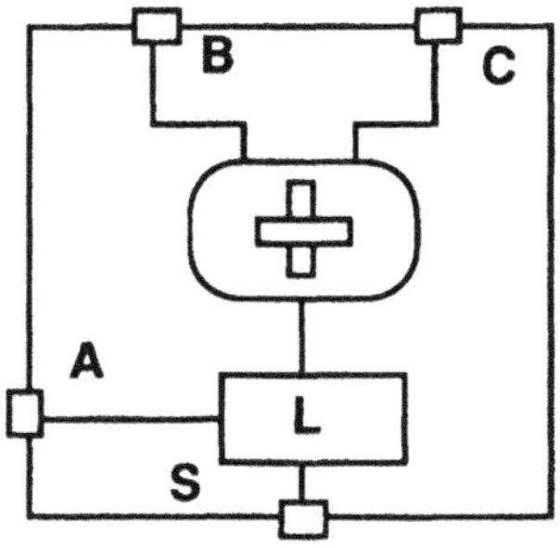

Fig 3.21 Synthesis View of Entity ADD

3.10.3.3. Variables

Variables are used to compute and/or keep intermediate results within a process. Some variables are optimized away during the elimination of common subexpressions. Variables used to keep track of information across process executions are mapped into latches or registers, and the others are transformed into wires. Variables are mapped into memory elements in the following circumstances:

- whenever a process defines paths through which a variable may be read before being updated, that variable must be latched;
- if the enclosing process contains a "**wait until**" statement, the memory element used to represent the variable will be synchronized with the condition of the "**wait until**" statement;
- in the case of processes with a sensitivity list (or with a "**wait on**" statement), the variable assignment statement should occur within a branch of a conditional statement. The relevant condition is used to control the memory element.

The number of latches required to store the VHDL variables may be optimized using so-called high-level synthesis techniques. The same techniques also apply to the communication between latches: various solutions ranging from one-to-one multiplex-based private connections to bus-based solutions may be considered.

3.10.3.4. Assignment Statements

Signal assignments are restricted to only one waveform element with no **after** clause. The general form of waveform in a signal assignment implies a tight (if not unrealistic) constraint that is difficult to implement with hardware. The consistency of such hardware with the simulation semantics of VHDL implies an infinite precision.

Sequences of sequential assignment statements may be mapped into a directed graph (DG). As the DG is built, common subexpressions may be pruned and constant expressions propagated. Static expressions, whenever possible, are evaluated and propagated. All these dataflow transformations simplify the resulting logic network while maintaining a consistency with the initial VHDL process statement.

Operations involved in the expressions are mapped into logic operators; signals and variables are mapped into wires or memory elements.

A cycle is produced in one of the following conditions:

- if a signal is both read and updated within the process;
- if the process defines at least one path through which a variable might be read prior to being updated;
- if a variable is used to store state information, i.e., any information whose lifetime spans many process invocations.

The signal or variable target of the assignment statement that caused the cycle to occur must be cast into a memory element. Due to possible critical racing problems raised by feedback loops in sequential circuits, the following restrictions apply to assignment statements:

- a signal must not be read before a "**wait until**" statement;
- signal assignment statements are not allowed before a "**wait until**" statement, because the value read from the signal according to the simulation semantics might deviate from the value produced by the corresponding logic network, as illustrated by the following example:

```
P1: process
begin
        A<=B or C;
        wait until CK;
end process ;
```

```
P2: process
begin
        wait until CK;
        A<=B or C;
end process ;
```

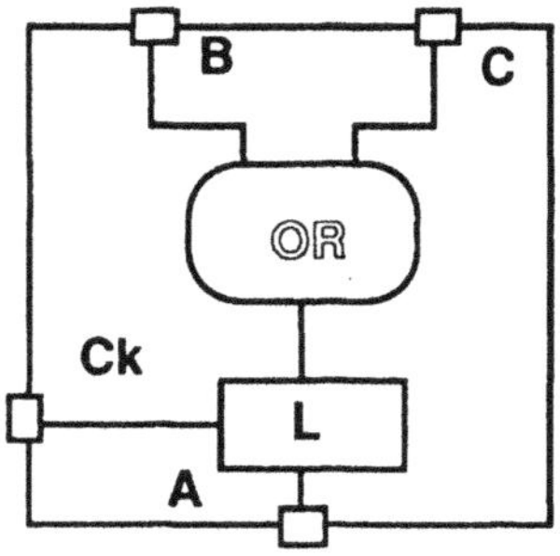

Fig 3.22 Synthesis View of P2

In the process P1, the value of 'B **or** C' is assigned to signal A before the wait statement is executed. In process P2, signal A is driven only when the condition of the wait statement evaluates to true. Process P2 reflects better the effects of the logic network. For process P1 to be consistent with the logic network, the initial value of 'B **or** C' should be latched in the register as an initial value. This value is not always easy to compute in the general case.

- If the enclosing process contains a "**wait until**" condition, then any variable cast into a memory element as a consequence of a cycle in the logic network shall be clocked by the condition of the wait statement.
- If the enclosing process does not contain a "**wait until**" statement, any assignment causing a loop in the logic network must occur within a conditional statement. In this case, the corresponding memory element will be governed by the condition causing the path containing the assignment to execute. Furthermore how would that initial value be latched in the register in a clean way?

Most of the restrictions stated earlier can be avoided by high-level synthesis, where operations are scheduled in distinct slices of time so as to prevent critical racing conditions from occurring.

3.10.3.5. Conditional Statements

Conditional statements comprise the **if** statements and the **case** statements. The DG keeps track of all the alternative values computed for signals or variables, along with the gating conditions (conditions in the IF or CASE statements). These values are connected to a multiplexer and the gating conditions used to trigger the multiplexer. Under certain circumstances, conditional statements are mapped into latches (see 3.10.3.2 and 3.10.3.3). Conditional statements may cause the multiplexer to be inserted as input to a device (wire or memory element) representing a variable or a signal. The size of the multiplexer for a particular variable or signal target depends on the set of branching paths of all of the conditional statements within which the target is assigned a value. In certain cases, multiplexers might not be generated, especially in two cases:

- if the target assigned by the conditional instruction is mapped into a register controlled by the condition of the statement;

```
P3 :  process (CK, B, C)
      begin
          if CK then
              S<=B and C ;
          end if ;
      end process ;
```

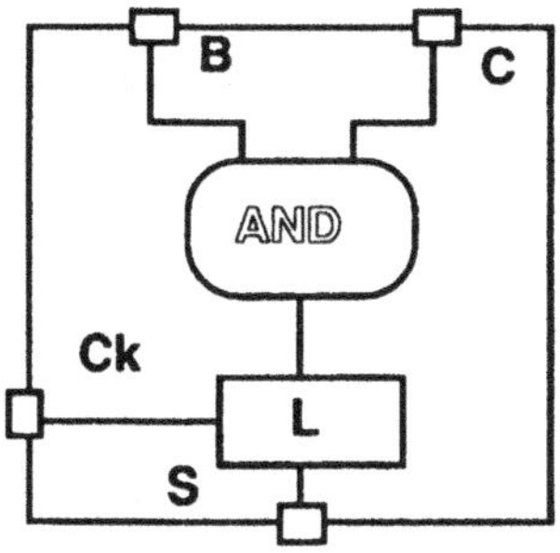

Fig 3.23 Synthesis View of P3

- if the condition of the conditional expression denotes a static expression whose value is known during the synthesis process. In this case, only the relevant path of the conditional statement is considered: there is no need for multiplexing.

3.10.3.6. Wait Statements

There are two forms of wait statements allowed:

- **wait on** <sensitivity list> ;

In a synthesis perspective, a process containing a "**wait on**" statement should be equivalent to a process with a sensitivity list. For that purpose, the wait statement must be the last statement of the process.

The "**wait on**" statement lists the signals to which the process is sensitive. Only these signals can possibly cause the process to update its output signals. Since a logic network is sensitive to all its inputs, it is mandatory that all the signals read by a process be listed in the list of the "**wait on**" statement.

- **wait until** <condition> ;

In a dataflow synthesis system, the "**wait until**" statement can be considered as a synchronization primitive, dictating the condition under which signals driven by the process acquire their new values. Thus, all of the signals driven within a process containing a "**wait until**" statement will be latched. For the corresponding logic network to perform in a fashion conforming to the VHDL simulation semantics, the **signal assignment** statements must occur **after** the wait statement. Furthermore, any reading from a signal must likewise occur after the "**wait until**" statement. In a dataflow synthesis system, only one wait statement is allowed within a given process: many wait statements would imply an FSM to sequence the wait statements according to the VHDL simulation semantics. In addition, the wait statement must not be embedded within any conditional or loop statement. In a high-level synthesis system, these restrictions can be removed.

After clauses are not allowed in wait statements, since the timing constraints implied are not generally easy to implement. In most of the existing synthesis systems, timing clauses are purely and simply discarded.

Because every signal driven within a process containing a "**wait until**" statement is mapped into a memory element, a good VHDL modeling practice for logic synthesis might be to refrain from driving many signals in a process containing a "**wait until**" statement.

```
entity COND1 is
        port(A,B,C: in BIT_VECTOR(1 to 5) ;
             CD,CK : BOOLEAN ;
             Z,P : out BIT_VECTOR(1 to 5)) ;
end COND1;

architecture BEHAVIOR of COND1 is
begin
  process
        variable TEMP,T2: BIT_VECTOR(1 to 5);
        constant N : INTEGER := 6;
  begin
        wait until CK=true;
        T2:=C;
        TEMP:= (A and B)nand C;
        if CD then
           TEMP:=A or B;
           T2:=A and B;
        end if;
```

```
        T2:=T2 xor C;
        Z<=TEMP or C;
        P<=T2 and C;
    end process;
end BEHAVIOR;
```

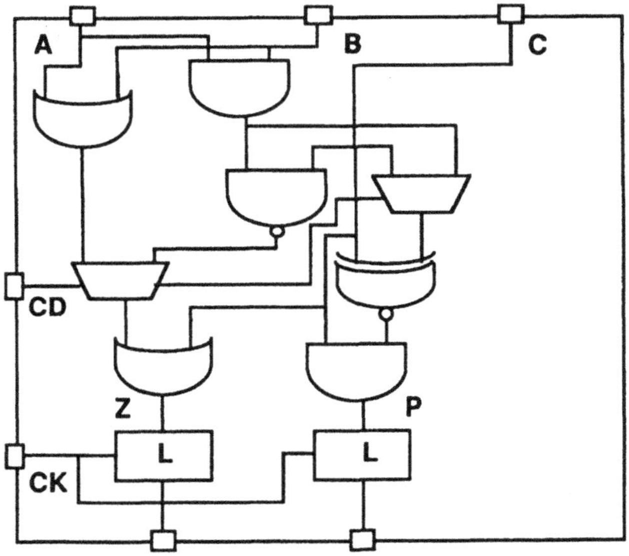

Fig 3.24 Synthesis View of Entity COND1

3.10.3.7. Subprogram Calls

Within a process, subprograms may be expanded. This solution is the simplest. The process obtained after expansion of all subprogram calls should respect the limitations defined for process statements. More complex solutions based on the definition of a stack or of communicating machines may be considered.

3.10.3.8. Loops

When handling **for** loops, at least two solutions may be considered by a dataflow synthesis system. In the first case, **for** loops will be unrolled; in the second case, an implementation based on a counterload device and some extra logic allows loops not to unroll and therefore saves hardware. In the first cases, the range of the loop subscript should be determined by static expressions. Furthermore, the process obtained after unrolling the loops should obey all the restrictions applying to VHDL processes accepted for synthesis purposes.

The other forms of loops are more difficult to handle in a dataflow synthesis process. However, they may be mapped into an FSM controlling a logic network. A high-level synthesis system can much more easily accept all forms of loops, including **next** and **exit** statements.

3.10.4. Synthesis View of VHDL Concurrent Statements

3.10.4.1. Process Statement

A VHDL process reads some signals and drives some other, possibly identical signals. A process statement may be mapped into a logic network. In VHDL, two kinds of process statements can be distinguished:

- Processes with a sensitivity list: the signals driven by a process with a sensitivity list may be updated only when an event occurs on at least one signal of the sensitivity list. On the other hand, a logic network is sensitive to any event occurring on its input signals. To maintain an equivalence between the simulation results and the effects achieved by the logic network, it is mandatory that any signal read by a process be listed in its sensitivity set.
- Processes with no sensitivity list: such processes must contain at least a wait statement. In fact, they must contain exactly one wait statement (see 3.10.3.6).

```
entity COND1 is
        port(A,B,C: in BIT_VECTOR(1 to 5) ;
             CD : in BOOLEAN ;
             Z,P : out BIT_VECTOR(1 to 5));
end COND1;
architecture BEHAVIOR of COND1 is
begin
   process(A,B,C)
        variable TEMP,T2: BIT_VECTOR(1 to 5);
        constant N : INTEGER := 6;
   begin
        T2:=C;
        TEMP:= (A and B)nand C;
        if CD then
            TEMP:=A or B;
            T2:=A and B;
        end if;
        T2:=T2 xor C;
        Z<=TEMP or C;
        P<=T2 and C;
   end process;
end BEHAVIOR;
```

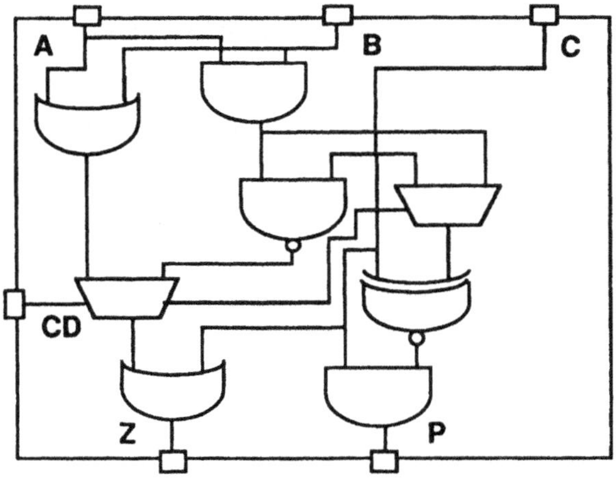

Fig 3.25 Synthesis View of Entity COND1

In VHDL, all the concurrent statements are equivalent to VHDL process statements. The mappings are therefore straightforward.

3.10.4.2. Concurrent Signal Assignment Statement

Concurrent signal assignments are equivalent to process statements containing corresponding sequential signal assignment statements. The logic network corresponding to a concurrent signal assignment statement may be derived from the equivalent process statement. The waveform element defining the right-hand side of the concurrent signal assignment statement should not contain any timing clause (see 3.10.3.4).

Consequently, a resolved signal assigned by a guarded signal assignment statement may be mapped into a register unless it is a guarded signal of the bus kind. In fact, the following VHDL descriptions are equivalent:

```
B1 : block(not CK'STABLE and CK='1')
begin
      S<=guarded B or C;
end block ;

B2 : block(not CK'STABLE and CK='1')
begin
      process(GUARD,B,C)
      begin
            if GUARD then S<=B or C ; end if ;
      end process;
end block ;
```

3.10.4.3. Component Instantiation Statement

Component instantiations are structural descriptions describing how parts of a given subsystem are interconnected. As such, they can be accepted by synthesis tools as a means provided to the designers to partially define the final solution: only the behavioral and dataflow parts of the initial description need to be synthesized; the structural description is viewed as the description of the remaining part of the final solution. However, if a component is configured into a design entity containing behavioral or dataflow descriptions, it can be expanded, and relevant behavioral or dataflow parts can be synthesized.

3.10.4.4. Generate Statement

Generate statements are means of describing highly regular structures. A synthesis tool must elaborate a given generate statement before synthesis is performed.

3.10.5. Open Issues

VHDL definition is mainly oriented towards simulation applications, and its use for synthesis raises many issues. These problems encompass all aspects of synthesis and module generation, including functional and timing aspects as well as technology-dependent issues.

3.10.5.1. Timing Constraints

A specification may state a timing constraint that is hard to express in VHDL in its actual form. These timing constraints are given in terms of min or max time that portions (loop, process, etc.) of the design may take to execute. User-defined attributes or assertion statements may be used to convey some of this control information.

3.10.5.2. Synchronous Design

VHDL does not propose a standard way to describe clocks. In fact they may be implemented by simple signals of type bit with the following statement:

```
CLOCK<= not CLOCK after CYCLE/2 ;
```

Here also, a user-defined attribute can be used to flag particular signals as clocks. But if clocks with multiple phases are needed, things become trickier.

3.10.5.3. Synthesis Validation

Synthesis is supposed to produce a design conforming to the input specification, at least with respect to I/O traces at the boundary of the design. But this conformity is to be defined and may require data abstraction and temporal abstraction. For combinational circuits, the synthesized solution must produce the same outputs as the specification delayed in time according to the propagation time associated with the hardware components.

For sequential circuits, things are more difficult. Transient states must be distinguished from stable states. Only the values corresponding to stable states must be checked for consistency. The specification must avoid over-specification to leave the door open for any consistent implementation. This must be addressed by appropriate methodologies.

A set of standard packages is currently specified by the VHDL Synthesis Interest Group that will address the problems raised by the use of VHDL in synthesis applications [SIG90].

3.11. CONCLUSION

This chapter addressed the use of VHDL in various fields. Issues such as abstraction, hierarchy, reusability, modularity, project management, and synthesis were covered, with emphasis on the relevant VHDL features. VHDL is shown to be quite a modern HDL, requiring methodologies inspired by software engineering practices. The use of VHDL in synthesis applications involves defining subsets with specific semantics and a set of conventions.

4. STRUCTURING THE ENVIRONMENT

4.1. CHOOSING A LOGIC SYSTEM

Digital integrated circuits compute binary information: in a stable state, each electrical node of a circuit may have one of two voltage values (0V or 5V in CMOS technologies) representing the two bit values 0 and 1. In hardware description languages, wires (signals) inside integrated circuits are thus usually (but not necessarily) modeled in a bit logic type.

4.1.1. Built-In Logic Type of Non-VHDL Simulators

Every IC designer is used to a given logic type. Before VHDL, this type was usually forced on the designer by the simulator and its hardware description language. The classical signal values that are handled by logic simulators are 0, 1, X (unknown), and Z (high impedance). All three hardware description languages that are compared with VHDL model signals with such a logic type:

- the M logic type has three levels (X, 0, and 1) and many strengths (initial, charged from 0 to 31, driven from 0 to 31, and supply);
- the Verilog logic type has three levels (X, 0, 1) and eight strengths (from high impedance to supply drive);
- the UDL/I logic type has four states (X, 0, 1, and Z).

In the first two languages, the high impedance can also be simply specified as a state (with character Z). Conflicts are resolved by the simulator with a built-in resolution function that is often based on strength comparisons or wired-X resolution. Other languages (and simulators) usually handle at least the four-level logic type (0, 1, X, and Z) with a wired-X resolution.

4.1.2. Modeling Logic Levels

4.1.2.1. The Predefined Type BIT

The designer who is used to such logic types is often very surprised (when not disappointed) to find in VHDL the unique predefined logic type:

```
type BIT is ('0', '1') ;
```

This type has neither unknown nor high-impedance states. Furthermore, it is not a resolved subtype, i.e., any signal that is declared with type BIT can only have one source (driver): an error occurs at compilation time if such a signal is connected to two gate outputs. Non-VHDL designers are more used to seeing X values appearing on such nodes. Thus, the common reaction of a new VHDL designer is to reject *a priori* the type BIT, which is found to be inadequate.

But nevertheless, this type BIT can be used in a wide range of models. For example, in a high-level model of a (synchronized) datapath, it is often useless and meaningless to propagate unknown values through combinational operators because these operators are described at a functional and not a structural level. Let us consider the following example of an adder:

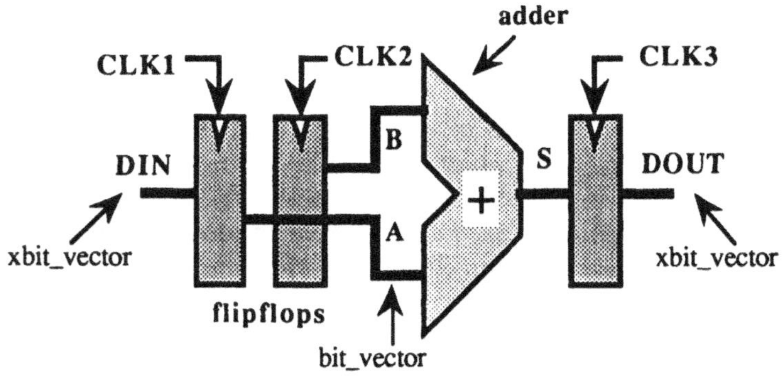

Fig 4.1 Use of BIT Type to Model Datapath

In this model, external signals DIN and DOUT are modeled with a user-defined type XBIT including an unknown state 'X' (type defined in 4.1.2.2), whereas all internal signals (A, B, and S) are modeled with the BIT type. Due to the safety of controller sequences, it may be assumed that datapath input registers should *never* latch unknown or unspecified values (otherwise it is a design error). Thus, instead of propagating unknown states inside the datapath, a conversion function from XBIT vectors to BIT vectors is performed inside the input registers. The resulting bit vectors may in turn be converted into integers to use the integer addition. Following is such a conversion function between XBIT and BIT types:

```
-- optimized BIT conversion function
function CONVERT_TO_BIT( S : XBIT) return BIT is
begin
        return BIT'VAL( XBIT'POS(S) - 1 ) ;
        -- type XBIT is ('X', '0', '1') ;
        -- an error is issued by the simulator when S = 'X' (BIT'VAL(-1) has no sense)
end CONVERT_TO_BIT ;
```

The designer may use a conversion function that includes an assertion to check the validity of input values:

```
-- conversion function using assertion
function CONVERT_TO_BIT( S : XBIT) return BIT is
begin
        case S is
         when '0' => return '0' ;
         when '1' => return '1' ;
         when 'X' => assert FALSE report "Unable to convert X to BIT" severity FAILURE ;
        end case ;
end CONVERT_TO_BIT ;
```

One of these conversion functions may then be used in the architecture of the registers:

```
use WORK.XBIT_PKG.all ;
entity REGISTER_ENTITY is
        port ( D : in XBIT ; CLK1 : in BIT ; Q : out BIT ) ;
end ;
architecture BEHAVIOR1 of REGISTER_ENTITY is
begin
        process
        begin
             wait until CLK1 = '1' ;
             Q <= CONVERT_TO_BIT(D) after 1 ns ;
             -- a simulation error or an assertion with severity FAILURE is raised when D is 'X'
        end process ;
end BEHAVIOR1 ;
```

After the adding operator, the output register performs a conversion from BIT vectors to XBIT vectors. In such an example, the model of the adding operator is assumed to be correct: the purpose of the simulation is more to check the controller behavior. In that case, any unknown state latched by one of the registers will cause the simulation to stop with an unambiguous message.

Debugging such a design is certainly faster than debugging a design that models the propagation of unknown values through the datapath.

4.1.2.2. The Unknown State: Modeling Lack of Information

Of course, it is often impossible to avoid the use of an unknown state. If the design includes conflicts between gate outputs that drive different logic values, it is an error if the effective value of such conflicts is set to '0' or '1': it should be specified as being *unknown* by the simulator (and the designer). Thus, any resolved subtype of type BIT cannot be used.

Declaring objects of type BIT may also hide some initialization problems: all signals of type BIT have as a default value the leftmost value of the enumeration type BIT, i.e., '0', which is not realistic. To avoid such troubles, the designer may want to specify that the signal state at the initialization is *unknown* to him.

It is important to notice that both cases do not request the specification of a *third* logic level, but instead introduce a state with the special meaning that any signal of this state has actually the level of one of the two (or more) other states, except that the simulator (and the designer) ignores which one. In this way, the unknown state models a *lack of information* on the real signal state.

In VHDL, the designer may define the following logic type:

```
type XBIT is ('X', '0', '1') ;
```

The character 'X' usually represents the unknown state. It is defined as the first enumeration value of the type, so that all the objects declared with type XBIT are initialized with value 'X'. Except for its use as a default value, the 'X' element has no other meaning due to the type declaration: it is just a character literal. Its meaning as an *unknown* state can only be specified by a user-defined resolution function and/or overloaded logic operators "and", "or", etc., which should be defined in the same package where XBIT is declared.

A typical resolution function on type XBIT is the following function:

```
type XBIT_VECTOR is array (NATURAL range <>) of XBIT ;

function WIRED_X (S : XBIT_VECTOR ) return XBIT is
        variable ONE, ZERO : BOOLEAN ;  -- := FALSE
begin
        for I in S'RANGE loop
            case S(I) is
             when '0' => if ONE then return 'X'; else ZERO := TRUE ; end if ;
             when '1' => if ZERO then return 'X'; else ONE := TRUE ; end if ;
             when 'X' => return 'X' ;
            end case ;
        end loop ;
end WIRED_X ;
```

This resolution function returns 'X' when both '0' and '1' are among the driving values or when at least one driving value is 'X': value 'X' now has a meaning as an *unknown state*. This meaning is somehow enforced by overloaded logical operator definitions (of course, the predefined logical operators are only defined on types BIT and BOOLEAN and cannot be used for XBIT):

```
function "and" (A, B : XBIT) return XBIT is
        type XBIT_TABLE is array (XBIT, XBIT) of XBIT ;
        constant TABLE : XBIT_TABLE := (('X','0','X'), ('0','0','0'), ('X','0','1')) ;
begin
        return TABLE(A,B) ;
end ;
```

The unknown state should be explicitly assigned to a given signal each time no supposition can be made on the value of this signal — not only before any explicit initialization or when the result of the resolution function cannot be decided, but also each time the control or input signals that are used to compute the value of the given signal are unknown.

Such an unknown state, although not predefined in VHDL, is actually used by every digital simulator.

4.1.2.3. The Uninitialized State

The previous 'X' state was modeling the lack of information on the real state of a signal at any time during the simulation (including the initialization step). It may be very useful for the designer to be able to differentiate the unknown state of a driven signal (one whose value has been evaluated at least once) and the state of an uninitialized signal (one whose value has never been evaluated during the simulation).

To achieve that purpose, it is necessary to introduce an uninitialized state in addition to the unknown state. For example:

```
type UXBIT is ('U', 'X', '0', '1') ;
```

Since it is defined as the leftmost enumeration value of type UXBIT, the 'U' state will be the default value of all signals declared with this type. Thus, if such a signal is never assigned during the simulation, it keeps the unknown state 'U'. The previous unknown state 'X' can only appear on a signal with an assignment. The designer is thus able to locate easily all the initialization problems and to identify *unused* nodes in the design. Unused signals often result from design mistakes and waste the memory space of the simulator.

4.1.2.4. The Don't-Care State: Synthesis Request

Somewhat close to the unknown state but used in a different context, the *don't-care* state is often introduced by synthesis or specification tools. It provides a means to specify that the designer literally does not care about the value of some inputs to a block: the output value(s) can be specified or computed without knowing the values of these entries.

In Verilog and in UDL/I, the character '?' is used as *don't-care* in the user-defined primitive syntax (see chapters 9 and 10). This symbol actually means: *for any input value*.... This meaning corresponds to a synthesis or specification semantics: the character '?' is *not* a simulation state and cannot be used in signal assignments or expressions. In other languages, the simulation semantics of such element (if defined) is usually the same as the semantic of any unknown state.

VHDL does not provide a predefined *don't-care* value. An example of its implementation in VHDL has been given in chapter 9 ("Verilog and VHDL").

4.1.3. Modeling Disconnection

4.1.3.1. The High-Impedance State

Besides the unknown state(s), another commonly requested and used state is the high-impedance state. This state is typically used to model tristate signals, such as buses in microprocessors: the high-impedance state is the state of the bus after the last tristate buffer (driver) disconnection.

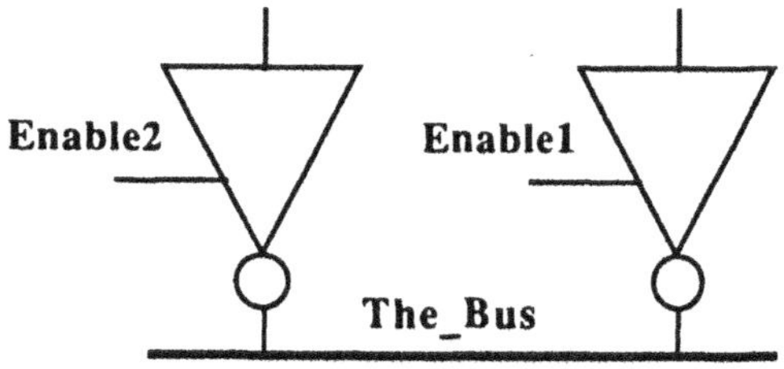

Fig 4.2 Modeling Tristate Buses

The high-impedance state is often designated as the 'Z' state. Thus, the designer who needs to model tristate buses could define the following type:

```
type ZBIT is ('X', '0', '1', 'Z') ;
```

As illustrated by the bus model, such a state is really meaningful only if the signal connects different gate outputs. Thus, tristate signals are usually declared with a resolved subtype (a resolution function is compulsory if the signal is assigned by more than one driver). A typical resolution function is the following:

```
type ZBIT_VECTOR is array ( NATURAL range <>) of ZBIT ;

function TRISTATE (S : ZBIT_VECTOR ) return ZBIT is
        type TABLE is array (ZBIT,ZBIT) of ZBIT ;
        constant RESOLVE : TABLE := (('X','X','X','X'), ('X','0','X','0'), ('X','X','1','1'), ('X','0','1','Z')) ;
        variable RESULT : ZBIT := 'Z' ; -- value returned when function is called without argument
begin
        for I in S'RANGE loop
            RESULT := RESOLVE(RESULT, S(I)) ;
            exit when RESULT = 'X' ; -- optimization
        end loop ;
        return RESULT ;
end TRISTATE ;
```

The result of this function is straightforward:

- all input (driving) values equal to 'Z' are ignored; if all driving values are equal to 'Z', then the result (the effective value) is 'Z';
- if all remaining driving values are equal to the same value ('0' or '1'), then the effective value is that common value;
- if one driving value is 'X' or if two driving values are equal to '0' and '1', then the effective value is 'X'.

Once again, the meaning of the enumeration value 'Z' is given by a resolution function. Many non-VHDL simulators use such a state to model tristate buses.

In VHDL, the high-impedance state can be introduced when the design includes a tristate bus or needs more generally to model driver disconnections, but it is useless when all signals are always driven by a gate output (in combinational blocks such as datapaths).

4.1.3.2. Use of VHDL Guarded Blocks

Another way to model tristate buses in VHDL is to use guarded signals of the **bus** kind. These signals are the targets of guarded assignments located inside guarded blocks. With such blocks, the physical disconnection of a tristate buffer from a bus is modeled by the VHDL mechanism of driver disconnection.

```
signal THE_BUS : TRISTATE ZBIT bus ; -- guarded signal of kind bus
...
ONE_TRISTATE_BUFFER : block (ENABLE = '1') -- guarded block
begin
        THE_BUS <= guarded ANY_SIGNAL after 2 ns ;
        -- guarded assignment
        -- when ENABLE /= '1', driver corresponding to this assignment is disconnected
end block;
```

Nevertheless, a resolution function must be defined for guarded signals. As the signal modeling the bus (THE_BUS) is declared to be of the **bus** kind, its state after the last driver disconnection is the value returned by the resolution function called without argument. In the previous model, the resolution function used was the TRISTATE function; thus the returned value was 'Z'. But the use of this value is not compulsory: the value could have been 'X' or '0' or '1'. A special state modeling the high impedance is not necessary when the bus is modeled by a guarded signal: VHDL driver disconnections model the bus disconnections.

4.1.4. Modeling Levels and Strengths

Actually, besides the tristate disconnection, the high-impedance value defined above also introduced the notion of *strength*. Indeed, the idea implied by the bus model is that a resistive level (i.e., the bus level when all tristate buffers are disconnected) is *weaker* than a driven level (assigned by a connected tristate buffer).

This notion of strength is almost compulsory to model devices at low levels of description (gate, switch). It is mainly used to resolve conflicts.

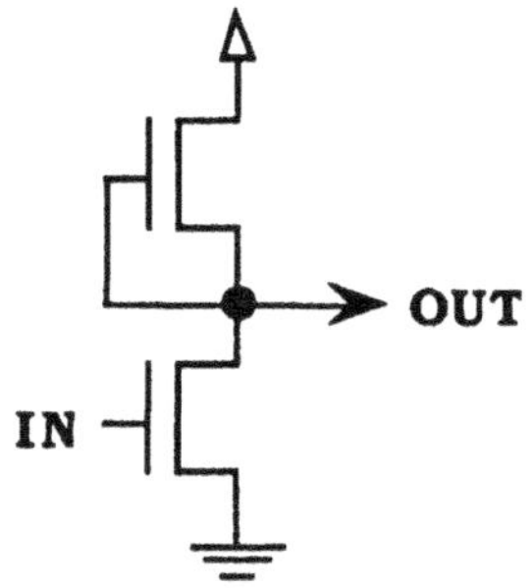

Fig 4.3 NMOS Inverter

In the example of the NMOS inverter of figure 4.3, a ZERO input will produce a *resistive* ONE, whereas a ONE input will produce a *driven* ZERO. Making the distinction between resistive and driven states is necessary to avoid the UNKNOWN state: a conflict between a resistive ONE and a driven ZERO produces a driven ZERO (and not the UNKNOWN value).

Some languages only handle three level values (X, 0, and 1), but specify for each level a range of strengths: HIGH IMPEDANCE, RESISTIVE, DRIVEN, and SUPPLY are examples of possible strength values (high impedance is indeed more a strength than a level). With such a logic type, each state is defined by a couple level-strength.

In VHDL, it is also possible to describe a logic type implementing the two notions of level and strength:

```
type LEVEL is (UNKNOWN, LOGIC0, LOGIC1) ;
type STRENGTH is (HIGH_IMPEDANCE, RESISTIVE, DRIVEN, SUPPLY) ;

type STR_BIT1 is record
        L : LEVEL ;
        S : STRENGTH ;
end record ;
-- number of states = 12
```

Another logic type using levels and strengths is the following:

```
type STR_BIT2 is ( 'X', '0', '1', 'W', 'L', 'H', 'Z' ) ;
```

The meaning that can be given to this enumeration type is the following: the first three states ('X', '0', '1') are defined, respectively, with logic levels unknown, logic 0, and logic 1 and all with a *strong* strength; the three next states ('W', 'L', 'H') are also defined with logic levels unknown, logic 0, and logic 1 but with a *weak* strength; and the last state ('Z') has no specific level but has the *weakest* strength. Such a logic system is often represented by the following diagram:

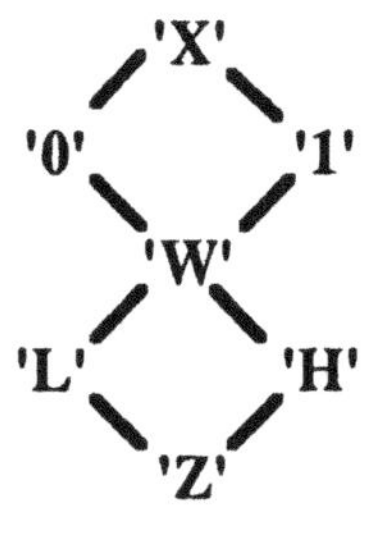

Fig 4.4 7-State Logic System

In both cases, a resolution function can be defined that is based on strength comparisons: a conflict between a weak (resistive) logic 0 and a strong (driven) logic 1 will result in a weak or strong logic 1.

Although both type definitions are correct, it is recommended to use the second kind of declaration, i.e., an enumeration type declaration. Indeed, digital simulators are more efficient with enumeration (scalar) logic types than with composite logic types: using enumeration types optimizes the simulation speed.

4.1.5. The 9-State STD_LOGIC_1164 Package

Since the standardization of VHDL by the IEEE in 1987, the VDEG (VHDL Design Exchange Group), later merged in the IEEE Model Standards Group, has been working on a standard VHDL package defining a multi-value logic system. The purpose of this logic system is to describe the interconnection datatypes used in modeling common TTL, CMOS, GaAs, NMOS, PMOS, and ECL digital devices (switched transistors are not addressed).

The motivation of this work was to define a multi-value logic system that could be supported by most of the VHDL simulator and synthesis tools. The use of a standard package to introduce this logic system avoids any language extension. On the other hand, tool builders can optimize the implementation of the package body if they want to support hardware accelerators or internal optimized types.

The STD_LOGIC_1164 package is just about to be submitted to the balloting vote (Spring 1992). Here is the version v4.2 [BIL91] of the package declaration (comments have been shortened):

```
package STD_LOGIC_1164 is
--------- Logic state system (unresolved) ------------------------------------------------
type STD_ULOGIC is ( 'U',  -- Uninitialized
                     'X',  -- Forcing Unknown
                     '0',  -- Forcing 0
                     '1',  -- Forcing 1
                     'Z',  -- High Impedance
                     'W',  -- Weak Unknown
                     'L',  -- Weak 0
                     'H',  -- Weak 1
                     '-'   -- Don't care     ) ;
--------- Unconstrained array of STD_ULOGIC for use with the resolution function ---------
type STD_ULOGIC_VECTOR is array (NATURAL range <>) of STD_ULOGIC ;
-------- Resolution function -------------------------------------------------------------
function RESOLVED( S : STD_ULOGIC_VECTOR) return STD_ULOGIC ;
--------- *** Industry standard logic type *** -------------------------------------------
subtype STD_LOGIC is RESOLVED STD_ULOGIC ;
--------- Unconstrained array of STD_LOGIC for use in declaring signal arrays ------------
type STD_LOGIC_VECTOR is array (NATURAL range <>) of STD_LOGIC ;
-------- Common subtypes ----------------------------------------------------------------
subtype X01    is RESOLVED STD_ULOGIC range 'X' to '1' ; -- ('X', '0', '1')
subtype X01Z   is RESOLVED STD_ULOGIC range 'X' to 'Z' ; -- ('X', '0', '1', 'Z')
subtype UX01   is RESOLVED STD_ULOGIC range 'U' to '1' ; -- ('U', 'X', '0', '1')
subtype UX01Z  is RESOLVED STD_ULOGIC range 'U' to 'Z' ; -- ('U', 'X', '0', '1', 'Z')
```

```
--------- Overloaded logical operators ------------------------------------------
function "and"  ( L : STD_ULOGIC ; R : STD_ULOGIC ) return  UX01 ;
function "nand" ( L : STD_ULOGIC ; R : STD_ULOGIC ) return  UX01 ;
function "or"   ( L : STD_ULOGIC ; R : STD_ULOGIC ) return  UX01 ;
function "nor"  ( L : STD_ULOGIC ; R : STD_ULOGIC ) return  UX01 ;
function "xor"  ( L : STD_ULOGIC ; R : STD_ULOGIC ) return  UX01 ;
function "not"  ( L : STD_ULOGIC                  ) return  UX01 ;
--------- Vectorized overloaded logical operators -------------------------------
function "and"  ( L, R : STD_LOGIC_VECTOR  )      return  STD_LOGIC_VECTOR  ;
function "and"  ( L, R : STD_ULOGIC_VECTOR )      return  STD_ULOGIC_VECTOR ;
function "nand" ( L, R : STD_LOGIC_VECTOR  )      return  STD_LOGIC_VECTOR  ;
function "nand" ( L, R : STD_ULOGIC_VECTOR )      return  STD_ULOGIC_VECTOR ;
function "or"   ( L, R : STD_LOGIC_VECTOR  )      return  STD_LOGIC_VECTOR  ;
function "or"   ( L, R : STD_ULOGIC_VECTOR )      return  STD_ULOGIC_VECTOR ;
function "nor"  ( L, R : STD_LOGIC_VECTOR  )      return  STD_LOGIC_VECTOR  ;
function "nor"  ( L, R : STD_ULOGIC_VECTOR )      return  STD_ULOGIC_VECTOR ;
function "xor"  ( L, R : STD_LOGIC_VECTOR  )      return  STD_LOGIC_VECTOR  ;
function "xor"  ( L, R : STD_ULOGIC_VECTOR )      return  STD_ULOGIC_VECTOR ;
function "not"  ( L    : STD_LOGIC_VECTOR  )      return  STD_LOGIC_VECTOR  ;
function "not"  ( L    : STD_ULOGIC_VECTOR )      return  STD_ULOGIC_VECTOR ;
--------- Conversion functions --------------------------------------------------
function TO_BIT ( S : STD_ULOGIC ; XMAP : BIT :='0' ) return  BIT ;
function TO_BITVECTOR ( S : STD_LOGIC_VECTOR ; XMAP : BIT :='0' )
                                                  return  BIT_VECTOR ;
function TO_BITVECTOR ( S : STD_ULOGIC_VECTOR ; XMAP : BIT :='0' )
                                                  return  BIT_VECTOR ;
function TO_STDULOGIC  ( B : BIT )                return  STD_ULOGIC ;
function TO_STDLOGICVECTOR ( B : BIT_VECTOR )     return  STD_LOGIC_VECTOR ;
function TO_STDLOGICVECTOR ( S : STD_ULOGIC_VECTOR )
                                                  return  STD_LOGIC_VECTOR ;
function TO_STDULOGICVECTOR (B : BIT_VECTOR)      return  STD_ULOGIC_VECTOR ;
function TO_STDULOGICVECTOR ( S : STD_LOGIC_VECTOR )
                                                  return  STD_ULOGIC_VECTOR ;
--------- Strength strippers and type convertors --------------------------------
function TO_X01   ( S : STD_LOGIC_VECTOR  )       return  STD_LOGIC_VECTOR ;
function TO_X01   ( S : STD_ULOGIC_VECTOR )       return  STD_ULOGIC_VECTOR ;
function TO_X01   ( S : STD_ULOGIC        )       return  X01 ;
function TO_X01   ( B : BIT_VECTOR        )       return  STD_LOGIC_VECTOR ;
function TO_X01   ( B : BIT_VECTOR        )       return  STD_ULOGIC_VECTOR ;
function TO_X01   ( B : BIT               )       return  X01 ;
---------------------------------------------------------------------------------
function TO_X01Z  ( S : STD_LOGIC_VECTOR  )       return  STD_LOGIC_VECTOR ;
function TO_X01Z  ( S : STD_ULOGIC_VECTOR )       return  STD_ULOGIC_VECTOR ;
function TO_X01Z  ( S : STD_ULOGIC        )       return  X01Z ;
function TO_X01Z  ( B : BIT_VECTOR        )       return  STD_LOGIC_VECTOR ;
function TO_X01Z  ( B : BIT_VECTOR        )       return  STD_ULOGIC_VECTOR ;
function TO_X01Z  ( B : BIT               )       return  X01Z ;
---------------------------------------------------------------------------------
function TO_UX01  ( S : STD_LOGIC_VECTOR  )       return  STD_LOGIC_VECTOR ;
function TO_UX01  ( S : STD_ULOGIC_VECTOR )       return  STD_ULOGIC_VECTOR ;
function TO_UX01  ( S : STD_ULOGIC        )       return  UX01 ;
```

```
function TO_UX01  ( B : BIT_VECTOR           )     return  STD_LOGIC_VECTOR ;
function TO_UX01  ( B : BIT_VECTOR           )     return  STD_ULOGIC_VECTOR ;
function TO_UX01  ( B : BIT                  )     return  UX01 ;
-------- Edge detection -----------------------------------------------------------------
function RISING_EDGE  ( signal S : STD_ULOGIC )    return  BOOLEAN ;
function FALLING_EDGE ( signal S : STD_ULOGIC )    return  BOOLEAN ;
-------- Object contains an unknown -----------------------------------------------------
function IS_X   ( S : STD_ULOGIC_VECTOR     )     return  BOOLEAN ;
function IS_X   ( S : STD_LOGIC_VECTOR      )     return  BOOLEAN ;
function IS_X   ( S : STD_ULOGIC            )     return  BOOLEAN ;
-----------------------------------------------------------------------------------------
end STD_LOGIC_1164 ;
```

The package body is defined as well, but it has not been added here. Of course, both package declaration and package body are in the public domain. Simulator and synthesis vendors are strongly advised to support this logic system package without modifying any type or function declaration. Actually, this package is already provided by many CAD vendors.

The enumeration type defined in this package has nine values: the uninitialized state 'U' (see 4.1.2.3), the strong unknown state 'X' (4.1.2.2), the strong logic levels '0' and '1' (4.1.2.1), the high impedance 'Z' (4.1.3.1), the weak unknown 'W' and weak logic levels 'L' and 'H' (4.1.4) and the don't-care '-' (4.1.2.4).

The package also defines a resolution function RESOLVED. This function uses the following two-entries table declared in the package body:

```
constant RESOLUTION_TABLE : STDLOGIC_TABLE := (
    -   U    X    0    1    Z    W    L    H    -          -    -
    ------------------------------------------------------
    (  'U', 'U', 'U', 'U', 'U', 'U', 'U', 'U', 'U' ), --  U  -
    (  'U', 'X', 'X', 'X', 'X', 'X', 'X', 'X', 'X' ), --  X  -
    (  'U', 'X', '0', 'X', '0', '0', '0', '0', 'X' ), --  0  -
    (  'U', 'X', 'X', '1', '1', '1', '1', '1', 'X' ), --  1  -
    (  'U', 'X', '0', '1', 'Z', 'W', 'L', 'H', 'X' ), --  Z  -
    (  'U', 'X', '0', '1', 'W', 'W', 'W', 'W', 'X' ), --  W  -
    (  'U', 'X', '0', '1', 'L', 'W', 'L', 'W', 'X' ), --  L  -
    (  'U', 'X', '0', '1', 'H', 'W', 'W', 'H', 'X' ), --  H  -
    (  'U', 'X', 'X', 'X', 'X', 'X', 'X', 'X', 'X' )  --  -  -
);
```

The reader may notice that the *don't-care* state has a simulation semantics (defined by the resolution function) that is very close to the simulation semantics of the high impedance state ('Z'). A second point is that the uninitialized state ('U') always wins the resolution.

Overloaded logic operators have the *natural* semantics. Many conversion functions are also provided ; some of them include assertions checking the validity of the current conversion (they can be used in the fashion presented in 4.1.2.1). Finally, classical edge-detection and testing functions are defined.

Most of the logic state values and/or functions that are implicit in other hardware description languages are predefined by this VHDL package. At this level of abstraction, the use of the standard package STD_LOGIC_1164 is thus strongly recommended in order to provide model interoperability on most of the simulator and synthesis platforms.

4.1.6. The 46-State Logic Value System

One other famous logic value system (although not standardized) is the 46-state logic system based on intervals between logic values. This logic system is presented in great details in [COE89].

The two main purposes of this logic system are to support switch-level simulation (which is out of the scope of the STD_LOGIC_1164 system) and to offer technology independence. The idea was also to define a logic type that optimizes the simulation performances and reduces the pessimism of the simulation results.

This logic system is based on two logic levels (0 and 1) and five strengths:

- Forcing strength (F) — direct connection to the power supply
- Strong resistive (R) — resistive connection to the power supply
- Weak resistive (W) — high resistive connection to the power supply
- High impedance (Z) — isolated node with capacitive charge
- Disconnect (D) — isolated node without capacitive charge (no level)

Thus, nine primary states are defined: F0, R0, W0, Z0, D, Z1, W1, R1, F1. Then the 46-state values are generated by taking each possible contiguous range or interval in these nine values. Figure 4.5 defines all these 46 logic values used in the logic system. The VHDL type below enumerates these states:

```
type LOGIC_46 is (
        U,  -- uninitialized value (unknown due to non-computation)
        D,  -- disconnect value (neither 0 nor 1)
        Z0, Z1, ZDX, DZX, ZX,
        -- states of strength Z
        W0, W1, WZ0, WZ1, WDX, DWX, WZX, ZWX, WX,
        -- states of strength W
        R0, R1, RW0, RW1, RZ0, RZ1, RDX, DRX, RZX, ZRX, RWX, WRX, RX,
        -- states of strength R
        F0, F1, FR0, FR1, FW0, FW1, FZ0, FZ1, FDX, DFX, FZX, ZFX, FWX, WFX, FRX, RFX, FX
        -- states of strength F
  );
```

Depending on the technology used, a given subset can be extracted from this 46-state type. For applications that are independent from the technology, only the five classical states can be used: uninitialized state U, forced unknown FX, forced zero F0, forced one F1, and high-impedance unknown ZX.

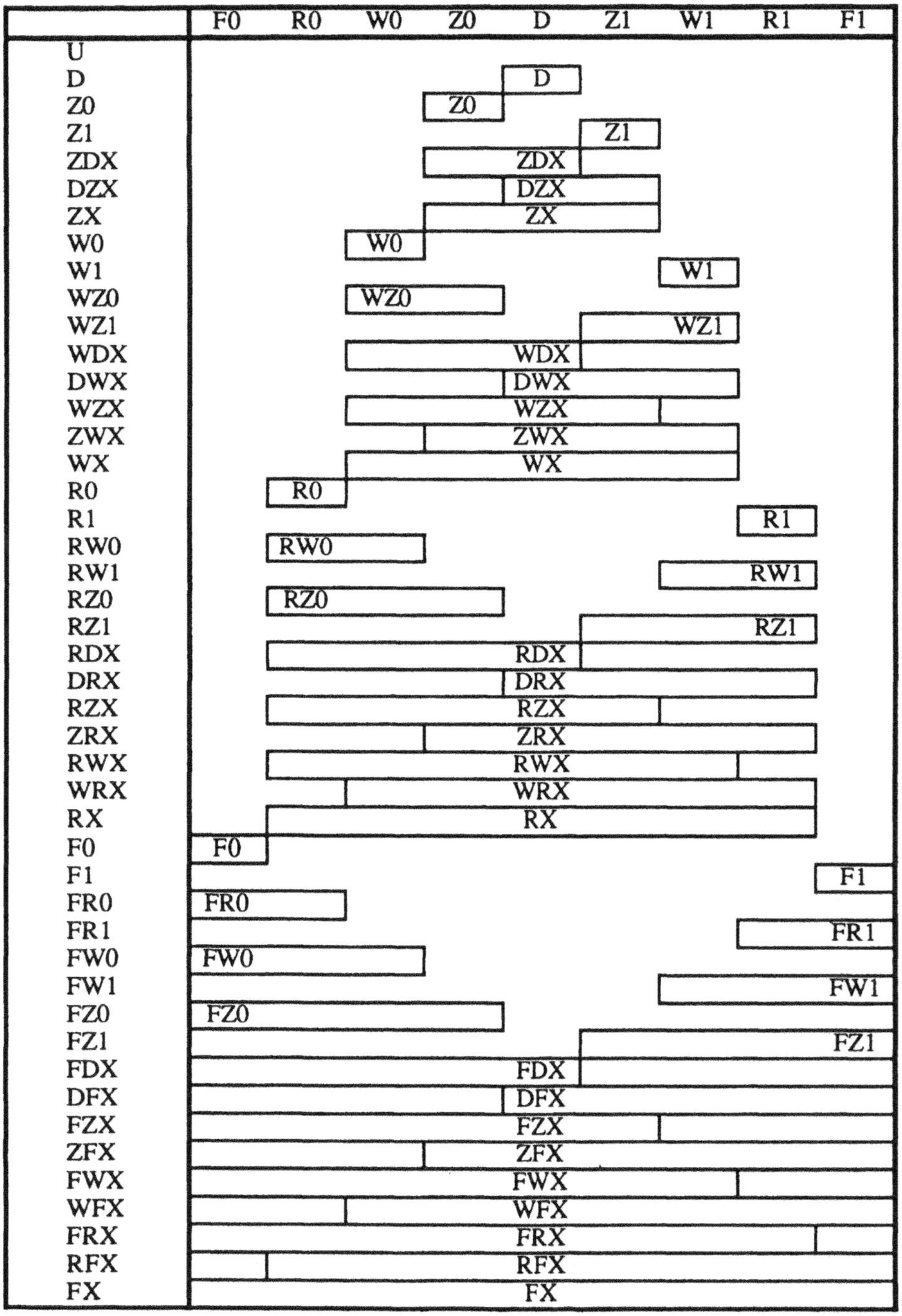

Fig 4.5 46-State Logic System

The main advantage of this 46-state type is that it provides more accuracy and less pessimism in the simulation results than the simple type with only one unknown value. Such a type is particularly well adapted for switch-level modeling, although VHDL is not very efficient at this level of abstraction.

4.1.7. Conclusion

The choice of a logic system strongly depends on the application. As a general rule, it is always better to use the simplest enumeration type that is sufficient for the application. Even the type BIT can be used in a wide range of models, especially at a functional level of modeling (which is also independent from the technology). A high-impedance state should be used when the model includes tristate buses. When it becomes necessary to introduce strengths (to distinguish a weak '1' from a strong '1'), then the package STD_LOGIC_1164 should be used: this is definitely a good investment for interoperability.

4.2. UTILITY PACKAGES AND LIBRARIES

VHDL enables the designer to define or simply reuse a complete modeling environment that can be specific to his or her application. The design of this environment is often unnecessary (and impossible) in other hardware description languages, which already provide only one extended logic type (with at least X and Z states) and miscellaneous components and subprograms. In VHDL, this number of predefined constructs is very small: the designer is almost obliged to define his own set of components, types, and subprograms to achieve efficient modeling. Hopefully, VHDL provides the designer with enough features to build his environment: libraries, packages, overloaded operators, user-defined types, and subprograms. Finally, once such an environment has been built, many designers of the same company only reuse it.

The notion of separate compilation is a main characteristic of VHDL: this language provides a means to store miscellaneous objects (constants, types, subprograms, components, signals, and so forth) in packages that are analyzed separately from the structural entities and stored in different libraries. Once a given package is defined, debugged, and compiled in a resource library (read-only library), any designer may use this package with maximum efficiency and security.

4.2.1. Logic System Package

Of course, one of the first packages that must be defined is the package that specifies the used multi-value logic system. The previous section presented a methodology to choose the appropriate logic system and several package source codes.

Some non-native VHDL simulators (QuicksimII from Mentor Graphics, for example) provide a VHDL logic package that implements the *built-in logic type of the simulator kernel.* Although a package body is given with this package (for portability purpose), the simulator will actually not use it: the VHDL type and associated resolution functions or logical operators are recognized by the simulator and directly mapped to its built-in logic type and functions. Simulation results are the same (they should be!), but the simulation using kernel constructs is much faster. This efficiency is another criteria to take into account when choosing a logic system.

4.2.2. Packages of Components

VHDL does not predefine any structural entity: even the basic gates (e.g., and, or, nand, nor) are not predefined by the language. Of course, such gates are so commonly used that it would be inefficient for each designer to redefine the same components.

A good design methodology will thus include the definition of packages of components. These components will then be available to each designer for structural descriptions. One typical example could be the package defining basic components corresponding to predefined logic operators:

```
package BASIC_GATES is
    component ORGATE
        generic ( N  :  NATURAL := 2   );
        port (    A  :  in BIT_VECTOR(1 to N) ;    Y  :  out BIT );
    end component;
    component ANDGATE
        generic ( N  :  NATURAL := 2   );
        port (    A  :  in BIT_VECTOR(1 to N) ;    Y  :  out BIT );
    end component;
    component NORGATE
        generic ( N  :  NATURAL := 2   );
        port (    A  :  in BIT_VECTOR(1 to N) ;    Y  :  out BIT );
    end component;
    component NANDGATE
        generic ( N  :  NATURAL := 2   );
        port (    A  :  in BIT_VECTOR(1 to N) ;    Y  :  out BIT );
    end component;
    component XORGATE
        generic ( N  :  NATURAL := 2   );
        port (    A  :  in BIT_VECTOR(1 to N) ;    Y  :  out BIT );
    end component;
end BASIC_GATES ;
```

In this package, components have a generic number of inputs. The type of the ports is simply the type BIT, but another package could have been defined for any other multi-value logic type. Of course, because these components do not contain any functional or timing information (which will be included in a corresponding entity and its architecture), they can be used in a wide range of structural models.

4.2.3. Technology-Dependent Packages and Libraries

The functional and timing information mentioned previously is included in the architecture of the corresponding entities. For the same reasons, all the basic entities and their architecture can be defined once and compiled in a resource library. Here is an example of an entity modeling an OR gate:

```
-- Logic entity
entity OR_ENTITY is
        generic ( N   :   NATURAL range 2 to 5 );
        port (   A    :   in BIT_VECTOR(1 to N);
                 Y    :   out BIT );
end OR_ENTITY;

-- Architecture
library CMOS;
architecture BEHAVIOR of OR_ENTITY is
        constant DELAY : TIME := CMOS.CMOS_PACKAGE.CMOS_OR_DELAY(N) ;
begin
        process
            variable RESULT : BIT ;
        begin
            RESULT := '0' ;
            for I in A'RANGE loop
                if A(I) = '1' then RESULT := '1' ; exit ; end if ;
            end loop ;
            Y <= RESULT after DELAY;
            wait on A ;
        end process ;
end BEHAVIOR ;
```

In this architecture, the delay used in the assignment depends on a table value CMOS_OR_DELAY(N), where the index N is the number of inputs of the entity. This table of delays is specified in another package CMOS_PACKAGE stored in the library CMOS. The last library is a technology-dependent library.

This example illustrates the way architectures may be written without being timing dependent. Here, the timing specifications are included in another package that is specific to the technology CMOS. Following is an example of such a package:

```
-- In library CMOS
package CMOS_PACKAGE is
        type DELAY_TABLE is array ( NATURAL range <> ) of TIME ;
        constant CMOS_OR_DELAY : DELAY_TABLE(2 to 5) ;
        constant CMOS_NOR_DELAY : DELAY_TABLE(2 to 5) ;
        ...
end CMOS_PACKAGE ;

package body CMOS_PACKAGE is
        constant CMOS_OR_DELAY : DELAY_TABLE(2 to 5) := (1.8 ns, 2.0 ns, 2.8 ns, 3.0 ns) ;
        constant CMOS_NOR_DELAY : DELAY_TABLE(2 to 5):= (1.3 ns, 1.5 ns, 2.3 ns, 2.5 ns);
        ...
end CMOS_PACKAGE ;
```

The effective timing values are specified in the package body; the package declaration only declares deferred constants. This VHDL mechanism allows the designer to change the CMOS timing values without having to recompile the package declaration, and thus all design units *using* this package declaration. Only the package body (which is invisible to other units) needs to be recompiled after a change of values.

4.2.4. Bit Arithmetic Package

Modeling datapath blocks in VHDL from scratch could be a very long and painful process: no arithmetic functions are predefined on type BIT_VECTOR. Many other hardware description languages handle bit vectors as unsigned integers: thus, direct integer operations can be performed on signals and buses. This is not possible in VHDL, which defines the type BIT as an enumeration type and not as an integer type.

In order to increase the efficiency of datapath modeling, it is important to dispose in the modeling environment of a user-defined package declaring classical arithmetic functions on bit arrays. Following is a package declaration defining such operations on type BIT (a similar package could have been defined on any other multi-value logic type):

```
-----------------------------------------------------------------------------------------
-- This package includes the definition of common arithmetic functions operating on bit vectors.
-- Conventions:
--      Any bit vector used as an argument or being the returned value of one of the
--      following functions is coding an integer value with the TWOS-COMPLEMENT notation.
--      The most significant bit (sign bit) of a bit vector is given by the 'LEFT attribute,
--      whereas its least significant bit is given by the 'RIGHT attribute.
-----------------------------------------------------------------------------------------

package TWOS_COMPLEMENT is
```

```
--------------------------------------------------------------------------------
-- Addition and subtraction of bit vectors
--------------------------------------------------------------------------------
function "+"(A,B : BIT_VECTOR) return BIT_VECTOR;
function "-"(A,B : BIT_VECTOR) return BIT_VECTOR;
--------------------------------------------------------------------------------
-- Warning: the two vectors must have the same length (A'LENGTH = B'LENGTH).
-- If both vectors have a length of n bits (A'LENGTH = B'LENGTH = N), then the
-- result of the operation is a vector of length N+1

--------------------------------------------------------------------------------
-- Two's complement of a bit vector (given thus the opposite integer value)
--------------------------------------------------------------------------------
function "-"(A :  BIT_VECTOR) return BIT_VECTOR;
function TWOS_COMP(A :  BIT_VECTOR) return BIT_VECTOR;   -- equivalent to "-"
--------------------------------------------------------------------------------
-- The range of the result is equal to (A'LENGTH - 1 downto 0).
-- Warning: This function does not accept A = "100..0" (Lowest integer)

--------------------------------------------------------------------------------
-- Multiplication of two bit vectors
--------------------------------------------------------------------------------
function "*"(A, B :  BIT_VECTOR) return BIT_VECTOR;
--------------------------------------------------------------------------------
-- If vector A has a length of n bits (A'LENGTH = N) and vector B a length of
-- m bits (B'LENGTH = M), then the result of the multiplication is a vector of
-- length N+M with the range (N+M-1 downto 0).

--------------------------------------------------------------------------------
-- Bit vector extension (sign propagation to the left)
--------------------------------------------------------------------------------
function EXTEND(A: BIT_VECTOR; L: POSITIVE) return BIT_VECTOR;
--------------------------------------------------------------------------------
-- This function copies the bit vector A into a bit vector of the given
-- length L (L > A'LENGTH), propagating thus the most significant bit
-- (sign bit) to the left.
-- Warning: it is assumed that the length L is greater than A'LENGTH.

--------------------------------------------------------------------------------
-- Bit vector reduction (with overflow detection and saturation)
--------------------------------------------------------------------------------
function REDUCE(A: BIT_VECTOR; L: POSITIVE) return BIT_VECTOR;
--------------------------------------------------------------------------------
-- This function copies the bit vector A into a bit vector of the given
-- length L (L < A'LENGTH), retaining only the L least significant bits.
-- If it is not possible (overflow or underflow), the resulting bit vector is
-- saturated to the highest ("0111...11") or the lowest ("1000...00") value
-- that can be coded with L bits.
-- Warning: it is assumed that the length L is less than A'LENGTH.
```

```
--------------------------------------------------------------------------------
-- Overflow detection
--------------------------------------------------------------------------------
function OVERFLOW(A: BIT_VECTOR; L: POSITIVE) return BIT;
--------------------------------------------------------------------------------
-- This function looks if it is possible to fit the bit vector A into a
-- bit vector of the given length L (L < A'LENGTH).
-- If it is not possible (overflow detection), the resulting bit is '1'.
-- Warning: it is assumed that the length L is less than A'LENGTH.

--------------------------------------------------------------------------------
-- Bit vector truncation
--------------------------------------------------------------------------------
function TRUNCATE(A: BIT_VECTOR; L: POSITIVE) return BIT_VECTOR;
--------------------------------------------------------------------------------
-- This function operates a truncation, i.e. it returns a bit vector of the
-- given size L (L < A'LENGTH) containing the L most significant bits (without
-- modification) of the bit vector A.
-- Warning: it is assumed that the length L is less than A'LENGTH.

--------------------------------------------------------------------------------
-- Bit vector rounding
--------------------------------------------------------------------------------
function ROUND(A: BIT_VECTOR; L: POSITIVE) return BIT_VECTOR;
--------------------------------------------------------------------------------
-- This function operates a rounding, i.e. it returns a bit vector of the
-- given size L (L < A'LENGTH) containing the L most significant bits of the
-- bit vector A, rounded by the addition of half the least significant retained bit.
-- Warning: it is assumed that the length L is less than A'LENGTH.

--------------------------------------------------------------------------------
-- Conversion of a twos-complement bit vector into the corresponding integer
--------------------------------------------------------------------------------
function VALUE(A: BIT_VECTOR) return INTEGER;

--------------------------------------------------------------------------------
-- Minimum bit vector length necessary to code the given integer
--------------------------------------------------------------------------------
function SIZE(I: INTEGER) return NATURAL;

--------------------------------------------------------------------------------
-- Conversion of an integer into the corresponding twos-complement bit vector
--------------------------------------------------------------------------------
function VECTOR(I: INTEGER; L: NATURAL) return BIT_VECTOR;
--------------------------------------------------------------------------------
-- This functions returns a bit vector of the given size L representing
-- the given integer I in the twos-complement notation.
-- Warning: it is assumed that the length L is greater than or equal to
-- the minimum bit vector length necessary to code the integer I. This
-- minimum length is given by SIZE(I).
```

```
end TWOS_COMPLEMENT;
```

As explained by the comments, this package declares operations on values of type BIT_VECTOR using the twos-complement notation (bit vectors are handled as signed integers). Another package could have been defined to declare unsigned arithmetic operations.

The proposed package includes all basic arithmetic operations (overloaded "+", "-", "*" operators), bit manipulation operators (shifting, rounding, saturation, etc.), and conversion functions.

The body of this package is listed below.

```
--------------------------------------------------------------------------------
-- Package Body
--------------------------------------------------------------------------------
package body TWOS_COMPLEMENT is

-- Size of machine word, usually 32
constant INTEGER_LENGTH : NATURAL := SIZE(INTEGER'HIGH);

-- Function returning TRUE if A codes - 2 ** (A'LENGTH - 1) (if A = "100..0")
function LOWEST_VECTOR(A : BIT_VECTOR) return BOOLEAN is
  alias AA: BIT_VECTOR(A'LENGTH - 1 downto 0) is A;
begin
  for i in 0 to A'LENGTH-2 loop
    if (AA(i) = '1') then
      return FALSE;
    end if;
  end loop;
  if (AA(A'LENGTH-1) = '0') then
    return FALSE;
  else
    return TRUE;
  end if;
end LOWEST_VECTOR;
--------------------------------------------------------------------------------

--------------------------------------------------------------------------------
-- Adder Primitives
function BITSUM(A,B,CIN: BIT) return BIT is
begin
  return (A xor B xor CIN);
end BITSUM;
--------------------------------------------------------------------------------
function BITCARRY(A,B,CIN: BIT) return BIT is
begin
  return ( (A and B) or ((A or B) and CIN) );
end BITCARRY;
--------------------------------------------------------------------------------
```

```
-------------------------------------------------------------------------------
-- Addition of two bit vectors
-------------------------------------------------------------------------------
function "+"(A,B : BIT_VECTOR) return BIT_VECTOR is
-------------------------------------------------------------------------------
alias AA : BIT_VECTOR(A'LENGTH - 1 downto 0) is A;
alias AB : BIT_VECTOR(B'LENGTH - 1 downto 0) is B;
variable SUM, INTER: BIT_VECTOR(A'LENGTH downto 0);
begin
  assert (A'LENGTH = B'LENGTH)
  report "Addition of two bit vectors having a different length!"
  severity FAILURE;
  if (A'LENGTH < INTEGER_LENGTH) then       -- optimization
     SUM := VECTOR(VALUE(A) + VALUE(B), A'LENGTH + 1);
  else
    INTER(0) := '0';
    for I in AA'REVERSE_RANGE loop
       SUM(I) := BITSUM(AA(I),AB(I),INTER(I));
       INTER(I+1) := BITCARRY(AA(I),AB(I),INTER(I));
    end loop;
       SUM(A'LENGTH) := BITSUM(AA(AA'LEFT),AB(AA'LEFT),INTER(A'LENGTH));
  end if;
  return SUM;
end "+";
-------------------------------------------------------------------------------

-------------------------------------------------------------------------------
-- Substraction of two bit vectors
-------------------------------------------------------------------------------
function "-"(A,B : BIT_VECTOR) return BIT_VECTOR is
-------------------------------------------------------------------------------
alias AA : BIT_VECTOR(A'LENGTH - 1 downto 0) is A;
constant NOTB : BIT_VECTOR(B'LENGTH - 1 downto 0) := not B;
variable SUB, INTER: BIT_VECTOR(A'LENGTH downto 0);
begin
  assert (A'LENGTH = B'LENGTH)
  report "Substraction of two bit vectors having different length"
  severity FAILURE;
  if (A'LENGTH < INTEGER_LENGTH) then       -- optimization
     SUB := VECTOR(VALUE(A) - VALUE(B), A'LENGTH + 1);
  else
    INTER(0) := '1';
    for I in AA'REVERSE_RANGE loop
       SUB(I) := BITSUM(AA(I),NOTB(I),INTER(I));
       INTER(I+1) := BITCARRY(AA(I),NOTB(I),INTER(I));
    end loop;
       SUB(A'LENGTH) := BITSUM(AA(AA'LEFT),NOTB(AA'LEFT),INTER(A'LENGTH));
  end if;
  return SUB;
end "-";
-------------------------------------------------------------------------------
```

```
--------------------------------------------------------------------------------
-- Two's complement of a bit vector (given thus the opposite integer value)
--------------------------------------------------------------------------------
function "-"(A: BIT_VECTOR) return BIT_VECTOR is
--------------------------------------------------------------------------------
alias AA: BIT_VECTOR(A'LENGTH - 1 downto 0) is A;
variable TWOSA : BIT_VECTOR(AA'RANGE) := AA;
variable FOUND_FIRST_ONE : BOOLEAN;
begin
 assert (not LOWEST_VECTOR(AA))
 report "Function TWOS_COMP (Unary -) called with 100..0!"
 severity FAILURE;
 for I in AA'REVERSE_RANGE loop
  case AA(I) is
   when '0' => if FOUND_FIRST_ONE then TWOSA(I) := '1'; end if;
   when '1' => if FOUND_FIRST_ONE then TWOSA(I) := '0';
               else FOUND_FIRST_ONE := TRUE; end if;
  end case;
 end loop;
 return TWOSA;
end "-";
--------------------------------------------------------------------------------
function TWOS_COMP(A: BIT_VECTOR) return BIT_VECTOR is  -- equivalent to "-"
begin
 return (-A);
end TWOS_COMP;
--------------------------------------------------------------------------------


--------------------------------------------------------------------------------
-- Multiplication of two bit vectors
--------------------------------------------------------------------------------
function "*"(A,B: BIT_VECTOR) return BIT_VECTOR is
--------------------------------------------------------------------------------
alias AA : BIT_VECTOR(A'LENGTH - 1 downto 0) is A;
alias AB : BIT_VECTOR(B'LENGTH - 1 downto 0) is B;
variable MULT: BIT_VECTOR(A'LENGTH + B'LENGTH - 1 downto 0) := (others => '0');
variable EXTENDEDA: BIT_VECTOR(A'LENGTH + B'LENGTH - 1 downto 0)
       := (others => AA(AA'LEFT));
variable INTER_SUM : BIT_VECTOR(A'LENGTH + B'LENGTH downto 0);
begin
 if (A'LENGTH + B'LENGTH <= INTEGER_LENGTH) then    -- optimization
   MULT := VECTOR(VALUE(A) * VALUE(B), A'LENGTH + B'LENGTH);
 else
   EXTENDEDA(AA'RANGE) := AA;
  assert (B'LENGTH >= 2) report "Multiplication with B too short (B'LENGTH < 2)!"
  severity FAILURE;
  for I in 0 to B'LENGTH - 2 loop
   if (AB(I) = '1') then
     INTER_SUM := MULT + EXTENDEDA;
     MULT := INTER_SUM(MULT'RANGE);
```

```
    end if;
    for J in EXTENDEDA'LEFT downto 1 loop
      EXTENDEDA(J) := EXTENDEDA(J-1);
    end loop;
    EXTENDEDA(0) := '0';
  end loop;
  if (AB(AB'LEFT) = '1') then
    INTER_SUM := MULT - EXTENDEDA;
    MULT := INTER_SUM(MULT'RANGE);
  end if;
 end if;
 return MULT;
end "*";
--------------------------------------------------------------------------------

--------------------------------------------------------------------------------
-- Bit vector extension (sign propagation to the left)
--------------------------------------------------------------------------------
function EXTEND(A: BIT_VECTOR; L: POSITIVE) return BIT_VECTOR is
--------------------------------------------------------------------------------
alias AA : BIT_VECTOR(A'LENGTH - 1 downto 0) is A;
variable RESULT : BIT_VECTOR(L-1 downto 0);
begin
 assert (L > A'LENGTH) report "Function EXTEND called with L <= A'LENGTH"
 severity FAILURE;
  RESULT(AA'RANGE) := AA;
 for I in A'LENGTH to L-1 loop
    RESULT(I) := AA(AA'LEFT);
 end loop;
 return RESULT;
end EXTEND;
--------------------------------------------------------------------------------

--------------------------------------------------------------------------------
-- Bit vector reduction (with overflow detection and saturation)
--------------------------------------------------------------------------------
function REDUCE(A: BIT_VECTOR; L: POSITIVE) return BIT_VECTOR is
--------------------------------------------------------------------------------
alias AA: BIT_VECTOR(A'LENGTH - 1 downto 0) is A;
subtype LM2 is BIT_VECTOR(L-2 downto 0);
variable RESULT: BIT_VECTOR(L-1 downto 0);
variable A_POSITIVE, A_NEGATIVE : BOOLEAN;
begin
 assert (L >= 2) report "Function REDUCE called with L = 1" severity FAILURE;
 assert (L < A'LENGTH) report "Function REDUCE called with L >= A'LENGTH"
 severity FAILURE;
 for I in L-1 to AA'LEFT loop
  if (AA(i) = '0') then
    A_POSITIVE := TRUE;
  else
    A_NEGATIVE := TRUE;
```

```
   end if;
  end loop;
  if (A_POSITIVE and A_NEGATIVE) then
   if (AA(AA'LEFT) = '0') then      -- Overflow
     RESULT(L-2 DOWNTO 0) := LM2'(others => '1');
     RESULT(L-1) := '0';
   else       -- Underflow
     RESULT(L-2 DOWNTO 0) := LM2'(others => '0');
     RESULT(L-1) := '1';
   end if;
  else
   RESULT := AA(L-1 downto 0);
  end if;
  return RESULT;
end REDUCE;
----------------------------------------------------------------------------

----------------------------------------------------------------------------
-- Overflow detection
----------------------------------------------------------------------------
function OVERFLOW(A: BIT_VECTOR; L: POSITIVE) return BIT is
----------------------------------------------------------------------------
alias AA: BIT_VECTOR(A'LENGTH - 1 downto 0) is A;
variable A_POSITIVE, A_NEGATIVE : BIT;
begin
  assert (L >= 2) report "Function OVERFLOW called with L = 1!"  severity FAILURE;
  assert (L < A'LENGTH) report "Function OVERFLOW called with L >= A'LENGTH!"
        severity FAILURE;
  for I in L-1 to AA'LEFT loop
   if (AA(I) = '0') then
     A_POSITIVE := '1';
   else
     A_NEGATIVE := '1';
   end if;
  end loop;
  return (A_POSITIVE and A_NEGATIVE);
  end OVERFLOW;
----------------------------------------------------------------------------

----------------------------------------------------------------------------
-- Bit vector truncation
----------------------------------------------------------------------------
function TRUNCATE(A: BIT_VECTOR; L: POSITIVE) return BIT_VECTOR is
----------------------------------------------------------------------------
alias AA: BIT_VECTOR(A'LENGTH - 1 downto 0) is A;
begin
  assert (L < A'LENGTH) report "Function TRUNCATE called with L >= A'LENGTH"
        severity FAILURE;
  return  AA(AA'LEFT downto AA'LEFT - L + 1);
end TRUNCATE;
----------------------------------------------------------------------------
```

```
--------------------------------------------------------------------------------
-- Bit vector rounding
--------------------------------------------------------------------------------
function ROUND(A: BIT_VECTOR; L: POSITIVE) return BIT_VECTOR is
--------------------------------------------------------------------------------
alias AA: BIT_VECTOR(A'LENGTH - 1 downto 0) is A;
variable HALF_LSB, TRUNCA: BIT_VECTOR(L downto 0);
variable SUM: BIT_VECTOR(L+1 downto 0);
variable RESULT: BIT_VECTOR(L-1 downto 0);
begin
  assert (L < A'LENGTH) report "Function ROUND called with L >= A'LENGTH"
        severity FAILURE;
  HALF_LSB(0) := '1';
  TRUNCA := AA(AA'LEFT downto AA'LEFT - L);
  SUM := TRUNCA + HALF_LSB;
  RESULT := SUM(L downto 1);
  return RESULT;
end ROUND;
--------------------------------------------------------------------------------

--------------------------------------------------------------------------------
-- Conversion of a twos-complement bit vector into the corresponding integer
--------------------------------------------------------------------------------
function VALUE(A: BIT_VECTOR) return INTEGER is
--------------------------------------------------------------------------------
alias AA: BIT_VECTOR(A'LENGTH - 1 downto 0) is A;
variable I: INTEGER := 0;
begin
  if (AA(AA'LEFT) = '0') then
    for J in AA'RANGE loop
      I := I * 2;
      if (AA(J) = '1') then
        I := I + 1;
      end if;
    end loop;
  elsif (not LOWEST_VECTOR(AA)) then
    I := -VALUE(-AA);
  else
    I := - 2 ** (A'LENGTH - 1);
  end if;
  return I;
end VALUE;
--------------------------------------------------------------------------------

--------------------------------------------------------------------------------
-- Minimum bit vector length necessary to code the given integer
--------------------------------------------------------------------------------
function SIZE(I: INTEGER) return NATURAL is
--------------------------------------------------------------------------------
variable INT: INTEGER := I;
```

```
variable N: NATURAL := 2;
variable POWER_OF_TWO: BOOLEAN := TRUE;
begin
 if (INT = 0) then return 1; end if;
 while (INT /= 1 and INT /= -1) loop
   N := N + 1;
   if (POWER_OF_TWO and (INT mod 2 = 1)) then
     POWER_OF_TWO := FALSE;
   end if;
    INT := INT / 2 ;
 end loop;
 if (I < 0 and POWER_OF_TWO) then  -- for I being negative and a power of two
   N := N - 1;
 end if;
 return N;
end SIZE;
--------------------------------------------------------------------------------

--------------------------------------------------------------------------------
-- Conversion of an integer into the corresponding twos-complement bit vector
--------------------------------------------------------------------------------
function VECTOR(I: INTEGER; L: NATURAL) return BIT_VECTOR is
--------------------------------------------------------------------------------
variable VECT: BIT_VECTOR(L-1 downto 0);
variable IVECT: BIT_VECTOR(L downto 0);
variable INTER: INTEGER := I;
begin
 assert (L >= size(I)) report "Function VECTOR called with L < SIZE(I)!"
 severity FAILURE;
 if (I >= 0) then
   for J in VECT'REVERSE_RANGE loop
     if (INTER mod 2 = 1) then  VECT(J) := '1';  end if;
      INTER := INTER / 2;
   end loop;
 elsif (I >= -INTEGER'HIGH) then
    IVECT := -VECTOR(-I,L+1);
   -- intermediate step necessary in the case of I = - 2 ** (L - 1)
   -- Ex: L=8, I=-128, size(-I) = size(128) = 9 > L!
     VECT := IVECT(VECT'RANGE);
 elsif (I = INTEGER'LOW) then
   for J in 0 to INTEGER_LENGTH-2    loop VECT(J) := '0'; end loop;
   for J in INTEGER_LENGTH-1 to L-1  loop VECT(J) := '1'; end loop;
 else
   assert FALSE report "Unexpected integer implementation (function VECTOR)!"
   severity FAILURE;
 end if;
 return VECT;
end VECTOR;
--------------------------------------------------------------------------------

end TWOS_COMPLEMENT ;
```

Using such a package, any datapath modeling becomes easier. Indeed, each combinational operator may be described at a functional level with very few statements. The use of this package also improves the overall security of the design. Following is a short example illustrating the use of the previous arithmetic package:

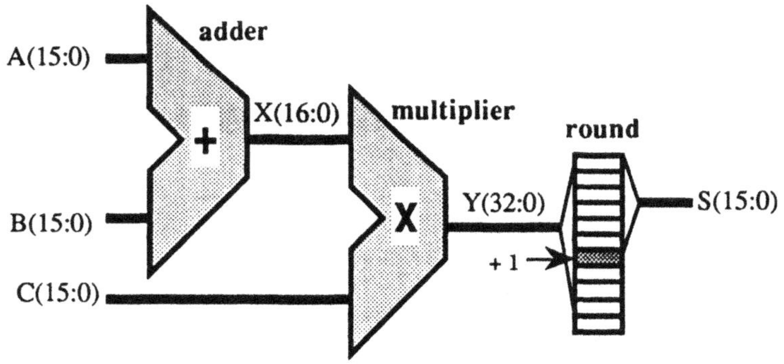

Fig 4.6 Datapath Operators

These combinational operators may be modeled with the architecture below:

```
library ARITH ;
use ARITH.TWOS_COMPLEMENT.all ;
architecture STRUCTURE of DATA_PATH is
        signal A, B, C, S : BIT_VECTOR(15 downto 0) ;
begin
        S <= ROUND( (A + B) * C , 16 ) after GEN_DELAY ;
end STRUCTURE ;
```

This architecture uses the overloaded addition on bit vectors ("+"), the overloaded multiplication on bit vectors ("*"), and a rounding function (ROUND).

4.2.5. Application-Specific Packages

Many other functional packages like the previous one can be defined. Some of them may be specific to the application. For example, the design of digital signal or image microprocessors may request packages implementing classical algorithms such as Fast Fourier or Discrete Cosine transforms, miscellaneous filter computations, and trigonometric functions.

Of course, it is not possible to give an exhaustive list of these packages. Each IC designer may need very specific entities and components, each system designer may need to simulate specific algorithms, and so on. VHDL provides libraries and packages to store, organize, and secure all these hardware and algorithmic models.

4.2.6. Timing Verification Packages

Specification or simulation models of ICs and PCBs often include timing constraint checks. These checks may be performed by concurrent procedure calls inside entity or architecture statement parts. These procedures are typical subprograms that should be written once (by one designer) and stored in a package that will be used by many designers. Following is the example of a procedure checking the setup time between a data signal and a reference clock.

```
-- Setup verification (on the rising edge of CLK)
procedure SETUP_CHECK (  signal CLK    : BIT;
                         signal DATA : BIT_VECTOR;
                         MIN_SETUP  : TIME              ) is
 variable T1, T2 : TIME := 0 ns;
begin
 loop
   wait on DATA, CLK;
   if DATA'EVENT then
     T1 := NOW;
   end if;
   if CLK'EVENT and CLK = '1' then
     T2 := NOW;
     assert (T2-T1) >= MIN_SETUP report "Minimum setup delay violation" severity ERROR ;
   end if;
 end loop;
end SETUP_CHECK;
```

Many different ways exist for writing such a procedure, but the skeleton of the previous one can be used for all similar subprograms, such as the hold time, the clock cycle time, or the pulse-width constraints check routines.

4.2.7. Conclusion

The general philosophy of VHDL is to let the designer define all the specific constructs he or she needs: abstract datatypes, subprograms, generic components, and entities. In a sense, VHDL can be seen as a kit to build its own hardware description language. As the set of predefined types and operators is very restricted, such a possibility becomes a necessity. Any modeling environment should thus include resource packages: general-purpose utility packages as well as application-specific packages. Both will significantly increase the efficiency of the design.

5. SYSTEM MODELING

5.1. INTRODUCTION

The design problem at the system level is to split an initial consistent specification into a set of interacting software and hardware subsystems. The specification is often expressed with specific formalisms. These formalisms generally define a small set of primitives (i.e., places and transition in a Petri-net) with well-defined semantics on which formal analysis methods (i.e., formal proof) can be applied to prove certain properties of the specification (such as liveness and safety). For example, to validate a communication protocol, the communicating nodes can be represented by a finite-state machine (FSM) specified in Petri-nets. The resulting description can be checked by formal methods to prove some properties of the protocol (no conflicts or deadlocks, for example).

The techniques used perform analysis on the set of reachable states and cannot apply on systems with a very large number of states. This is why system simulation is often needed.

This chapter suggests ways to implement system specification formalisms in VHDL so that they can be simulated.

5.2. THE FSM WITH A SINGLE THREAD OF CONTROL

An FSM with a single thread of control represents a controller view of the system. It is completely specified by two functions. The first determines the next state of the FSM as a consequence of the current state and the input patterns. The second describes the outputs or commands of the FSM in a similar way. These two functions are often presented in state-transition graphs. At any time, the behavior of the FSM is completely determined by its current state and the input patterns.

The traffic light controller will serve as an example to illustrate the various ways to describe FSM in VHDL.

5.2.1. The Traffic Light Controller

The traffic light controller example was first proposed by Mead and Conway in their best seller entitled Introduction to VLSI Systems [MEA80]. Its purpose is to control traffic lights at a crossroad junction between a farm road and a highway.

Priority is given to the highway road as long as there is no car on the farm road: the lights are green on the highway and red on the farm road. Things remain the same until one or many cars are detected on the farm road. Then lights switch to yellow and next to red on the highway. After that, the lights on the farm road turn to green. This situation remains unchanged until no car passes on the farm road, but for no longer than a time period (MEDIUM_TIMEOUT).

Then, the lights on the farm road switch to yellow and then to red while the highway lights turn back to green. The farm road is sensed again only after another delay (LONG_TIMEOUT), and the process is repeated. SHORT_TIMEOUT is the duration of the yellow lights. Figure 5.2 below gives a state-transition diagram of the traffic light controller.

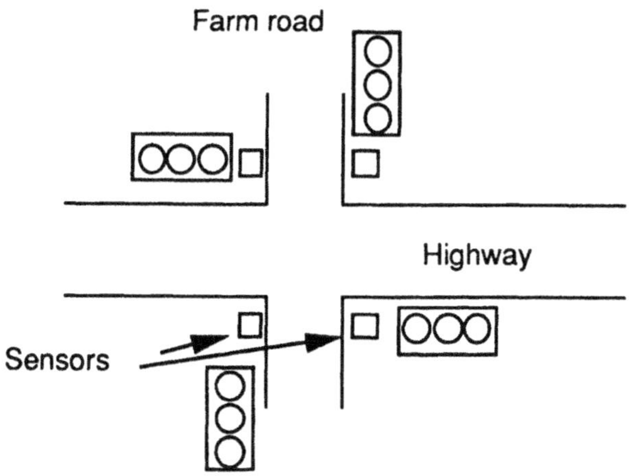

Fig 5.1 Traffic Light Problem

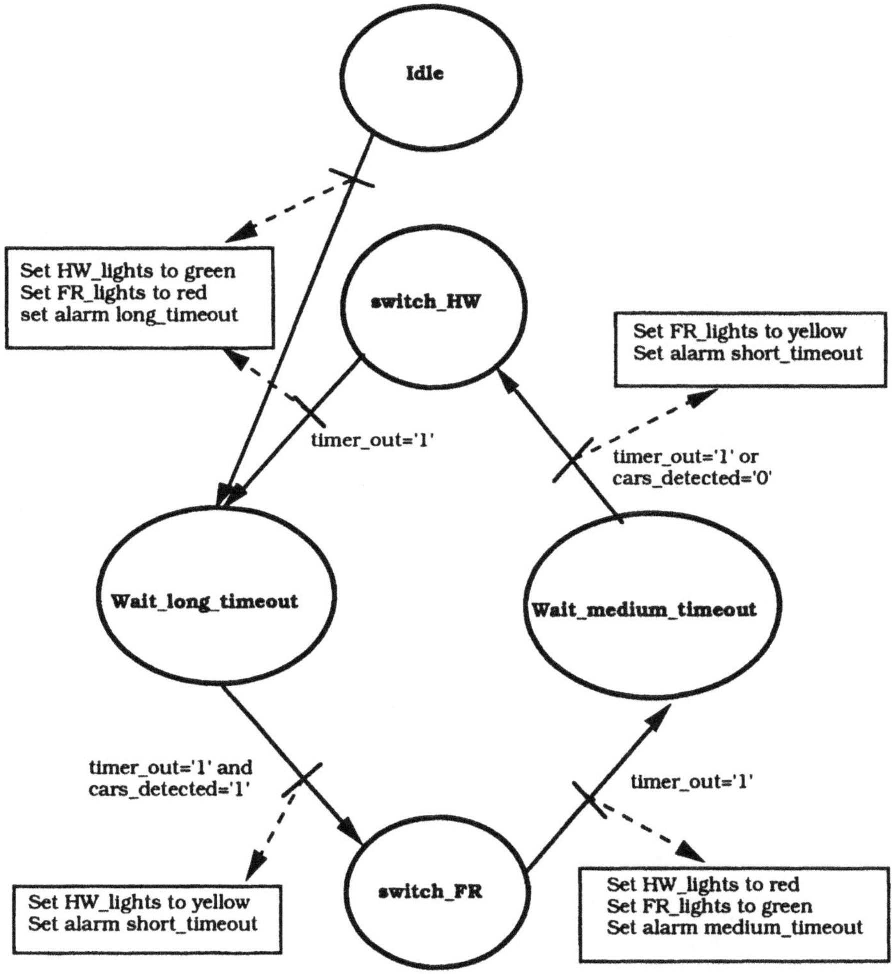

Fig 5.2 Traffic Light State Diagram

5.2.2. The FSM with One Process

The set of states is described by an enumeration type. The status of the traffic lights is also represented by an enumeration type. The FSM is modeled by a case statement within the process. Notice the judicious choice to use a variable to store the state information.

```
package TRAFF is
        type THREE_COLORS is (GREEN,YELLOW, RED) ;
        type CONTROLLER_STATES is ( IDLE, WAIT_LONG_TIMEOUT, SWITCH_FR,
                                    WAIT_MEDIUM_TIMEOUT, SWITCH_HW ) ;
end TRAFF;

use WORK.TRAFF.all;
entity TRAFFIC_CONTROLLER  is
        port (   CARS_DETECTED          :    in BIT;
                 FR_LIGHTS, HW_LIGHTS  :    out THREE_COLORS ) ;
        constant SHORT_TIMEOUT     :    TIME := 20 SEC ;
        constant MEDIUM_TIMEOUT    :    TIME := 180 SEC ;
        constant LONG_TIMEOUT      :    TIME := 360 SEC ;
 end TRAFFIC_CONTROLLER ;

 architecture FSM of TRAFFIC_CONTROLLER is
        signal TIMER_OUT : BIT;
 begin
        process (TIMER_OUT,CARS_DETECTED)
            variable CURRSTATE : CONTROLLER_STATES ;
        begin
          case  CURRSTATE is
            when IDLE=>
                HW_LIGHTS<=GREEN;
                FR_LIGHTS<=RED;
                TIMER_OUT<='0','1' after LONG_TIMEOUT;
                CURRSTATE:=WAIT_LONG_TIMEOUT;
            when WAIT_LONG_TIMEOUT=>
                if CARS_DETECTED='1'  and  TIMER_OUT='1' then
                    HW_LIGHTS<=YELLOW;
                    TIMER_OUT<='0', '1' after SHORT_TIMEOUT;
                    CURRSTATE:=SWITCH_FR;
                end if ;
            when SWITCH_FR=>
                if TIMER_OUT='1' then
                    HW_LIGHTS<=RED;
                    FR_LIGHTS<=GREEN;
                    CURRSTATE:=WAIT_MEDIUM_TIMEOUT;
                    TIMER_OUT<='0','1' after MEDIUM_TIMEOUT;
                end if;
            when WAIT_MEDIUM_TIMEOUT =>
                if TIMER_OUT='1' or CARS_DETECTED='0' then
                    FR_LIGHTS<=YELLOW;
                    TIMER_OUT<='0','1' after SHORT_TIMEOUT;
                     CURRSTATE:=SWITCH_HW;
                end if;
            when SWITCH_HW=>
                if TIMER_OUT='1' then
                    FR_LIGHTS<=RED;
                    HW_LIGHTS<=GREEN;
                    TIMER_OUT<='0','1' after LONG_TIMEOUT;
```

```
                CURRSTATE:=WAIT_LONG_TIMEOUT;
            end if;
        end case;
      end process;
end FSM;
```

Another way to model an FSM is to use two process statements: one for the next state function and the other for the output function.

A third way to model an FSM in VHDL is to use guarded blocks. The state could be represented by a signal and the next state function by a set of concurrent guarded assignment statements. A guarded block may represent one or more transitions. In the first case, the guard condition is equal to the transition condition:

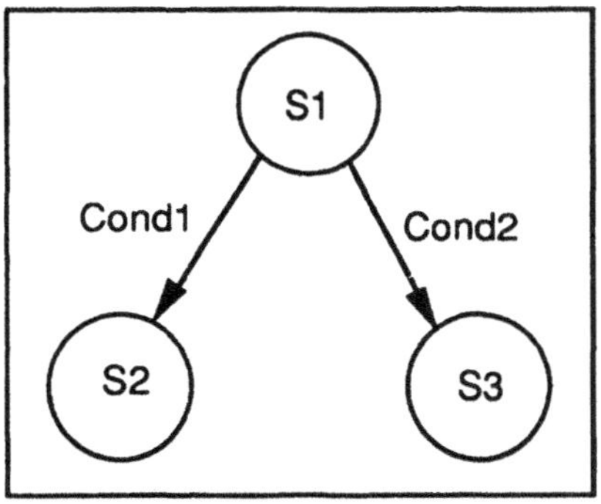

Fig 5.3 FSM Transitions

```
TR1 : block (STATE=S1 and COND1)
begin
      STATE<= guarded S2;
end block TR1 ;

TR2 : block (STATE=S1 and COND2)
begin
      STATE<= guarded S3;
end block TR2 ;
```

In the second case, the block statement may encompass all the transitions from a given state value, such as

```
TR1 : block (STATE=S1)
begin
      STATE<= guarded S2  when COND1
                  else S3 when COND2
                  else STATE;
end block TR1 ;
```

The outputs of the automaton can also be described with guarded block statements. Two problems must be addressed:

- First, the FSM transitions are described with a set of concurrent assignment statements to the same STATE signal: this signal must be resolved. Furthermore, the drivers of **guarded** assignments must be disconnected when the **guard** conditions are false: the STATE signal must be a **guarded** signal. When no transition is active, the STATE signal should keep its value, so it must be of the **register** kind.
- The FSMs considered here have a single thread of control, so at any time, at most one transition is active. This condition can be checked in the resolution function associated with the STATE signal. In case there is an active transition driver, its value must be returned as the next value of the state. This resolution function may be written as

```
function FSM_RESOLVE(ARG : STATE_ARRAY) return STATE_TYPE is
begin
      assert (ARG'LENGTH = 1)
           report "Error! multiple active transitions"
           severity ERROR ;
      return ARG(ARG'LEFT);
end FSM_RESOLVE;
```

Notice that this function assumes that its parameter contains exactly one value. When no driver is active, the resolution function will not be called because the STATE signal is of the **register** kind.

An advantage of the above approach is that descriptions can be generated automatically from graphical formalisms. But it uses concurrent statements to simulate a sequential FSM, which is not very efficient. We will see later that this approach is appropriate to implement hierarchical FSMs where a state is mapped into another FSM.

One limitation of simple FSMs is that they do not reflect concurrent activities. The next proposed formalism overcomes this difficulty by allowing parallel activities to be expressed.

5.3. MULTIPLE THREADS OF CONTROL: PETRI-NETS

An FSM with multiple threads of control may be defined with a Petri-net formalism: a set of places and a set of transitions interconnected by arrows. The previous formalism gives a controller view of the system. It is close to implementation. The Petri-net gives a dynamic view of the system. It models all the concurrent activities and reflects all the access conflicts.

Places are represented by circles and are characterized by two state values depending on whether or not they contain a token. When they contain a token, they are said to be "marked". The state of the system is defined by the set of all the marked places. A Petri-net with N places has N(N-1)/2 possible states.

Transitions are depicted by rectangles and may be associated with conditions or actions. They are connected to places by arrows. Arrows can only connect places to transitions and vice versa. The places linked to a transition can be classified into two subsets: the first subset comprises places standing at the destination of arrows originating from the transition and are known as output places; the second set defines the input places to the transition.

A transition is said to be "firable" or active when all of its input places are marked and the optional associated condition is true (as shown in figure 5.4). "Firing" a transition consists in removing the marks from its input places, setting a mark in each of its output places, and executing the associated actions. The behavior of a Petri-net is obtained by continually firing all the active transitions. It is split into many activities performing in parallel, transitions providing the synchronization mechanism.

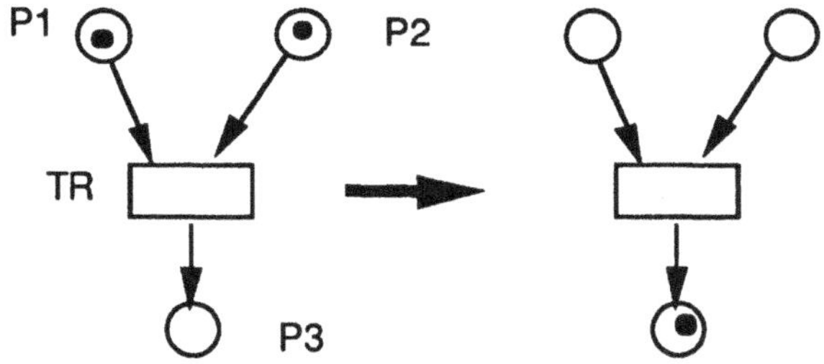

Fig 5.4 Firing a Transition

Conflicts arise when many active transitions share the same place as input. An example is given in figure 5.5.

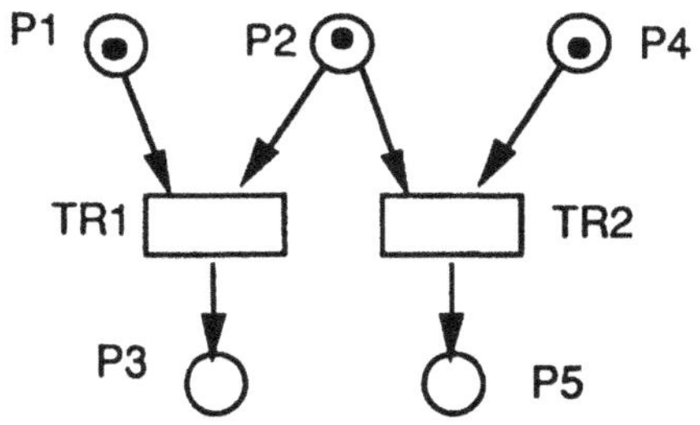

Fig 5.5 Access Conflict between TR1 and TR2

Such conflicts require a mutual exclusion to be implemented to solve the conflict. The pure Petri-net semantics does not specify the choice rule when many transitions are firable and share some input places. In the proposed VHDL translation, such situations will be flagged as errors.

5.3.1. VHDL Translation of Petri-Nets

Here follows a straightforward VHDL model of Petri-nets that lends itself well to automatic generation by a translation tool.

5.3.1.1. Places

Places are represented by guarded signals of the bit type and of the **register** kind . Marked places are associated with signals with values '1'. Because guarded signals are resolved signals, a resolution function is needed. In extended Petri-nets, places may be allowed to have more than one mark. Then they may be implemented by signals of the integer type. Here places are allowed to hold at most one mark.

```
function PETRI_RESOLVE(ARG:BIT_VECTOR) return BIT is
begin
        assert(ARG'LENGTH<=1 )
            report "Conflict: Place already busy..."
            severity ERROR;
        if ARG'LENGTH=1 then
            return ARG(ARG'LEFT) ;
        else
            return '0';
        end if;
end PETRI_RESOLVE ;

subtype PLACE is PETRI_RESOLVE BIT ;
```

This resolution function ensures that there is no access conflict in a place. The places of figure 5.4, before firing the transition, are declared as:

```
signal P1, P2 : PLACE register :='1';
```

5.3.1.2. Transitions

Each transition is described by a guarded block statement. The guard condition checks whether or not the transition is firable. The effects of the transition are achieved through guarded assignment statements. These effects must take place only when the transition is active and should not extend beyond that period. That is why signals of the **register** kind are used to represent the places. This ensures that places connected to a given transition are not affected by that transition when it is not active: the relevant driver is disconnected as a result of the guard condition becoming false. Here follows the VHDL text proposed for describing transition TR of figure 5.4:

```
TR :  block (P1='1' and P2='1')
      begin
          assert not (GUARD and P3='1')
              report "trying to put a mark in a non-empty place!" severity ERROR;
          P1<= guarded '0';   -- removes the mark from P1
          P2<= guarded '0';   -- removes the mark from P2
          P3<= guarded '1';   -- puts the mark in P3
      end block;
```

The assertion statement checks that no mark already exists in P3 just before the firing of TR.

5.3.1.3. The Bus Arbiter Example

The bus arbiter (shown in figure 5.6) is used to monitor bus access requests from two units. Any unit may request the bus by a rising edge on BUS_REQ1 (for unit 1) or BUS_REQ2 (for unit 2). Depending on the bus status (free or busy), the controller grants the bus (BUS_GRANT1 or BUS_GRANT2). When the grant signal rises, the requesting unit uses the bus and sets signal BUS_BUSY to '0' to notify end of use. Then the controller may process other requests.

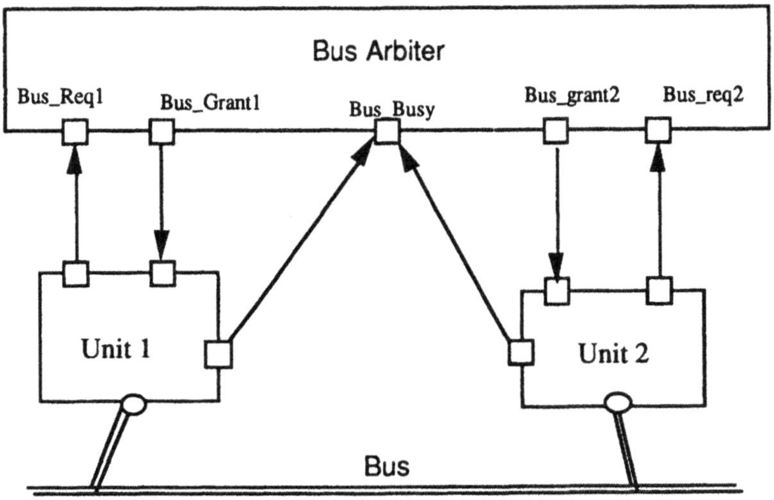

Fig 5.6 Bus Arbiter

Figure 5.7 gives a Petri-net description of the bus arbiter. Notice the conflict between SELECT1 and SELECT2. This conflict reflects the situation where the two units ask for the bus at the same time. The solution retained here is to give priority to UNIT1. In order to do that, an additional condition on SELECT2 requires place REQUEST_1 to be empty. This condition defines an extension to the bare Petri-net formalism known as Petri-net with inhibitor arrows, where an inhibitor arrow tests for the absence of a mark in a place.

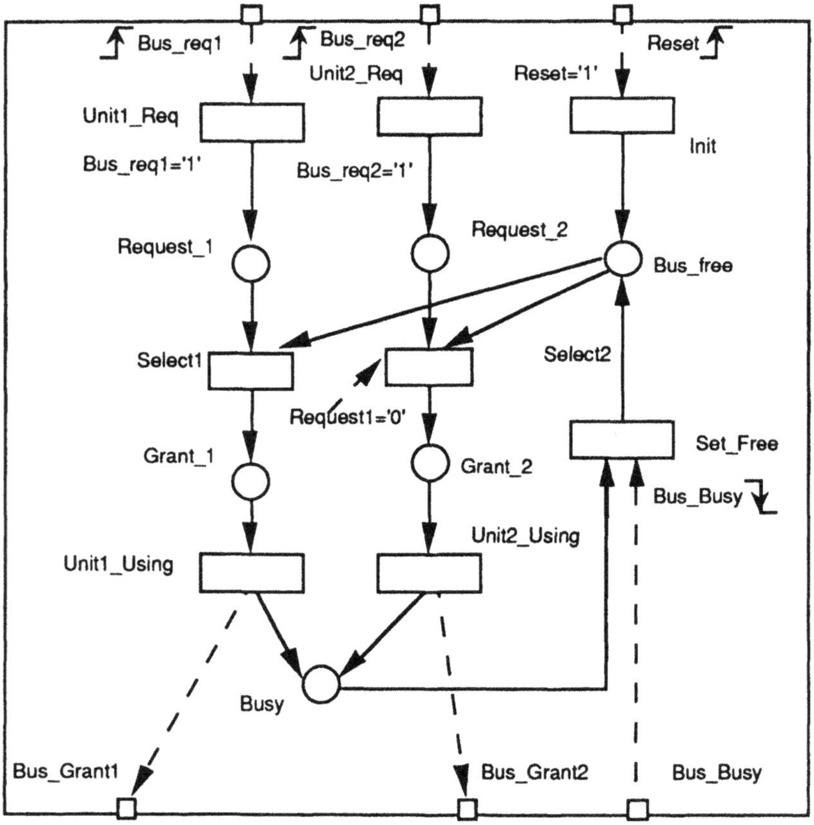

Fig 5.7 Petri-Net Description of the Bus Arbiter

Using such a formalism, transitions SELECT1 and SELECT2 could be represented as shown in figure 5.8:

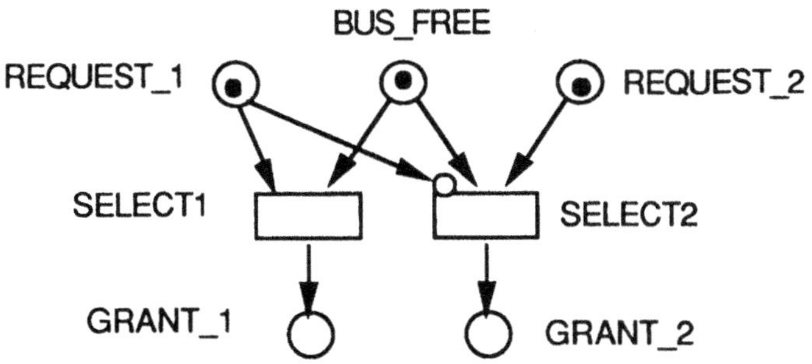

Fig 5.8 Mutual Exclusion with Inhibitor Arrow

Transition SELECT2 can be implemented in VHDL as:

```
SELECT2 : block( REQUEST_2='1' and BUS_FREE='1' and
                 REQUEST_1='0' )-- Inhibitor arrow
        begin
            assert not (GUARD and GRANT_2='1')
                report " Putting a mark in a non-empty place!"
                severity ERROR;
            BUS_FREE <= guarded '0';
            REQUEST_2 <= guarded '0';
            GRANT_2 <= guarded '1';
        end block SELECT2;
```

and transition SELECT1 as:

```
SELECT1 : block ( BUS_FREE='1' and REQUEST_1='1' )
        begin
            assert not (GUARD and GRANT_1='1')
                report " Putting a mark in a non-empty place!"
                severity ERROR;
            BUS_FREE <= guarded '0' ;
            REQUEST_1 <= guarded '0';
            GRANT_1 <= guarded '1';
        end block SELECT1 ;
```

Assuming type PLACE is defined as above in package UTIL_PKG, the corresponding VHDL text is straightforward:

```
entity ARBITER is
    port (  BUS_REQ1,BUS_REQ2,RESET, BUS_BUSY : in BIT ;
            BUS_GRANT1,BUS_GRANT2 : out BIT) ;
end ARBITER ;
```

```
architecture DATAFLOW of ARBITER is
        use WORK.UTIL_PKG.all;
        signal REQUEST_1,REQUEST_2,GRANT_1,GRANT_2,
                READY, BUS_FREE, BUSY : PLACE register ;
begin
   INIT : block( not RESET'STABLE and RESET='1')
        begin
            BUS_FREE <= guarded '1';
        end block;

   UNIT1_REQ : block( not BUS_REQ1'STABLE and BUS_REQ1='1' )
        begin
            REQUEST_1 <= guarded '1';
        end block;

   UNIT2_REQ : block( not BUS_REQ2'STABLE and BUS_REQ2='1' )
        begin
            REQUEST_2 <= guarded '1';
        end block;

   SELECT1 : block ( BUS_FREE='1' and REQUEST_1='1' )
        begin
            assert not (GUARD and GRANT_1='1')
            report " Putting a mark in a non-empty place!"
            severity ERROR;
            BUS_FREE <= guarded '0' ;
            REQUEST_1 <= guarded '0';
            GRANT_1 <= guarded '1';
        end block SELECT1 ;

assert not (REQUEST_1='1' and REQUEST_2='1')
report "Conflict detected: priority given to Unit 1"
severity NOTE;

  SELECT2 : block( REQUEST_2='1' and BUS_FREE='1' and
                REQUEST_1='0' )-- Inhibitor arrow
        begin
            assert not (GUARD and GRANT_2='1')
            report " Putting a mark in a non-empty place!"
            severity ERROR;
            BUS_FREE <= guarded '0';
            REQUEST_2 <= guarded '0';
            GRANT_2 <= guarded '1';
        end block SELECT2;

   UNIT1_USING : block ( GRANT_1='1' )
        begin
            GRANT_1 <= guarded '0' ;
            BUSY <= guarded '1' ;
            BUS_GRANT1<='1' when guard else '0' after 5 ns;
        end block UNIT1_USING ;
```

```
    UNIT2_USING : block ( GRANT_2='1' )
        begin
            GRANT_2 <= guarded  '0'  ;
            BUSY <= guarded '1' ;
            BUS_GRANT2<='1' when guard else '0' after 5 ns;
        end block UNIT2_USING ;

    SET_FREE : block ( not BUS_BUSY'STABLE and BUS_BUSY ='0' )
        begin
            BUSY <= guarded '0' ;
            BUS_FREE <= guarded '1';
        end block SET_FREE;
end DATAFLOW ;
```

For action on signals that are not of the **register** kind (the ports BUS_GRANT1, BUS_GRANT2), notice the use of conditional signal assignment statements to restrict the action of the transition to when it is active.

5.3.1.4. Adding Delay to Transition

Petri-nets can be annotated with delays. Associated with a transition, a delay represent the lapse of time it takes for a transition to be fired. The resulting VHDL description will add these delays to the signal assignment statements associated with the transitions. The transition of figure 5.9 follows:

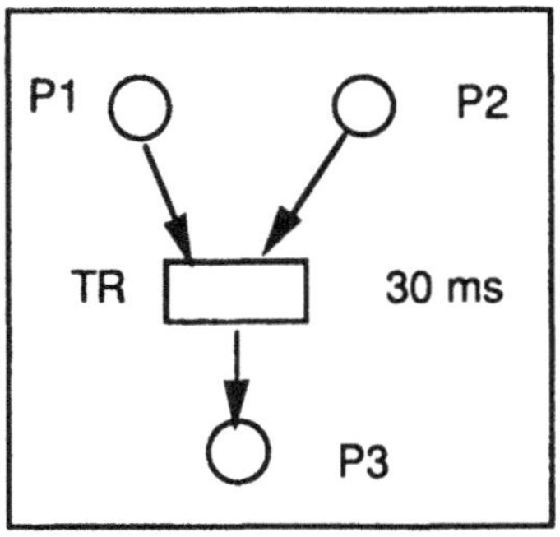

Fig 5.9 Transition Annotated with a Delay

This can be modeled with

```
TR : block(P1='1' and P2='1')
    begin
        P1<='0' after 30 ms;
        P2<='0' after 30 ms;
        P3<='1' after 30 ms;
    end block;
```

5.3.1.5. Conclusion

Petri-nets extend the descriptive power of FSMs by adding the concept of concurrency. Still, for very complex systems, they do not support step refinement methods by which a system is incrementally described by the concept of macro states (state charts) and macro transitions (structured nets).

5.4. HIERARCHY: STATE CHARTS AND S-NETS

5.4.1. State Charts

State charts overcome the limitations of FSMs with a single flow of control by adding two concepts for handling complexity: concurrency and hierarchy. This is achieved by allowing a given state to be refined as a set of concurrent FSMs. This refinement can be performed recursively on the states of the embedded FSM states.

A macro state can be mapped into a block statement enclosing the translation of the FSMs.

When a macro state is entered, the embedded FSMs can be started from default or an explicit initial state or can resume from the state they were in when the macro state was previously exited. This latter alternative implements a history mechanism.

A macro state can be exited under one of two situations:

- asynchronously, when triggered by an explicit transition;
- synchronously, when all the embedded FSMs are exited.

The FSM model with block statement presented earlier can easily be extended to support state charts and their various synchronizations.

Each macro state is described by a guarded block statement enclosing its internal FSM. A signal of enumeration type local to this guarded block stores the current state of this FSM. The guard condition identifies the macro state. This mapping is illustrated by the following example (figure 5.10).

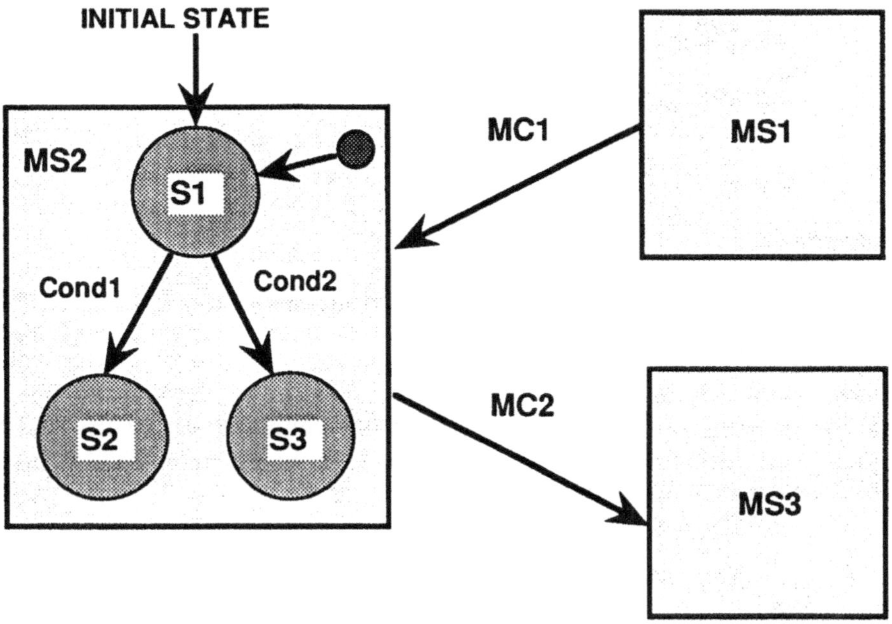

Fig 5.10 State Chart

A possible VHDL translation can be deduced as:

```
B1 : block (MSTATE = MS1)
begin
        MSTATE<= guarded MS2 when MC1
                   else MSTATE ;
end block B1;

B2 : block (MSTATE=MS2)
        type STATE_TYPES is (S1, S2, S3) ;
        signal STATE : STATE_TYPES ;
begin

        MSTATE<= guarded MS3 when MC2
                   else MSTATE;
        -- Macro state transition

        BI1 : block (GUARD and not GUARD'STABLE)
        begin
             STATE<= guarded S1 ;
             -- implicit initial state
        end block;
```

```
BI2 : block (GUARD and GUARD'STABLE
    -- equivalent to MSTATE=MS1 and MSTATE'STABLE
    and STATE=S1)
begin
    STATE<=guarded S2 when COND1
              else S3 when COND2
              else STATE;
end block;

end block B2 ;
```

Notice the utility of using signals to hold the state information. This state information must be accessed by other block statements implementing the embedded FSMs in order to switch them off when the macro state is asynchronously exited.

In the example previously depicted, the FSM enclosed by macro state MS2 is an assigned implicit default state S1. Any time macro state MS2 is entered, this FSM must resume execution from state S1. For that, STATE must be set to S1 as soon as MS1 is entered. This is achieved by the block statement

```
BL : block (GUARD and not GUARD'STABLE)
begin
    STATE<=guarded S1 ; -- implicit initial state
end block;
```

In case an embedded FSM must resume from its previous state, the above block statement must simply be removed from the description. Indeed, since all internal state transitions include the GUARD condition of the block representing this FSM, all these internal transitions are invalidated when the macro state changes. Furthermore, as signals are of the **register** kind, they keep their last value when all guard conditions become false. Consequently, the embedded FSM resumes from its previous internal state value.

The macro state is exited by simply setting the macro state signal to the new macro state. This can occur asynchronously, e.g.,

```
BL : block(ASYNCHRONOUS_CONDITION)
begin
    MSTATE<=guarded NEW_MSTATE;
end block;
```

If the macro state must exit synchronously when all the embedded FSMs are exited, the corresponding block statement for a macro state containing two FSM looks like the following:

```
BL : block(STATE1=EXIT_STATE1 and STATE2=EXIT_STATE2)
begin
    MSTATE<=guarded TARGET_MSTATE;
end block;
```

5.4.2. Structured Nets

Structured nets introduce hierarchy in Petri-nets by refining a transition (then called macro transition) into another Petri-net. This extension is straightforward. Firing the macro transition is the same as executing the corresponding sub-Petri-net.

5.5. CONCLUSION

In this chapter we have seen that many system formalisms can be mapped into VHDL. The suggested mapping can be automatically produced by software utilities from graphical entries. VHDL block statements provide a straightforward way of implementing FSMs, Petri-nets, structured graphs, and structured nets. Other system formalisms such as VAL/VHDL increase the capabilities of VHDL in the field of abstract specification, hierarchical development of design, and validation [AUG91].

6. STRUCTURING METHODOLOGY

6.1. STRUCTURING THE MODELING EFFORT

Many different ways of structuring VHDL modeling are available to the designer. Even if, at the end of the elaboration process of the design hierarchy, all the VHDL hierarchy is translated into a set of nested blocks by the compiler, source code structuring will lead to the important properties of the description.

However, what is really expected from an ideal HDL structuring?

- It must provide a *good readability* that allows people outside a given team to be able to read and easily understand the goal of the design.
- It must also be *reusable*. This means that another team must be able to use and benefit from this description as part of its own design. To be reusable by people outside the original team, the description must have a good readability but also must be as general as possible. This generality must always remain related to the needs of the user, but a description that is too application-specific will never be reusable.
- A good HDL description has to be *easy to refine*. A common method in a top-down approach to the design is to refine the description step by step. Ease of change can be measured by the ease of going from one step to another and the length of the design cycle time.
- An appreciable feature of an HDL is the ability to *facilitate the debugging phase*. Ease of debugging can be seen as the ease of diagnosing the bug and correcting the description.
- Another aspect of an ideal HDL is the necessity of being able *to write it quickly*. This property is often opposed to reusability, but there are some cases where it is important to provide prototypes quickly.

We will now review some VHDL features and show how they can be used to make a description satisfy one or more of the preceding properties.

6.2. WHAT ARE THE POSSIBILITIES OF VHDL?

6.2.1. Block Statement

For a VHDL beginner, at the first glance, the block statement can appear to be the simplest way of structuring a description. Actually, this is the case in very small modeling, but drawbacks quickly arise when using it in real modeling. Hierarchy is not the only goal of blocks in VHDL.

The block statement has two other functionalities, which are more important than hierarchy:

- It allows certain declarations to be encapsulated in its declarative part, in order to share them between all local statements. The declarations local to the block are seen from the entire set of concurrent statements of the block, but do not encumber the enclosing declarative region.
- It offers the possibility (in its guarded form) of factorizing certain signal assignments. This short-writing facility is very much appreciated when certain signals are synchronized by another signal.

When used to structure a description, the block statement has a unique advantage: it allows quick and concise writing, but the drawbacks are numerous.

To simplify, we can say that a block is equivalent to a component that has been instantiated once. If we need two instantiations of the same component, the only solution is to duplicate the block source code. Even if this cut/paste operation seems easy, the number of code lines will increase dramatically, the readability will decrease, and the debugging will become difficult. If a design error occurs in the block, how can the correction of all the duplicated code of this block be ensured?

The block is a statement and not a design unit. This implies that it is not possible to share it between designers. It is impossible to store blocks in a library in the same way that are stored entities. A block is by its very nature not reusable.

Can the block statement be considered as an initial means of structuring the design in order to quickly test it before translating it into a design entity?

No, this strategy is not a good one; the task of translating a block into a design entity is not easy. The problem is due to the scope of the block. Concurrent statements inside the block can access every object of the enclosing declarative region. An example is given in figure 6.1, where the signals E1, E2, and S can be used inside the block. When transformed into a design entity, this signal should be bound to ports. This translation is difficult, and errors can be made during this operation. In fact, a block is not a good way of initializing the hierarchy of a design. Such a description is difficult to refine in term of components.

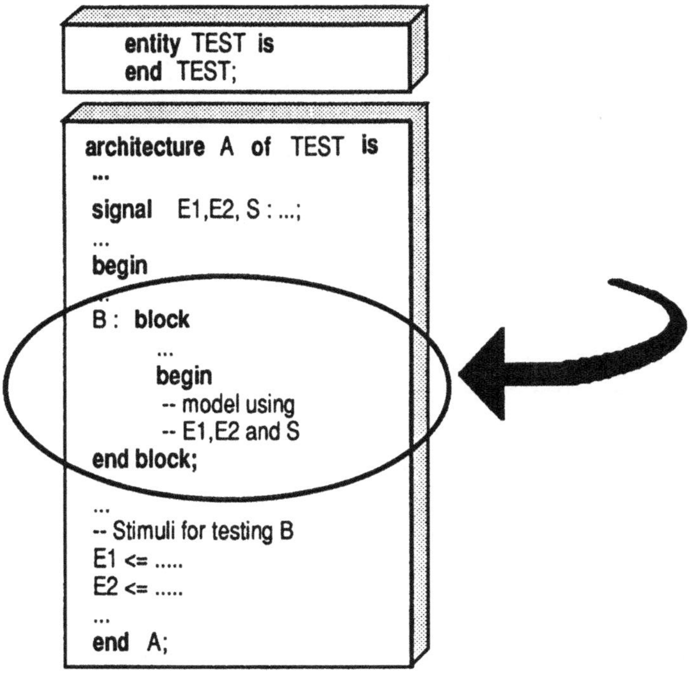

Fig 6.1 Example of Design Using a Block as Hierarchy

To summarize, a block has to be used for its other functionalities, declaration encapsulation and signal assignment factorization (guarded block), and not as a way of explicitly structuring a design.

Note: The block statement is also very useful for allowing automatic translation from another HDL to VHDL, but this kind of translation is not expected tò produce a reusable or readable VHDL source code.

6.2.2. Design Entity and Component Instantiation

A design entity is defined (LRM § 1.1) as an entity declaration, an external view of a black box, together with a corresponding architecture body describing the hierarchy and/or behavior of this black box.

It is possible to make a successive decomposition of a design entity into components that are bound (using the configuration mechanism) to other design entities. The complete design consists of a collection of design entities. The binding of indications between a design entity and a component is ensured by the configuration mechanism. Figures 6.2 and 6.3 show examples of such a binding, the former using a configuration specification, the latter using a configuration design unit. These two mechanisms are conceptually equivalent but provide different structuring properties.

A configuration specification is very easy and concise but makes the links between the entity and the component dependent on the architecture: to change the configuration, it is necessary to recompile the architecture. In the case of a large design, this kind of recompilation can be cumbersome.

On the other hand, writing a configuration design unit is a more flexible way of expressing the link between the components of the design and the entities to use as models, but its syntax is not very exciting. Creating this kind of design unit when the number of components is small or when the links between components and entities do not offer other alternatives can also appear cumbersome.

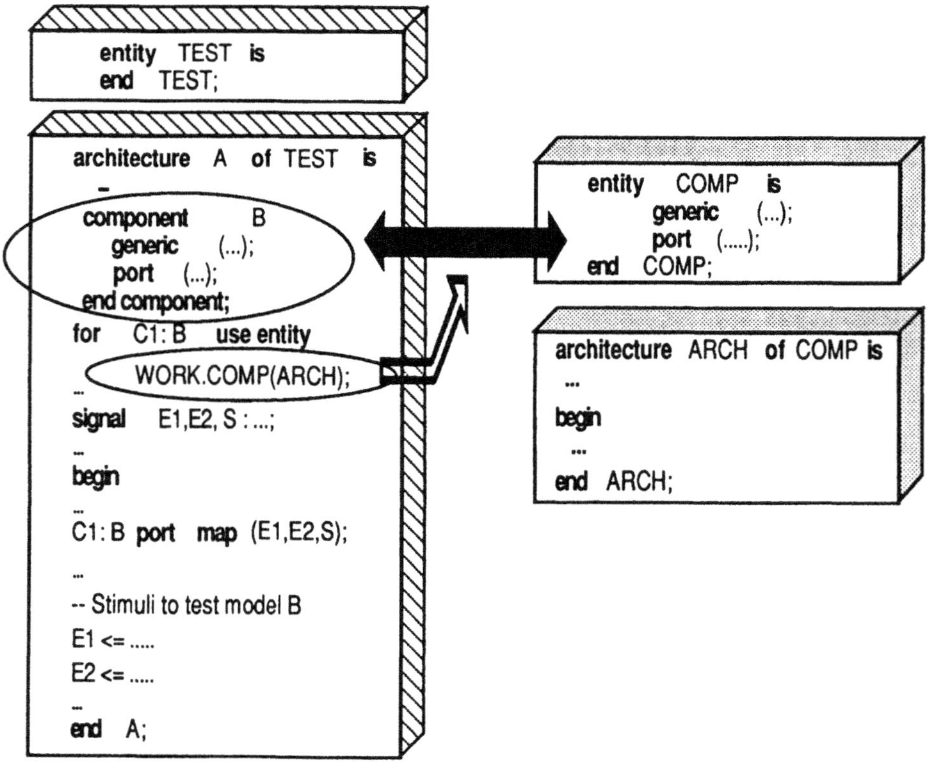

Fig 6.2 Binding Component and Design Entity Using a Configuration Specification

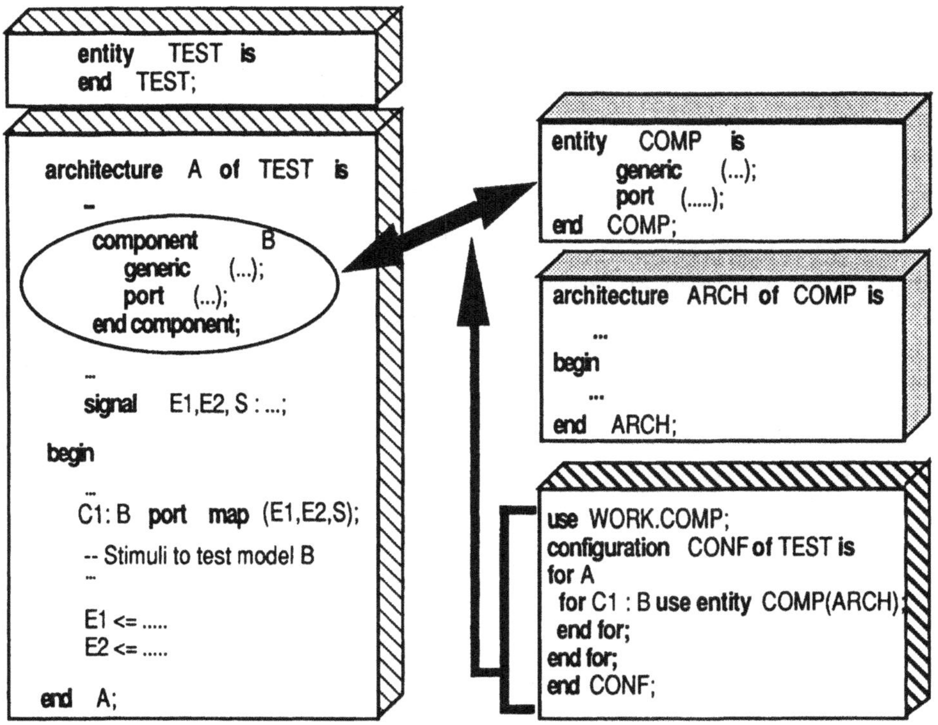

Fig 6.3 Binding Component and Design Entity Using a Configuration Design Unit

For a given design entity, the component is the only way to make successive decompositions in other design entities.

The advantages of decomposing a given design in design entities is well known. Every design entity is composed of two design units (the entity declaration and the architecture body). Each part can be compiled separately with the guarantee that all consistency checks with the other elements of the design have been performed. Any update of the architecture body has no influence on the entity declaration or on any architecture body using components bound to the previous entity. So, it is possible to debug the architecture body without recompiling the whole design. In addition to acting as a debugging facility, this feature of VHDL reduces the design cycle time. In the top-down approach, the designer must frequently replace an architecture body by another, more accurate one. This step of the design (refining) can be carried out (using this mechanism) without any useless recompilation.

Of course, another advantage of such a separation between entity and architectures is to increase readability. It is possible to have a global view of the interface of a model (the entity declaration) without being lost in the details of architecture implementation that remain hidden in the architecture body. A good practice is to attach some comments to ports and generics of the entity declaration in order to make their use clearer.

Another point is that reusability often means generality. If you create and test a general component (in VHDL, *general* can be *generic*), you have a much greater chance of reusing it. The configuration mechanism not only creates links between formal and actual ports, but also between formal and actual generic parameters.

Writing a component declaration, especially if there are numerous ports and generics, can appear boring as well as redundant in terms of entity declaration. In fact, redundancy is not the general case. Component declaration consists in writing "what is needed" for the design, whereas entity declaration describes "what is available". The same scheme is used when making a PCB: you need a given circuit, with given ports and characteristics, but you have to use what the databook proposes. At this level, it is possible to make the binding between the ports you have and those you need (figure 6.4).

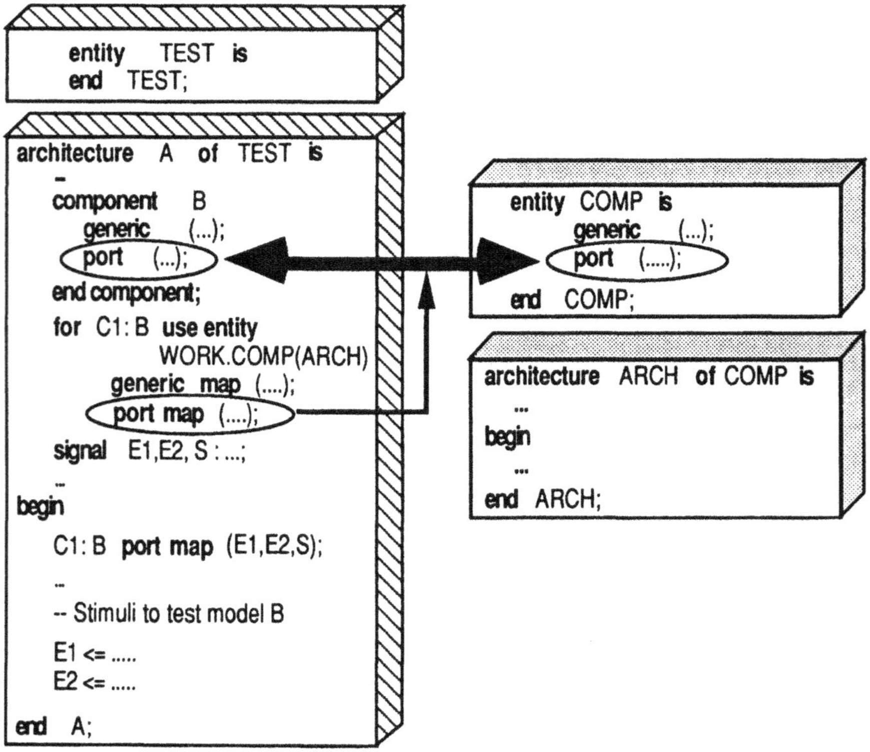

Fig 6.4 Binding Component and Design Entity Ports

This binding consists in giving the mapping of entity port names and component port names. In this mapping, certain ports can be left open (not connected) or can be associated with default values. In many cases where the same designer has written the entity and is using it, the number and names of the component and entity ports are identical, and this port map clause is therefore not useful.

The same binding is available for the parameters in figure 6.5.

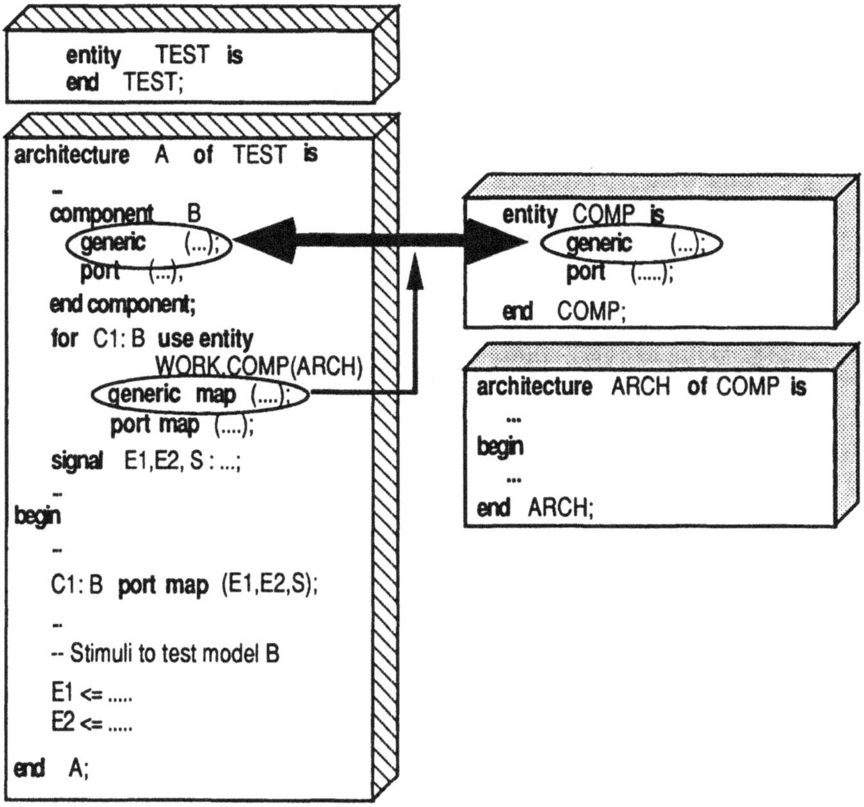

Fig 6.5 Binding Component and Design Entity Generics

At a second level and for each instantiation of a component, the binding between actual (signals or ports you are working with) and formal ports of the component has to be specified (as shown by figure 6.6).

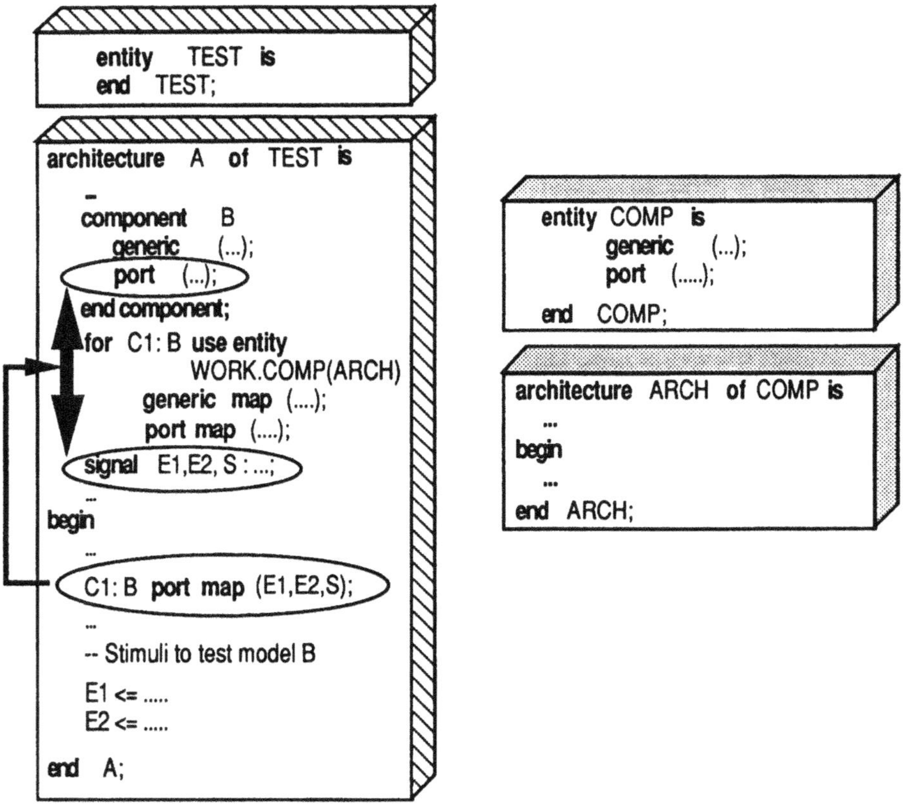

Fig 6.6 Binding Actual Signals and Formal Ports of the Component

Of course, the same mechanism is available for linking the current values of parameters (local constants, for example) to the formal generics of the component. Figure 6.7 illustrates this point.

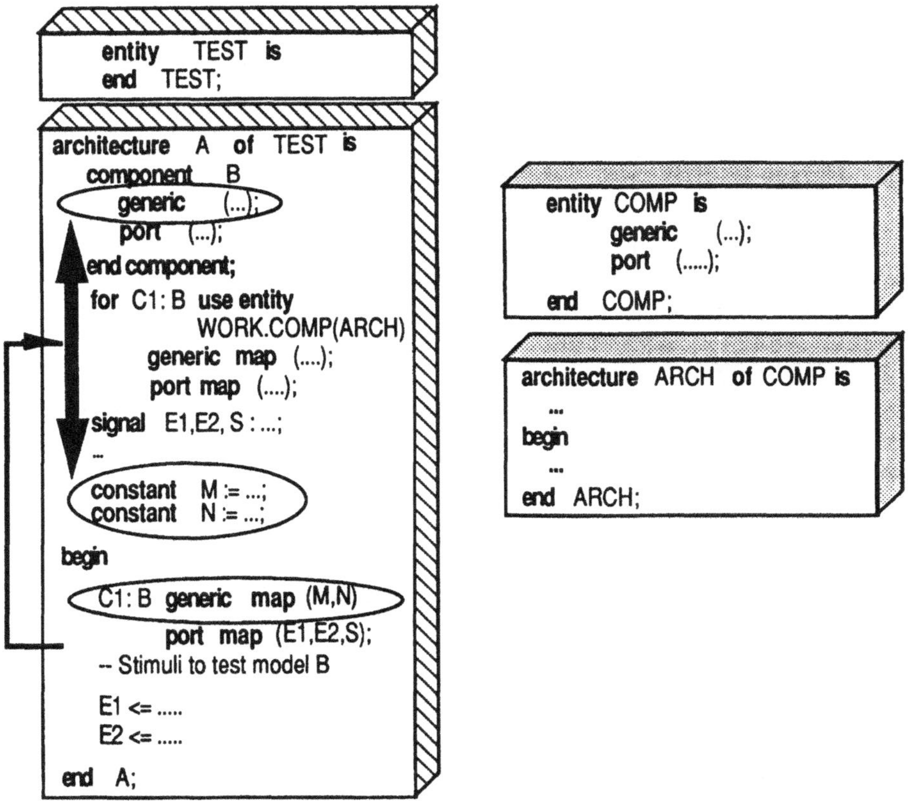

Fig 6.7 Binding Parameter Values and Formal Generics of the Component

This binding mechanism in two steps is very rich and powerful. This power makes it very cumbersome to use in simple cases where the binding could be implicit. Declaring, instantiating, and configuring a component lead to a large number of source code lines. The designer sees the component declaration as a boring copy of the design entity. A full-page editor with cut/paste commands makes it easy, but when the declared component is only used once, the number of code lines can appear to be unnecessarily large.

This is often the case in the upper levels of a top-down approach. Unlike real components, the difference between "what we need" and "what we have" does not exist. The designer divides his description into elements that are "what he needs". No reference to a given library of predefined design entities is made; he writes them as he thinks they should be. The binding between components and deduced design entities is implicit.

Even in this case, this structuring has its own advantages. If a given component has to be instantiated in another part of the design, the structuring strategy is ready to do this. The design entity is stored in a library and can be used by everybody having access to it, and if necessary, the component declaration can be shared by putting it in a package of components. All this structuring leads to reusability and generality.

6.2.3. Design Entity and Package of Components

The package of components is the most powerful structuring in VHDL. As shown in figure 6.8, component declarations are put together in a package declaration. No package body is necessary if the package declaration only contains component declarations.

This structuring strategy offers all the advantages of the previous one "design entity and component instantiation" and is more comfortable to use. Even if a package of components can appear cumbersome and boring to write, the writing of the part of the design that uses it will be simplified.

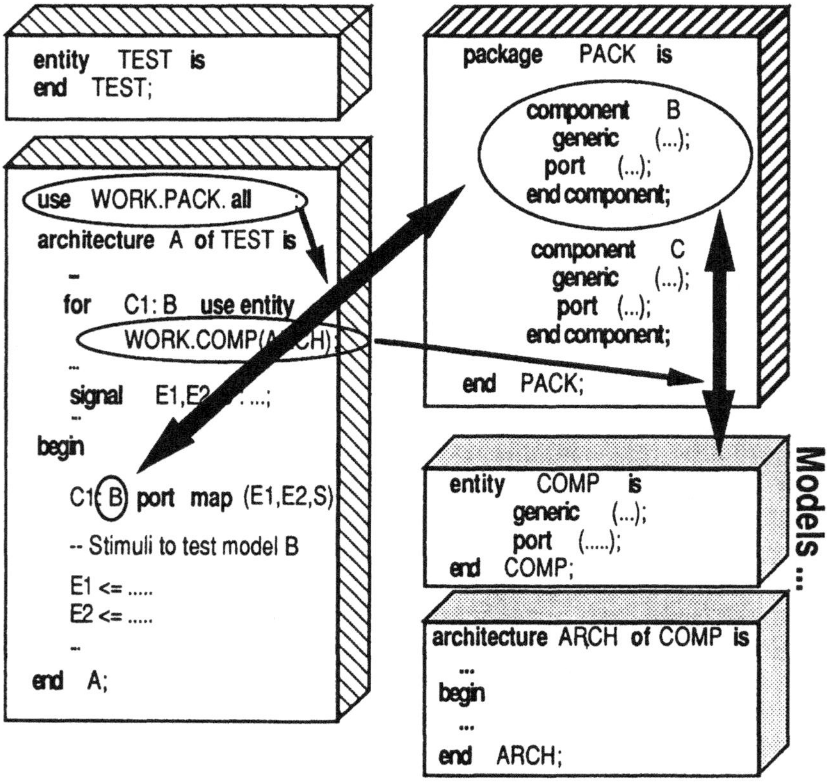

Fig 6.8 Binding of a Component Exported by a Package

When referencing a package of components using a **use** clause (like other packages), the designer saves the burden of component declarations and only has to instantiate and configure them. Therefore, readability and short writing can be achieved in the design.

This possibility of the language has other attractive properties. The available components are declared in a unique place, and all these declarations can be shared by designers. The content of this set of components can be easily seen at a glance. Prolific comments in this package declaration ensure a good reusability of the components.

Furthermore, this structuring allows a powerful flexibility of the design. As mentioned above, the package of components presents "what we have". For example, at a gate level, the package can contain the whole "library" of gates available for standard cell design. The design entities modeling the real cells can easily be compiled in other VHDL libraries. A good organization might be to use a specific library for a given technology. It would thus be possible, using a simple configuration unit compilation, to compare the behavior of a design in different technologies.

It is possible to make a comparison of how to organize a structural description (in blocks, components, and packages of components) and the possibilities of structuring a functional one (in subprograms and packages). Figure 6.9 summarizes this duality.

Using a block statement means putting a source code in line to achieve a given functionality. It works, and is easy to write, but if this portion of code has to be run twice or more, it has to be duplicated. This duplication is a drawback. It is not only cumbersome to duplicate a source code, but if some updates have to be made in this description, this correction has to be written twice or more, with all the bug possibilities that this implies.

The answer to this problem in a functional description is to write the portion of code in a subprogram. The component is the dual solution in a structural description. A subprogram call corresponds to component instantiation. This solution is much more flexible than the first, but does not solve the reusability problem. If another person wishes to use your subprogram (your component), this is not possible if it is nested in the structure.

Both for functional and structural hierarchy descriptions, VHDL suggests the package mechanism for this purpose. The subprogram (the component) is placed in the package, from where it can be exported to everybody referencing the package. This is the most powerful way to produce a reusable code.

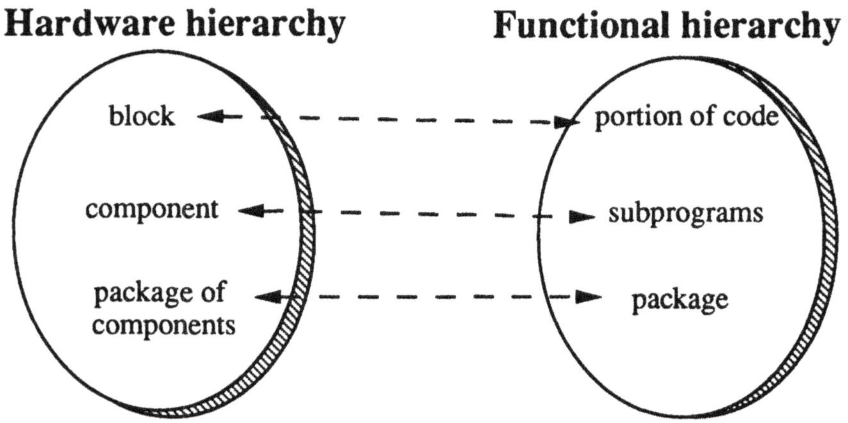

Fig 6.9 Duality in the Use of Components and Subprograms

Note: The goal of this comparison is to show a certain similarity of use between subprograms and components. The essence of a component is very different from that of a subprogram. For example, a subprogram only influences the execution when called, whereas a component is always connected and active. Outside the scope of this comparison is the concurrent procedure call statement, where the iterativity of the equivalent process associated with the reusability of the subprogram makes a powerful trade-off between structural and functional description.

6.2.4. Process Statement

If a part of a design can be described as a list of sequential statements, it is very natural to code it as a process. For all its conciseness, this way of structuring a description has two main drawbacks:

The first is the reusability of such a part of the design. Here again, we face one of the drawbacks of structuring in blocks. If two parts of the design are very similar and can be described by almost the same process, two processes have to be written. There is no way of putting processes in packages for reuse. We will see next that concurrent procedure call statements of procedures stored in packages are close in concept to a "package of processes", but face some restrictions and imply a careful definition of the interface.

The second disadvantage is perhaps the most important. A process is allowed to use any port of the entity and any signal of the enclosing architecture. This coupling makes it very difficult to evaluate the interface of a process with its environment. A glance at its sensitivity list (if it exists) does not solve the problem: a process can use signals that are not on its sensitivity list.

The only way to know the interface of complex processes is to list them and to note all the signals used. This is a painstaking task that has to be systematically performed when deciding to replace a process by a component instantiation in a common top-down approach (refining). The readability of such a structuring is closely related with programming style: processes with multiple wait statements can be difficult to read. Furthermore, the lack of a clear interface with the environment makes it mandatory to explore the entire code to extract the external view of the process.

The sequentiality of processes can make them appear easy to debug. Many debugging tools are proposed in order to set breakpoints within the source code. Other debugging facilities which are well known in programming languages are also proposed. Here also, a bad programming style can have heavy consequences for the debugging phase.

6.2.5. Concurrent Procedure Call

A concurrent procedure call is a concurrent statement and thus a process. The two main advantages of this specific process over an explicit one correspond to the two main drawbacks of the latter.

The reusability problem is solved by the very nature of the procedure. If two parts of the same design have very similar behavior, two different concurrent calls (with perhaps some differences in their parameter parts) can solve the problem. Maintainability is made easier because the algorithm only exists in one place: the procedure body.

Since the interface list of a procedure is very clear, it is very easy to evaluate at a glance the interaction of this part of the design with the environment. Readability is improved, and the transformation of a concurrent procedure call to a component instantiation is one of the easiest things to do in a conventional top-down approach (refining).

Using a concurrent procedure call solves the main drawbacks of using a process but introduces some new limitations.

First, the use of very useful predefined attributes for signals such as STABLE, QUIET, or DELAYED are not allowed in procedures because they define (are) signals. This can lead to the impossibility of expressing a given behavior inside such a procedure.

Second, the sensitivity list of the equivalent process of such a procedure is automatically constituted by all the signals appearing in the actual part.[12]

---

[12] In [9.3], the LRM says exactly: "If there exists a primary that denotes a signal in the actual part of any association element in the concurrent procedure call, and that signal is associated with a formal parameter of mode in or inout, then the equivalent process statement includes a final wait statement with a sensitivity clause that contains the longest static prefix of each signal name appearing as a primary in an actual part and associated with such a formal parameter; otherwise, the equivalent process statement contains a final wait statement that has no explicit sensitivity clause, condition clause, or timeout clause".

Control of the sensitivity of such a process by the designer (excluding some actual parameters from it, for example) must be achieved by replacing the process iterativity by an infinite loop within the procedure. The process never reaches its end, and convenient wait statements can be put inside the loop. Such tricks overcome the problem but do not help to increase the readability of a VHDL design.

Let us take the example of the design of a procedure concurrent call of a classical front-edge register. The interface of this model consists of one input data D, one input clock CLK, and one output Q, all of them being of type BIT. At each rising edge of CLK (but only at this moment), output Q changes to the value of input D.

This behavior is very simple to translate into a procedural form:

```
procedure LATCH (signal D, CLK : BIT; signal Q : out BIT) is
begin
if CLK='1' then
  Q <= D;
end if;
end LATCH;
```

When writing the following concurrent call with actual parameters D1, CLK1, and Q1, which are local signals

```
LATCH (D1, CLK1, Q1);
```

the compiler transforms this statement into its equivalent process:

```
process
begin
LATCH (D1, CLK1, Q1);
wait on CLK1, D1;    -- Here is the problem
end process;
```

The resulting design is sensitive to events on D. In this case, the behavior remains correct (it can be the reverse in other examples); but this kind of event, if D is always changing its value, can lead to an unnecessarily high CPU consumption. A reasonable choice is to write the concurrent procedure to control the sensitivity clause. Since the sensitivity clause of the wait statement of the equivalent process is automatically deduced, the only possibility is to mask it by a new one using an infinite loop.

The following code satisfies the behavior of the latch device and avoids useless activations of the process. However, the readability is not improved by this trick.

```
procedure LATCH (signal D, CLK : BIT; signal Q : out BIT) is
begin
loop
      if CLK='1' then
            Q <= D;
      end if;
      wait on CLK;
end loop;
end LATCH;
```

Note: A good idea *that does not work* could be to control the sensitivity clause of the wait statement of the equivalent process by removing the reserved word signal of parameter D in the procedure interface:

```
procedure LATCH (signal D : BIT; CLK : BIT; signal Q : out BIT) is
```

However, this does not solve any problem. As seen in the previous footnote, the sensitivity clause of the wait statement of the equivalent process is made up of all signals appearing in the actual parameter part, which is always the case for D. This change in class of a formal parameter (from signal to constant) has no consequences on this clause.

All these subtle differences do not simplify the writing of such structuring and can make the debugging phase laborious. Beginners must be very careful when coding concurrent procedures.

6.3. TO SUMMARIZE

At the beginning of this chapter, we enumerated the main characteristics of a good structuring: ease of reading, reusing, debugging, refining , and writing. In the light of the previous paragraphs, we can classify the different features of the VHDL structuring according to their properties in the modeling. Figure 6.10 summarizes this table.

	block	component	pack. of comp	process	conc. proc. call
easy to read	no	yes	yes	no	yes
easy to reuse	no	yes	YES	no	yes
easy to debug	no	yes	yes	yes **	no
easy to refine	no	yes	yes	no	yes
easy to write	yes	no	no *	yes	no

* but make easier the writing of the design using it
** depends on programming rules

Fig 6.10 Modeling Characteristics Depend on Structuring Features

Of course, no structuring feature brings together all the advantages without any drawbacks. But what is not shown by figure 6.10 is that the properties of a description (readability, reusability, etc.) do not have the same importance within a given design.

The relative importance of such a property towards another one depends on the design style chosen. Some examples can easily illustrate this point:

- Designs that can only be achieved by a team of designers have to be easy to read in order to facilitate integration.
- When writing utilities or libraries of models, the reusability aspect of a description is a prime necessity.
- Testing is a very important phase in complex design, and good debugging properties greatly reduce the design cycle time in these cases.
- When coding is a first step of a global top-down approach methodology, a good point is to use structuring features that are easy to refine.
- If the purpose of the design is to quickly carry out a prototype, features that make writing easier have to be privileged. Here, we must be careful. Cases where speed of coding is important are not so common. The experience of programming languages shows that the time spent to first write a design is often a very small part of the design life cycle (which includes debugging, testing, and maintaining). Nevertheless, a good hardware description language must provide features to quickly test an idea, even if once the idea is validated it must be written again with care. And VHDL is also a good language for prototyping.

7. TRICKS AND TRAPS

7.1. MODELING TRAPS

7.1.1. Simulation Time and User Time

Simulation time and user time can be distinguished in the VHDL world. For beginners, this can lead to very surprising behavior of VHDL descriptions.

Here we will point out the basic difference between user time and simulation time. The following process (using package TEXTIO) is supposed to wait until the user enters an integer, wait for ten minutes, and then repeat this operation.

```
process
        variable L : LINE;
        variable I : INTEGER;
begin
READLINE (INPUT, L);
READ (L, I);
wait for 10 min;
end process;
```

And this is definitively what the process does, with its own time. But the user has to enter an integer as soon as the previous one is entered. He does not see the expected ten minutes between each input.

Nevertheless, the simulation date (function NOW) has been incremented with these ten minutes. User time and simulation time are decoupled, even if they have synchronization points as keyboard inputs. Actually, no VHDL statements manage user time, also called "real-life time" (that is the work of the user's boss: "time is money").

Note that "real time" is not part of the VHDL world. Thus, if a simulation describes the first second of run of a system, no designer expects this simulation to be executed in exactly one second. Real-time systems can be described in VHDL, but they will not be simulated in real time.

7.1.2. Signal and Driver

7.1.2.1. Signal Current Value Update

Signal current value update is the first trap for the beginner. After a sequential signal assignment without delay (zero-delay assignment), the current value of the target signal *is not* the result of the right expression. This result is only stored in the target signal driver as a scheduled future value in zero femtoseconds.

If we want a process to be suspended forever when the product of two signal current values (B and C) is greater than a given constant MAX, the following example does not work.

```
process
begin
A <= B*C;
if A>MAX then
  wait;
end if;
wait on B,C;
end process;
```

Quite obviously, the test A>MAX refers to the value of A corresponding to the previous values of B and C. This process will perform one iteration more than expected before waiting forever.

The current value of a signal is only updated on synchronization points: **wait** statements. After these synchronization points, the simulation time (function NOW) can be moved forward. In the case of processes with an explicit sensitivity clause (and so without explicit **wait** statements), the current values of target signals are updated just before the **end process** keywords, but conceptually just after the implicit **wait** statement.

When no **wait** statements seem necessary but the updated values of signals are required, it is possible to create synchronization points. The goal of such a statement is only to provoke the updating of the current values of zero-delay assignment target signals. One form of this synchronization statement is

```
wait for 0 ns;
```

All the current values of signals whose drivers are associated with the current process are updated. There are not many tricks in VHDL, but this statement is one of them. When it appears (and it is very common) in VHDL descriptions, it does not mean that one must wait for a delay, but that one must update the current values of signals. The previous example can be corrected by using this statement:

```
process
begin
A <= B*C;
wait for 0 ns ; -- or fs, or sec... that is the trick
if A>MAX then
  wait;
end if;
wait on B,C;
end process;
```

Now, the value of A compared to MAX is always the last result of the B*C expression.

Note: It is important to note that the use of a signal (instead of a local variable) to store only an intermediate result of a process is very expensive; this will be discussed in paragraph 7.2.3. In the previous example, we assume signal A has another functionality such as communication with other processes. In this case, the solution of using a local variable to store the result B*C in order to compare it with MAX is also valid but decreases the readability of such a code. Why this new local variable? What is its use? The preceding solution (**wait for** 0 ns) is quite common and is easily translated by the reader as the update of target signal drivers. If the designer finds this trick nasty, he can declare a procedure whose sole statement is "**wait for** 0 ns" and whose name is UPDATE_DRIVERS! At any rate, one should always expect a trade-off between efficiency (adding a variable) and readability (only the signal).

A direct consequence of the signal current value update mechanism previously seen is that the two following processes are equivalent. This equivalence can seem amazing to the beginner, but signal assignment *is not* variable assignment, even in the sequential domain.

Then, the process

```
process
begin
A <= B;
C<= A;
wait;
end process;
```

is equivalent to the process:

```
process
begin
C<= A;
A <= B;
wait;
end process;
```

7.1.2.2. Uncontrolled Driver Filtering

Another very common mistake, especially for beginners, is to forget that a signal assignment is implicitly inertial.

For example, assuming MAX equals 3 and PERIOD equals 12 ns, the following process should generate the waveform shown by figure 7.1.

```
process
begin
for I in 0 to MAX-1 loop
   S <=     '0' after I*PERIOD, '1' after I*PERIOD+(PERIOD/3);
end loop;
wait;
end process;
```

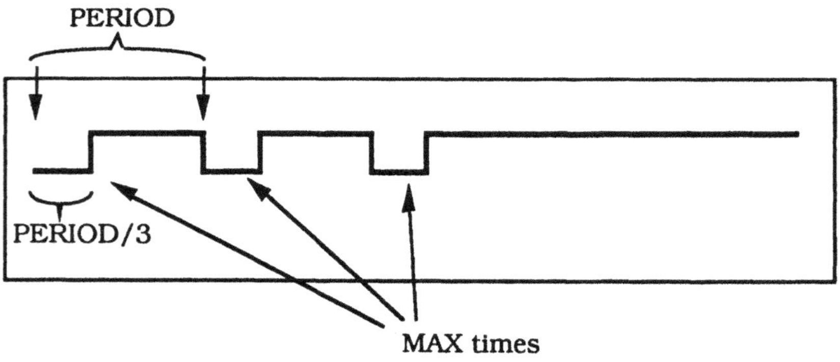

Fig 7.1 Expected Waveform

This is obviously not the case. The inertial signal assignment filters the values of the driver of S at each iteration. All transactions whose date is less than I*PERIOD are erased from the driver at the Ith iteration. The generated waveform shown in figure 7.2 is very different from the expected one.

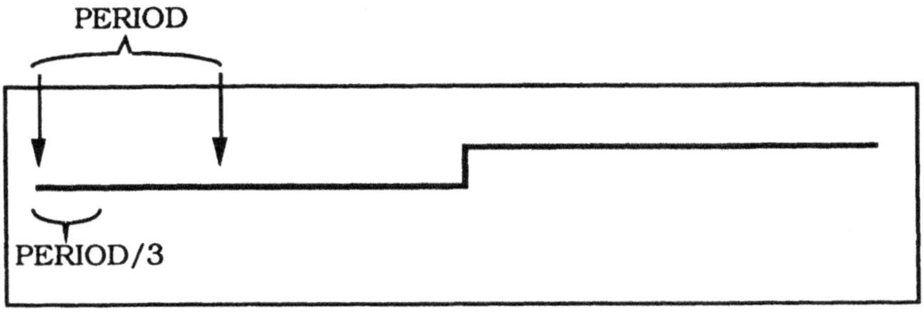

Fig 7.2 Generated Waveform

The correct process can be written only by adding the reserved word **transport**.

```
process
begin
for I in 0 to MAX-1 loop
   S1 <= transport '0' after I*PERIOD, '1' after I*PERIOD+(PERIOD/3);
end loop;
wait;
end process;
```

Note: Another way to write this example is to make the time progress after each signal assignment. Both kinds of signal assignment (inertial or transport) give the same result in this case, because no filtering is performed.

```
process
begin
   for I in 0 to MAX-1 loop
      S2 <= '0','1' after PERIOD/3;
      wait for PERIOD;
   end loop;
   wait;
end process;
```

7.1.2.3. Uncontrolled Driver Erase

The two forms of signal assignment, conditional and selected, are very short writings of powerful processes. Nevertheless, some aspects of their syntax can be very confusing.

The designer often wants to make some assignments if some conditions are true and to preserve the present signal values in other cases.

Using the conditional form (for example, but the same problem can occur with the selected form), the **else** clause seems convenient, but the result is different from the expected one.

In the following example, we assume that S is a previously defined signal of type BIT and that COND1 is a predefined boolean expression including signals.

```
S <= '1' after 10 ns when COND1 else S;
```

This simple statement can be interpreted as follows: signal S receives the value '1' after a delay of 10 ns if the value of the expression COND1 is true; if COND1 is false, there is no change in S. This second part of the interpretation is completely wrong.

If the condition is false:

- A transaction is performed on S. The attribute S'TRANSACTION switches.
- The driver of S is erased of its previous values. It is not the signal S itself that is assigned to S, but only its current value. All waveforms previously stored in the S driver are lost.

This behavior is very different from what the designer thinks by "no change in S", and errors can occur.

The only way to properly solve this problem in VHDL'87 is to write an explicit process. Assuming S_1 and S_2 are signals involved in expression COND1, the previous example becomes

```
process
begin
if COND1 then
  S <= '1' after 10 ns;
end if;
wait on S_1, S_2;
end process;
```

Although this code takes longer to write, it does not generate any transaction on S (does not modify the driver of S) if the condition COND1 is false.

Conditional and selected assignments would always be translated in an explicit process to avoid "no change" problems.

7.1.3. Process Iterativity

When no explicit sensitivity list appears after the reserved word process, the execution will immediately continue with the first statement after the last statement is executed. This illustrates the iterativity of the process.

A common mistake of the VHDL beginner is to forget the sensitivity clause *and* wait statement in the sequence of statements.

Since every process is (at least) activated once during initialization, this process will loop forever. The symptom seen by the user is a high CPU consumption without any results. Certain implementations detect this loop, and a meaningful message is displayed; others leave it up to the designer .

The following process shows this behavior:

```
MY_NAND : process
constant DELAY : TIME :=50 ns;
begin
        OUT_1 <= IN_1 nand IN_2 after DELAY;
end process MY_NAND;
```

Of course, to avoid such a pitfall, the right process must be written:

```
MY_NAND : process
constant DELAY : TIME :=50 ns;
begin
        OUT_1 <= IN_1 nand IN_2 after DELAY;
        wait on IN_1, IN_2;
end process MY_NAND;
```

Or, if we suppose that the constant DELAY is declared in an enclosing declaration area:

```
MY_NAND : OUT_1 <= IN_1 nand IN_2 after DELAY;
```

7.1.4. Resolution Functions Properties

A resolution function is a user-defined function that computes the resolved value of a resolved signal. The term "user-defined" indicates that this function is written by the user, but the resolution function is implicitly invoked during the simulation cycle. The user does not control when this function call happens.

Beginners usually think that the resolution function is only called when there is a conflict and, therefore, when there are at least two sources with different values. This is a false idea that can lead to a trap. Resolution functions are called even if the sources have the same value and even if there is only one source. At the gate level, if two sources are providing '1', the conflict is not obvious; but what can de said about two sources at the system level providing the same value as REQUEST_TO_THE_BUS? In some guarded blocks, a resolution function can also be called when all sources are disconnected. The designer must not make any assumption about when the resolution function is called.

Furthermore, the user does not handle the way in which the single input parameter used by such a function represents the multiple sources. The order of this collection of inputs is implementation dependent and can vary during simulation.

This implies that a resolution function whose result depends on the order of the parameters can yield different results even if called with the same set of input values. This breaks the determinism of VHDL and produces non-predictable and non-portable resolution functions.

Nowadays, it seems impossible to ask the compiler to verify that resolution functions have the right property, and the LRM does not ask for it. But the user has to be very careful; it is very easy to write non-commutative or non-associative functions, and the non-deterministic behavior of a program is very difficult to debug. The next example shows a simple example of such a function (the **nand** operation is not associative).

```
function BAD_IDEA (SOURCES : BIT_VECTOR) return BIT is
        variable RESULT : BIT :=SOURCES(SOURCES'LEFT);
begin
        for I in SOURCES'LEFT+1 to SOURCES'RIGHT loop
          RESULT := RESULT nand SOURCES(I);
        end loop;
        return RESULT;
end BAD_IDEA;
```

Note: No function can have side effects in VHDL, except one: NOW. It is possible to call it twice (or more) with the same parameter values (none) and to obtain (of course) different results.[13] Much discussion preceded its introduction into VHDL (should a model depend on the simulation time?). The use of this function to compute the result in a resolution function can also be a good way to break the determinism of VHDL.

7.1.5. Transactions and Events

Let us consider two processes working on signals A, B, R, and M of type NATURAL. A first process, DIVISION, is performing an operation on signals A and B and will wait for a new value of A. The second process, MAIN, waits for the result of one of the previous operations (the modulo) and then performs a new assignment on signal A.

The following code does not conceptually translate the behavior previously described:

```
DIVISION : process
begin
R <= A / B;
M <= A MOD B;
wait on A, B ;
end process DIVISION;
```

[13] But do not forget that you can also call it twice and get the same result; between two wait statements, the time does not advance.

```
MAIN : process
begin
wait on M;
A <= M;
end process MAIN;
```

These two processes are waiting for each other on events (i.e., on value changes) and not on transactions (i.e., on performed operations).

The consequences are easy to point out: if signal A takes the value 6 and signal B the value 2, signal M will be computed by the process DIVISION as 0. This value does not represent a change (an event) with regard to the initial value of signal M (a natural is initialized by 0), and so the process MAIN will still wait for a new value of signal M. It will not assign a new value to signal A, and then the two processes will wait forever.

Of course, the correction of such a trap is easy: replacing **wait for** M (respectively, **wait for** A) by **wait for** M'TRANSACTION (respectively, **wait for** A'TRANSACTION) will solve the problem.

Nevertheless, and especially when working on high-level datatypes (here INTEGER is a higher-level datatype than BIT, for example), the designer has to clarify if the synchronization mechanism works on changes (events) or on operations (transactions).

7.1.6. Overloading of Predefined Operators

Overloading is defined for enumeration literals and subprograms.

It can be very useful to redefine some predefined operators. In the next example, we modify the integer division to avoid an error when dividing by zero. A specific treatment occurs in this case, but the standard predefined division is used in all other cases.

The following code, however natural it may be to write, does not work. When overloading a subprogram, the visibility rules make the original operator not directly accessible.

```
function "/"(A,B : INTEGER) return INTEGER is
      begin
      if B=0 then
            if A = 0 then return 0;
            elsif A>0 then
                  return INTEGER'RIGHT;
            else
                  return INTEGER'LEFT;
            end if;
      else
            return A/B;   -- A beautiful recursive call indeed!
      end if;
end "/";
```

The hidden predefined operator "/" still exists and is accessible using dot notation. LRM 6.3 says: "An expanded name denotes an entity declared in a package if the prefix denotes the package and the suffix is the simple name, character literal, or operator symbol of an entity whose declaration occurs immediately within that package."

When a subprogram SUB with same parameter type profile is defined in two different visible packages, P1 and P2, the notations WORK.P1.SUB or WORK.P2.SUB allow the appropriate one to be selected. But in which library and package are the predefined operators declared?

The answer is simple: predefined operators on a given datatype are implicitly declared when this datatype is declared: in our case, the package STANDARD in library STD.

In the previous example, the dot notation STD.STANDARD."/" will allow access to the standard integer division. The previous example can be easily rewritten:

```
function "/"(A,B : INTEGER) return INTEGER is
        begin
        if B=0 then
            if A = 0 then
                return 0;
            elsif A>0 then
                return INTEGER'RIGHT;
            else
                return INTEGER'LEFT;
            end if;
        else
            return  STD.STANDARD."/"(A,B);
        end if;
end "/";
```

But another case exists and cannot be solved by LRM 6.3: the overloading of a predefined operator in a package where the type is also declared.

The following paragraph (LRM 7.3) is therefore applicable: "Two declarations that occur immediately within the same declarative region must not be homographs, unless exactly one of them is the implicit declaration of a predefined operation. In such case, a predefined operation is always hidden by the other homograph. Where hidden in this manner, an implicit declaration is hidden within the entire scope of the other declaration; the implicit declaration is visible neither by selection nor directly."

An example can illustrate this situation:

```
package LOG is
        type BIT4 is ('x', '0', '1', 'z') ;
        function "="(A,B : in BIT4) return BOOLEAN ;
end LOG ;
```

```
package body LOG is
        function "="(A,B : in BIT4) return BOOLEAN is
        begin
                -- The access to the implicit "=" function is impossible
                -- This code is incorrect in VHDL
                if LOG."="(A, 'x') or LOG."="(B, 'x') then
                        return FALSE ;
                else
                        return LOG."="(A, B) ;
                end if ;
        end ;
end LOG ;
```

In this specific case, access from the package body (and also from other packages) is impossible: the implicit "=" function is defined as soon as the type BIT4 is declared and in the same declarative region. The next line declares, in the same declarative region, the explicit "=" function, which hides the previous one. The only solution is to define the equality without using the predefined one (through the use of tables, for example).

Note: To declare a new function (EQUAL) between the type BIT4 declaration and the explicit function "=" does not solve any problem. When writing its body within the package body, the use of the implicit "=" operator will also be prohibited. The package body knows all the declarations of the package declaration.

7.1.7. Kinds of Predefined Attributes on Signals

There are two kinds of predefined attributes on signals: functions and signals. To choose the appropriate kind for a given description can be difficult and may lead to destabilizing traps.

If we want to create a guarded block to perform the assignments of signals X and Y when a given signal S of type BIT rises from '0' to '1', the following description seems simple enough to write:

```
B1 : block (S'EVENT and S='1')
begin
        X <= guarded Z1;
        Y <= guarded Z2;
end block B1;
```

Assignments of X and Y are supposed to be only carried out on the rising edge of S; but in fact, they are executed on the high state of S. The trap is real, and the description does not reflect at all what we want to model.

Where is the problem? The kind of predefined attribute EVENT, a function, is guilty.

From the previous description, an implicit GUARD signal is created. The value of this signal (S'EVENT **and** S='1') is re-evaluated each time an event on S occurs. By definition, the value of the function S'EVENT each time an event on S occurs is TRUE. Then, the previous expression is reduced to (S='1'). In our description, each time an event occurs on signal Z1 or Z2, the value of the GUARD signal is checked (but not re-evaluated!), and if this value is true, which is the case during the entire period where S equals '1', the assignment of X or Y is performed. This guarded block is not sensitive to edges *but to levels*.

The correction is easy. Instead of the 'EVENT function, we need the dual signal. Events on this signal will lead to the re-evaluation of the value of the GUARD signal. Unfortunately, there is no predefined attribute of the signal kind that offers the same functionality, but one with the opposite value exists. The signal 'STABLE, as shown in figure 7.3, takes the value TRUE each time no event occurs on the signal.

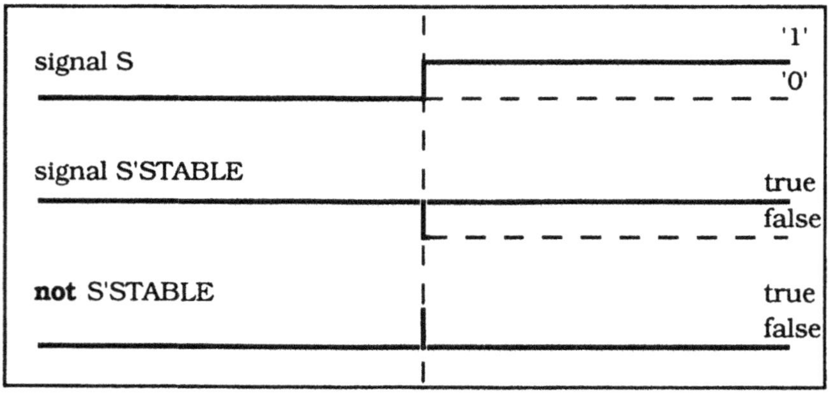

Fig 7.3 Use of Signal S'STABLE

Using this signal, the following description is appropriate:

```
B2 : block (not S'STABLE and S='1')
begin
      X <= guarded Z1;
      Y <= guarded Z2;
end block B2;
```

Note: If A is a boolean type signal, using the previous explanation, it is easy to conclude that the concurrent assignment

```
A <= B'EVENT;
```

is equivalent to

```
A <= TRUE;
```

which is not what was intended!

7.1.8. Concurrent Procedure Call Statements

Never forget to put the reserved word **signal** to transmit a signal as an actual parameter in a concurrent call statement. If you forget to do so, the behavior of your description will be surprising, and the bug will be difficult to localize.

The next example illustrates this situation. This VHDL code designs and tests a behavioral description of a latch register instantiated as a concurrent procedure call.

```
package PACK_COMP is
        procedure REGISTRE(CK : in BIT; signal  D : in BIT; signal Q : out BIT) ;
end PACK_COMP;

 package body PACK_COMP is
        procedure REGISTRE(CK : in BIT; signal D : in BIT; signal Q : out BIT) is
        begin
        wait until CK='1';
              Q <= D;
        end REGISTRE;
end PACK_COMP;

 use WORK.PACK_COMP.all;
entity TEST is
end TEST;

 architecture SIMPLE of TEST is
        signal CLOCK, INPUT, OUTPUT : BIT;
begin
        CLOCK <= not CLOCK after 10 ns;
        INPUT <= '0', '1' after 5 ns, '0' after 25 ns,  '1' after 50 ns;
        C1 :REGISTRE(CLOCK, INPUT, OUTPUT);
end SIMPLE;
```

Except for some "smart" implementations which detect an until clause without signals within its condition, this "latch register" model does not work as previously expected. The equivalent process statement of a concurrent procedure call consists of two sequential statements. The first is the sequential procedure call with the same procedure name and actual parameter part that appear in the concurrent procedure call. The second is a wait statement with a sensitivity clause in which appears every signal name present in the actual part. Signals CLOCK and INPUT are present in this sensitivity list.

The formal parameters of the concurrent procedure are of either the constant or signal kind. If the reserved word **signal** is not used, these parameters are considered as constants. This is what happened to CK in the previous example. The object class of formal CK inside the procedure is constant, and the statement **wait until** CK='1'; appears as a nice trap because no signal is used in the condition and the list of signals of the implicit **wait on** statement is empty. This statement is equivalent to a wait forever (**wait**;).

A good practice to avoid such errors, which can be very difficult to trap in large designs, is to explicitly specify the class (constant or signal) of each formal parameter.

7.1.9. Sensitivity List Close Control Subtleties

The more explicit your sensitivity list, the more efficient will be the running of your code.

Implicit features are very pleasant to use, but can also be treacherous. Even the very simple wait statement must be handled with care:

- The implicit feature we are relying on may disappear

Supposing we are modeling a synchronous counter that is sensitive to the rising edge of clock CLK. We can write:

```
wait until CLK='1';
```

This sequential statement stops the process until a rising edge of CLK. If we need a description *also* sensitive to any event on a reset signal, the temptation is great to update this code to

```
wait on RESET until CLK='1';
```

And this code is no longer sensitive to the events on CLK!

An implicit sensitivity list of signal has been built in the first case because no explicit sensitivity list was expressed. So

```
wait until CLK='1';
```

was understood as

```
wait on CLK until CLK='1';
```

This implicit list consists of the signals that are present in the condition. But this implicit list is only built when the explicit one is empty, which is not the case in the second example:

```
wait on RESET until CLK='1';
```

This statement waits on an event on RESET in order to evaluate the condition. The implicit list has disappeared.

- An implicit feature may not be what we need

If we desire a counter able to count each rising edge of a clock CLK when the ENABLE signal is high, the following statement seems convenient:

```
wait until (CLK='1' and ENABLE='1');
```

But the process will go on every rising edge of ENABLE when CLK is high, which is not what we need. The implicit has transformed the previous statement in

```
wait on CLK, ENABLE until (CLK='1' and ENABLE='1');
```

but we only needed:

```
wait on CLK until (CLK='1' and ENABLE='1');
```

7.1.10. File Subtleties

7.1.10.1. Portability of Text Files

This package must be provided by a simulator vendor. The package is supposed to export types and primitives to read and write text files on the host machine. The primary goal for this package was to provide in the language the possibility of writing outputs that could be compared from one implementation to the other (test suites, for examples).

Problems may come from the fact that the notion of "text file" is not precise. It is only assumed that such a file is to be read or written by a text-editor.

One typical problem arises when a text file produced by an implementation on a given machine is ported for comparison purposes on another machine. It may happen that terminators (i.e., marks indicating end of line) are not the same, and thus a text file for one system will become a binary file for another.

Consequently, the same VHDL text may produce different text files on different operating systems, even with the *same* root of simulator!

7.1.10.2. File Open/Close Concept

Another issue with files is the problem of closing them. VHDL deliberately does not provide open or close primitives to prevent the use of files as a means of communication between processes. Signals, and only signals, are provided for the purpose of synchronization. You cannot, for example, open a file in a process, close it, and reopen it from another process. However, a given procedure can declare a file object, write in it, and exit. The scope of the declaration of the file is thus lost. If now we reenter the procedure, VHDL does not state which physical file is to be opened for writing: is it another one (for systems that handle different versions), the same at the end (append), or the same (scratch the old one)? Actually, different implementations make different choices. The result is that different text files may be produced by the same VHDL text on different implementations!

7.1.11. Constant Evaluation

If a constant of a VHDL text is initialized using a user-defined function, it will be rejected by some implementations. For example:

```
constant SIZE : SIZE_TYPE := COMPUTE_OBJECT_SIZE(OBJECT) ;
```

This is forbidden by a strict interpretation of the LRM (the result of a user-defined function is not said to be "globally static"). However, there is no real reason to reject a user-defined function call from the list of globally static values. This is why many implementations allow this kind of initialization. This seems to be a minor issue, but an implementation that forbids user-defined function calls at elaboration time also forbids the structure to be parameterized at the last moment (for instance, by reading the content of a file): generic arguments and arguments of generate statements should be globally static.

Another philosophical issue on a given tool is to check whether or not it is allowed to read the keyboard from within a function (a function should not have side effects, but this particular case is not explicitly forbidden in the LRM).

If your implementation forbids one of these two features, you will not be able to change generics at each run of your model. This might involve many recompilations!

If, on the contrary, your implementation allows them, you have to document internally the use of these features in a model, because its portability is questionable.

To test both of these features, just create a new entity and a package:

```
package P is
        function FN return INTEGER ;
        constant C : INTEGER ;
end P ;

use STD.TEXTIO.all ;
package body P is
        function FN return INTEGER is
            file F : TEXT is in "STD_INPUT" ;  -- the keyboard
            variable L : LINE ;
            variable I : INTEGER ;
        begin
            READLINE(F,L) ;  -- possible error
            READ(L,I) ;
            return I ;
        end FN ;
        constant C : INTEGER := FN ;  -- possible error
end P ;

entity TEST_STATIC is
end TEST_STATIC ;

use STD.TEXTIO.all, WORK.P.all ;
architecture ARCH of TEST_STATIC is
        component BENCH
            generic (N : POSITIVE) ;
        end component ;
begin
        B : BENCH generic map (FN) ;  -- possible error
end ARCH ;
```

The result of this test is easy to evaluate: either it compiles or it does not!

7.2. MODELING TRICKS

7.2.1. Name Identifier Policy

7.2.1.1. General Considerations

VHDL offers many features to write very large descriptions in terms of lines of codes. The mechanisms of encapsulation (design units, deferred constants, etc.) allow real teamwork. These two considerations imply a reality: you have to use part of a code written by other people, and your code has to be easily readable by any member of your team.

Name identifier policy is important to produce a good readability quality to your code.

The multiplicity of very powerful full-page editors with cut-paste commands allow the use of long and significant identifier names with underscores between words.

Identifiers such as MY_BUFFER_OUTPUT are self-documented, even though the use of multiple comments is also a good practice. Identifiers such as A, B, or C must be reserved for the writing of some VHDL book examples.

Now it is time to think about the people who will have to read your description — and not only to pity them.

7.2.1.2. The Problem of Port Names

Readability is a wish; VHDL constraints are a reality.

One well-known problem in using VHDL to model existing hardware devices is that the names of their ports cannot always be considered as legal VHDL identifiers. Certain device port names begin with a digit, conflict with VHDL reserved words, or contain special characters. The designer must use other names, and this does not help to clarify the interface with such a device. For the designers of this language, remove these naming constraints will lead to large lexical and parsing problems.

To solve this problem, some vendors provide a syntactic comment-based mechanism that is out of the range of the language. A better practice is to create user-defined attributes of the type STRING. Such attributes will be associated with each device port and will contain the exact name of the existing hardware device port. In VHDL, it is possible to read the values of these attributes and to write them in a file; debugging tools allow them to be examined. Readability is not as good as allowing non-constrained identifier names, but the possibility of handling such a name in VHDL (writing it in a file, for example) is very useful, and no language allows this to be done with identifier names.

7.2.2. Writing up Black Boxes

Two kinds of black boxes exist in VHDL. Software black boxes are package declarations, and hardware ones are entity declarations. The goal of these two classes of design units is to describe the external view of something — in other words, to separate declaration and implementation.

This is an important notion in object-oriented programming. The declaration usually defines the name of the object as well as the operations that could be performed on it. Separating this definition from the implementation of the operations provides many benefits. Encapsulation is made easier and favors good readability of the source code. Readability is good because the objects we are dealing with are well defined, and it is possible to hide many implementation details. With some dozens of source code lines, it is usually possible to give to the designer all the information he needs about the package or the model he will use. It is also very common that the implementation of such a package or model takes some hundreds of source code lines.

Some recommendations will nevertheless help to make the readability even better:

- It is a good practice to write comments to specify the semantics of the objects declared in this design unit. Self-documentation is a nice policy but is limited in its power of explanation.
- Especially for package declaration, it is important only to declare the objects exported by the package. Subprograms or types local to a package must be declared in the corresponding package body (and so be hidden to the package or model user).
- When using a model, nothing is more frustrating than to see unexpected assertion errors or warnings. When they concern the interface of the model (generics value consistency or port timing checks, for example), such verifications are better shown in the entity declaration. They can be concurrent assert statements or any passive process.[14] Errors or warnings related to generics or parameters checks must be considered as possible "outputs" of the model and documented accordingly.

---

[14] If the source code of such passive processes is too long, it is preferable to write it as a concurrent procedure call (in the entity declaration) in order to let it be visible without making it too cumbersome. The fact that a check is done is important for the model user, but the way to do it can (must) be hidden.

Entity (or package) declarations must be taken as the "directions for use" of the model (or package) itself. All useful information must be found there, but only it.

7.2.3. Source Code Optimization

There are many ways to code the same functionality in VHDL. This variety is not specific to HDL; conventional programming languages offer the same range. We saw in previous chapters that readability, reusability, or debugging capability depend on programming style. Performance in terms of CPU time and used memory can also depend on some modeling tricks, and some of them are well known by software people.

Since the performance of VHDL tools is often a sensitive matter, it is useful to remember certain practices:

- If several objects depend on the result (evaluation) of an expression, it is desirable to store them (in a variable, if possible) in order to avoid useless computations.[15]
- Use of subprograms is a good practice to avoid code duplication and to encapsulate data. A subprogram call has a cost, in term of memory (context storage) and CPU time (context restoration). Of course, this cost is commonly negligible in comparison with the convenient structuring it provides. However, there are some specific exceptions. For example, when an algorithm is based on a loop that calls a subprogram not used by other parts of the application, putting the subprogram code "in line" can lead to a great sparing of CPU time. In the same spirit, using recursive subprograms is not a very good idea if CPU time performances are an issue. Recursive codes, because of their stacked contexts, are always more expensive than the iterative form.
- As seen earlier, the use of procedures in place of components can be very efficient: remember that a concurrent procedure call is

  ```
  LBL : PROC(S1,S2,S3);
  ```

  and can be functionally equivalent to a component instance:

  ```
  LBL : COMP port map (S1,S2,S3);
  ```

  The central difference is that the functionality of the procedure is written in its body, whereas the functionality of the component comes from a (explicit or implicit) configuration. The inside of the procedure is, of course, purely sequential and cannot bear any structure. But should this

[15] Optimistic people often like to think that their compiler is well optimized for such cases. This may be true.

not be a limitation (e.g., leaf components), it is far faster to compile and test such a description, and the compiler has more chances to optimize the code.

- VHDL offers a large range of datatypes. When a complex algorithm has to be written, performances depend on the choice of adapted datatypes. For example, arithmetic operations will be more efficient when working on integers than on bit vectors. Of course, some accuracy constraints can imply the use of low-level types (bit true computation, for example), but in some cases the choice remains open.
- A signal is expensive. It requires the simulator to create a driver and to schedule all future values of the signal. The expense in term of size of generated code and simulator task is high. The use of a signal when a variable can fit is to be prohibited. In a first approach, a signal must only be used if it has a hardware meaning as interprocesses communication.
- Stimuli and trace usually imply file operations. They are commonly not very efficient in VHDL, and when a complex stimulus has to be used, a good strategy can be to write it in VHDL. Of course, when a great number of different scenarios of stimuli have to be envisaged, using files remains a more flexible solution.
- Earlier we saw the importance of close control of the sensitivity list. In the same spirit, it can be very efficient to split a VHDL description into processes that are only sensitive to a minimal number of signals. In processes with a large sensitivity clause, conditional statements commonly try to know which signal is active. This can even lead to the memorization of previous values of such signals. In this case, efficiency means partitioning in simple sensitivity clause processes.
- Use the **wait until** form of the wait statement only when necessary; as seen before (and assuming CLK is a signal), the statement

```
wait until CLK='1';
```

has the semantics of

```
loop
    wait on CLK;
    exit when CLK='1';
end loop;
```

which means that, each time CLK changes, the expression has to be checked, and the loop is actually executed in some way. The **wait on** and **wait for** forms are much cheaper to implement, because they are static and most of the work is done (once) at compilation time. If you need in many processes[16] such a "**wait until** CLK = '1'; ", an efficient practice is to create an intermediate signal and a process:

```
signal CLK_MOVES_TO_ONE : BOOLEAN;

process
begin
    if CLK ='1' then
        CLK_MOVES_TO_ONE<= not CLK_MOVES_TO_ONE;
    end if;
    wait on CLK;
end process;
```

Then **wait on** CLK_MOVES_TO_ONE[17] must replace the previous **wait until** CLK='1'! This trick guarantees that CLK='1' will be evaluated once only, and that all other wait statements will be efficient. Unfortunately, VHDL'87 has no shortcut to write this statement (no equivalent concurrent statement). If you need to handle many signals in this way, use a concurrent procedure call instead of a process to factorize the code:

```
procedure UPDATE_MOVES_TO_ONE ( signal THE_SIGNAL : XBIT;  -- see 4.1.2.2
                                signal MOVES_TO_ONE : inout BOOLEAN) is
begin
  -- Infinite loop to avoid being sensitive on MOVES_TO_ONE
  loop
      if THE_SIGNAL ='1' then
          MOVES_TO_ONE <= not MOVES_TO_ONE ;
      end if;
      wait on THE_SIGNAL;
  end loop;
end UPDATE_MOVES_TO_ONE ;
```

or this one, where the use of a local variable, faster to read, allows for an out mode for MOVES_TO_ONE, easier to use in the general case:

```
procedure UPDATE_MOVES_TO_ONE (signal THE_SIGNAL : XBIT;
                               signal MOVES_TO_ONE : out BOOLEAN) is
variable LOCAL : BOOLEAN;
begin
      if THE_SIGNAL ='1' then
          LOCAL := not LOCAL;
          MOVES_TO_ONE <= LOCAL ;
      end if;
      wait on THE_SIGNAL;
end UPDATE_MOVES_TO_ONE ;
```

---

16 And this is commonly the case when describing a synchronous system.

17 CLK_MOVES_TO_ONE will change each time CLK becomes '1', so the value itself of the boolean is irrelevant. We just use the event.

There is no need for an infinite loop in this example. MOVES_TO_ONE is an output port, and therefore is not involved in the sensitivity list.

7.2.4. Range of an Array

Especially in subprograms, it is important never to make any assumption about the range of an array transmitted as a parameter. This will give a more general code and will make the risk of constraint errors disappear. The VHDL beginner well knows this error, which classically occurs when indexing with zero an array whose index has been declared as positive, or, on the other hand, when using the element N of an array defined from 0 to N-1.

Making no suppositions on the range of array means a specific programming methodology with a wide use of predefined attributes. Resolution functions are commonly written using such a methodology: their single input parameter is an unconstrained array.

In the following example, a function to convert BIT_VECTOR into STRING makes no supposition concerning the size constraints of the BIT_VECTOR argument.

```
function IMAGE (P : BIT_VECTOR) return STRING is
        variable STR : STRING (1 to P'LENGTH);
        type BIT_TO_CHARACTER_TYPE is array (BIT) of CHARACTER;
        constant BIT_TO_CHARACTER : BIT_TO_CHARACTER_TYPE:=('0','1');
begin
        for I in P'RANGE loop
            STR(I):=BIT_TO_CHARACTER(P(I));
        end loop;
        return STR;
end IMAGE;
```

7.2.5. Manipulating "Abstract Datatypes"

One of the most powerful features of VHDL is that it provides a means to construct user-defined types in order to support abstract data and to encapsulate them with their related subprograms: the packages. Especially in system modeling, this possibility is widely used.

But VHDL does not provide any mechanism to protect the structure of these datatypes from undesirable manipulations, and the result may destroy the consistency of the modeling.

To illustrate this purpose, we will present the example of a package modeling a FILO (First In Last Out Stack). This package exports two procedures (PUSH and PULL).

```
package FILO is

        subtype FILO_SIZE is INTEGER range 1 to 10;       -- Range of the filo
        subtype ELEMENT_TYPE is REAL;                     -- Type stored in the filo

        type ELEMENT_ARRAY_TYPE is array (FILO_SIZE) of ELEMENT_TYPE;
        type    FILO_TYPE is record                       -- Data representing the filo
                INFO : ELEMENT_ARRAY_TYPE;                -- Contents of the filo
                P_START : FILO_SIZE;                      -- Starting address of the filo
                P_END : FILO_SIZE;                        -- Ending address of the filo
        end record;

        -- Procedure to add one element to the FILO
        procedure PUSH (STACK : inout FILO_TYPE; ELEMENT : in ELEMENT_TYPE);
        -- Procedure to get one element of the FILO
        procedure POP (STACK : inout FILO_TYPE; ELEMENT : out ELEMENT_TYPE);

end FILO;
```

As the structure of the stack itself is given in the package declaration, a designer can decide to update it without calling the related procedures. In that case, its own algorithms will manage the values of the starting and ending address of the FILO. This has some negative effects:

- Proposed procedures PUSH and PULL are supposed tested and safe to use. This may not be the case with the designer's procedures.
- Nothing will ensure the consistency of use of P_START and P_END between the package procedures and that of the designer. What will be the result if they are both used to operate on the same FILO?
- For debugging and maintainability reasons, it is not acceptable to multiply the algorithms that perform the same task on the same object.

Other, more complex examples using very sophisticated abstract datatypes can be easily found. Types using access types are among these, with their need to properly deallocate the objects after use.

Indeed, when a package is representing an object (here, a FILO), the structure of the object itself can (and often must) be ignored by the designer. The only information he needs is the meaning of the types he manipulates (a FILO, an element of a FILO) and the functionality (the semantics) of the related subprograms (to push a data into the FILO or to pop it). Ignoring the structure of the object will automatically lead him to adopt a correct policy by using the related subprograms.

7.2.6. Accessing the Value of Output Port

7.2.6.1. Describing the Problem

Especially when modeling simple hardware devices, it can appear very restricting not to be allowed to read output port values (i.e., to place them on the right-hand side of assignment statements).

For example, when describing with timing a simple nand gate, it can be necessary to give a different delay depending on which transition happens. When using a multi-valued logic system, the number of different delays (transitions) can be significant, but even when using a bit type, the transition from 0 to 1 must commonly be distinguished from the transition from 1 to 0. The knowledge of the previous output value is necessary to determine which transition (if any) is performed.

The following code, even if "spontaneous", is not correct! The designer attempts to read the output port value.

```
entity NAND2 is
        generic (FROM_0_TO_1, FROM_1_TO_0 : TIME);
        port (A, B: in BIT; O: out BIT);
end NAND2;
architecture INCORRECT of NAND2 is
begin
        process
        begin
        if (A nand B) /= O then              -- Incorrect
            if O='0' then                    -- Incorrect
                O <= A nand B after FROM_0_TO_1;
            else
                O <= A nand B after FROM_1_TO_0;
            end if;
        end if;
        wait on A,B;
        end process;
end INCORRECT;
```

What is the fundamental reason for this restriction on output port value use?

Resolution functions can be applied on output signals. Reading their value, in these cases, means reading the resolved value and not only the contributing value of your component. In other words, this is a way to enter information (the resolved value depends on other signal values in the network) using an output port. VHDL does not allow such an exception to its consistency.

Therefore, turnarounds must be found.

7.2.6.2. Using Buffer Ports

Buffer ports have been introduced in VHDL to solve the above kind of problem. They are a restricted form of output ports: the compiler ensures that buffer ports cannot be "wired together". In others words, a signal of buffer mode is updated by at most one source. For this kind of port, the contributing value and the external network value are identical.

Nevertheless, the concept of buses is very familiar in hardware, and it is often not possible to define an entity declaration with buffer ports whose outputs will never be wired together. From a practical viewpoint, buffer ports are not often used in VHDL.

7.2.6.3. Using Bidirectional Ports

For reading purposes, transforming an output port into a bidirectional port seems very easy in VHDL: the reserved word **out** becomes **inout**. In fact, this practice has to be discouraged.

Since this modification appears in the external view (entity declaration) of the model, declaring an output port as bidirectional just for reading purposes gives false information outside the design unit: as would declaring as bidirectional the data of a ROM. This is conceptually not admissible. Moreover, with the resolution function mechanism, this port (thought of as an input by the designer) can import a resolved value.

The designer puts this port as bidirectional in order to be able to read it, but he no longer knows what he is reading!

7.2.6.4. Using Local Variables or Signals

The use of a local[18] variable or signal to store the previous state of the output port is very popular. Although this is boring and has no hardware reality (there is no storage functionality in a nand gate), it is the best way to turnaround output port restrictions.

A correct architecture of the previous example is the following:

---

[18] Some authors speak of a "temporary" variable or signal. Because the wording "temporary" for a signal sounds inconsistent with the very nature of a signal (a signal exists all along the simulation), we prefer the term "local".

```
architecture CORRECT of NAND2 is
begin
      process
          variable PREVIOUS_STATE : BIT;
      begin
      if (A nand B) /= PREVIOUS_STATE then
          if PREVIOUS_STATE='0' then
                O <= A nand B after FROM_0_TO_1;
          else
                O <= A nand B after FROM_1_TO_0;
          end if;
          PREVIOUS_STATE := A nand B;
      end if;
      wait on A,B;
      end process;
end CORRECT;
```

7.2.7. Configuration Checks

The configuration mechanism plugs an actual entity into a given component. While the LRM is very clear about the dependencies between units, based on the actual use of the names of other units, it is not clear at all about the dependencies between the units used in the structure. For example, if the architecture A uses the couple entity/architecture B in a configuration, should B exist before A is compiled? or only the entity declaration of B? or nothing?

The ISAC issue #5 explains the problem very clearly, and identifies four approaches (*soft, semi-soft, hard, hard-even-in-architectures*) corresponding to the different solutions. This book is not the place to comment in greater detail on this issue, but the consequence is as follows: if one wants to port a VHDL description together with a "make" file, or to append all VHDL texts in the same file, he should take care of specifying the more constraining order: compile used entities and architectures before the place they are used. This order might be different from that suggested by the automatic recompilation utilities of the original platform.

7.3. PITFALLS

7.3.1. Testing the Existence of a File

The pitfall here is that there is no trick. No VHDL87 primitives allow this check, and furthermore, no error retrieval mechanism allows this case to be treated. If we write in a non-existing file, the latter is created, and if you read a non-existing file, a simulation error occurs. The simulation will stop; there is no way to treat this error.

7.3.2. Files to Communicate between Processes

The only way to communicate between processes in VHDL is through signals. The LRM describes in detail the mechanism to update the current value of signals. This mechanism, based on delta delays, signal drivers, and resolution functions, has to answer this question: when should a signal be updated, and what is its new current value?

Therefore, good properties of a model, such as concurrency, determinism, and even causality, are safe, unless if somebody short-circuits this mechanism. Language designers carefully put some constraints on the use of certain sensitive objects of the VHDL world. To avoid having information enter a model through an output port, restrictions on reading these ports have been determined. Since the modification of access type values could not be controlled, this type is prohibited for all objects of class signal. Communication between processes using variables was also banished, and global variables do not exist in VHDL'87.

In the same vein, files could be a way to communicate between processes without using any signals. This could lead to zero-delay communication paths, which destroy determinism and concurrency. That is why the LRM specifies that a given file can be read or written, but not both. Two kinds of files are coexisting in VHDL: input files, which usually contain stimuli or a description of a ROM, and output files, which usually record some traces.

Nevertheless, there is a way to overcome this restriction. Two different VHDL files can represent the same external file. Using this "trick", it can seem possible to build zero-delay communication paths between processes and to break the determinism of the language. But this exercise is no longer relevant to the VHDL world, but rather to the operating system the VHDL simulator is running on — and more precisely, its file management system.

A VHDL hacker may be tempted to use the operating system's facilities to work around the fact that, in principle, interprocess communication is not possible through files. Typically, UNIX's pipes may be seen by the VHDL compiler as a file, and a process could write in it while another process could read from it.

Fortunately, this will not work in most cases. This hacker is thinking of VHDL processes as UNIX processes, but the main difference is that VHDL processes are not code-generated to support unsynchronized inputs/outputs. In other words, the compiler produces a code where, most of the time, the file I/O *must* be completed before any other activity can take place, even in other processes. If a process is expecting an input from a pipe, it will *not* give control to the kernel for activation of the process supposed to write in the same pipe. The simulation is therefore deadlocked.

7.3.3. Non-Generic Entity and Generic Component

If we declare a generic component, it is impossible in VHDL87 to configure a non-generic model. No syntactic form allows this feature.

Assume the following component declaration:

```
component EXAMPLE is
        generic (N : INTEGER);
        port (A, B : in BIT; C : out BIT);
end component;
```

How is it possible to instantiate a non-generic entity in a given case? If N=4, I know that the entity CASE4 with the following declaration has to be configured. But how do you tell this to the compiler?

```
entity CASE4 is
        port (X,Y : in BIT; Z : out BIT);
end CASE4;
```

A syntactic form like

```
C4 : EXAMPLE generic map (4) use entity CASE4(ARCH) port map (.....
```

will be appropriate — but it is not VHDL'87 code.

Always remember that it is only possible to configure a generic entity on a generic (or not) component.

7.4. DESIGNER COMMANDMENTS

time?

- Simulation time is *not* user time. Real time is not part of the VHDL world, even if VHDL is a good candidate to simulate ("in batch") real-time systems.

event?

- The designer must always specify if his or her synchronization mechanism works on changes (events) or on operations (transactions).

driver?

- When considering a signal, always imagine the state of its drivers.
- Be sure your signal assignment is not unfortunately erasing previous values in the driver.
- Be aware that your concurrent signal assignment can erase your driver values.

process?

- Be aware that the very nature of a process is iterative.
- Always closely control the sensitivity list of a process.

subprogram?

- Always write associative and commutative resolution functions.
- Be alert when overloading predefined operators; you can hide the old ones.
- Never forget to put the reserved word **signal** to transmit a signal as an actual.
- Never directly access an abstract datatype: use the related subprograms.
- Never confuse relational operator and assignment symbols.

file?

- Always remember that the notion of text file is not portable.
- Note that the open/close semantics is not well defined in the LRM.
- Never try to use files to communicate between processes.
- Never try to test the existence of a file: this is not possible in VHDL.

attribute?

- Always choose the right kind of predefined attribute: signal or function?
- Never make assumptions about the range of an array: make wide use of predefined attributes.

constant?

- Be aware of what is allowed in constant declaration expressions: are function calls accepted by the implementation?

port?

- Never try to read the output port value: use local variables.
- Be conscious of buffer port restrictions: can you really use them?
- Never transform output ports into bidirectional ports solely to read their value.

configuration?

- Never try to configure a non-generic entity on a generic (or not) component.
- Plan the compilation order of your design to fit the configuration approach of your platform (soft, semi-soft, hard, etc.).

documentation?

- Always use self-documented identifier names.
- Always document the entity and package declarations as they were the "instructions for use" of the model or the package.
- Never put the local object of a package in the package declaration.
- Always make checks on generics and ports appearing in an entity declaration.
- Always think about the people who will have to read your VHDL description — and do not just pity them.

optimization?

- Never forget that software source code optimization tricks are also usable in VHDL.
- A signal is always more expensive than a variable. Do you really need a signal?
- Always be sure that the datatype you are manipulating is consistent with the problem you solve and the level of description you use.
- Always remember that file operations are expensive in term of performances.

tired?

- Never think that the designer commandments are too numerous to be remembered — just read them again.

8. M AND VHDL

8.1. INTRODUCTION

The purpose of this chapter could have been simply defined as *a comparison between two hardware description languages: M and VHDL.* But such a title would have been confusing, because these two languages cannot be compared feature by feature.

M is the description language of Lsim, a Mentor Graphics logic simulator (originally the Silicon Compiler Systems simulator), whereas VHDL is a standardized description language, and therefore is supposed to be portable on many different simulators.

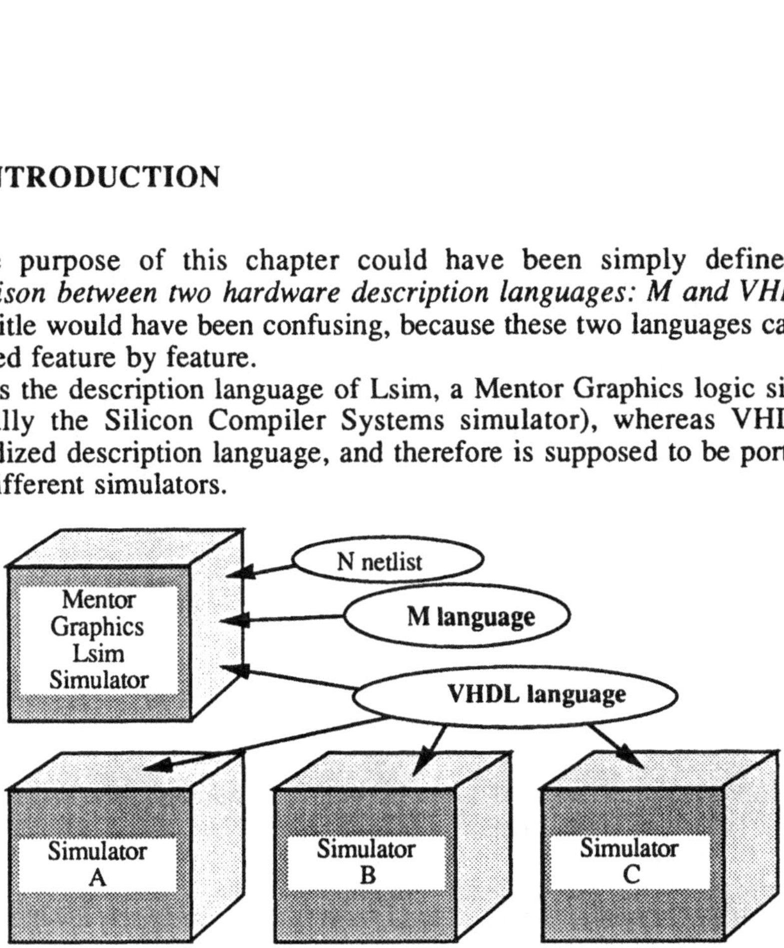

Fig 8.1 Languages and Simulators

M is built from the C programming language and includes features directly linked to the Lsim simulator. VHDL is mainly inspired by the Ada language and only assumes a specified set of features from the simulator.

Furthermore, VHDL is defined by a Language Reference Manual (LRM) published by the IEEE [IEE87]. Any simulator using VHDL is assumed to behave like the canonical event-driven simulator described in this reference manual (elaboration, initialization, and simulation cycles are explained in detail in the LRM). Thus, the implementation of VHDL on a given simulator can be checked against this reference manual.

On the other hand, speaking of M is equivalent to speaking of the implementation of M on Lsim, this simulator being the only *reference*. There is no language reference manual for M, but rather a Lsim user's manual presenting the M language as well as the N netlist format and all the simulator and graphical tools commands [MEN89]. Actually, Lsim is a multi-kernel simulation environment, allowing, for example, mixed-mode simulations of M models and SPICE netlists, M models being executed by an event-driven kernel.

Being aware of this main distinction between both languages is fundamental. As M is the language of the Lsim simulator, it will be sometimes difficult to decide if a given feature belongs to M or to Lsim. In that case, it could make no sense to try to find an equivalent feature in VHDL.

In order to list the equivalencies between the two languages, we will proceed step by step, from the obvious to the more questionable points. The different concepts handled in this chapter are here expressed more often with words taken from the VHDL Language Reference Manual. This is not an ideological choice between the two languages, but due to the standardization process, the VHDL concepts have now a common meaning for a large number of designers.

8.2. DESIGN UNIT

According to the VHDL Language Reference Manual, the design units are constructs that may be separately analyzed and inserted into a design library.

A VHDL design library is an implementation-dependent storage facility: in many simulation environments, each library is mapped to a filesystem or directory (via an index file not specified in the language). M files can be grouped by related subjects into UNIX directories, but there is no larger library semantics in the M language than in the C language.

In VHDL, a design unit is either an *entity declaration*, a *configuration declaration*, a *package declaration* (three primary units), an *architecture body*, or a *package body* (two secondary units). One or more design units can be put together in a design file, which is used as analyzer input. But from the compilation point of view, only the design unit is relevant.

In M, the fundamental design unit is the *module*, which is stored in a *file* (containing only one module). Since M is built from the C language, it is possible to include files containing functions, or to call from the module external functions stored in different files. As in C, from the compilation point of view, the only relevant object is the file.

First of all, design units from both languages can be split into two families:

- the VHDL entity, architecture, and configuration, as well as the M module, are closely related units in the sense that they represent hardware objects: they will be designated as *hardware design units*;
- the VHDL package and package body, as well as M functions files, can group constants, types, and subprograms declarations: they will be called *software design units* (VHDL packages may also declare hardware objects such as signals and components, but this feature will not be considered here).

8.2.1. Hardware Design Unit

Here is a typical example of an M module skeleton:

```
MODULE SHIFT_REGISTER (NB_BIT, LENGTH)
 int NB_BIT ;
 int LENGTH ;
{
 IN LOGIC DATA_IN [NB_BIT] ;
 IN LOGIC CLOCK ;
 OUT LOGIC DATA_OUT [NB_BIT] ;

 MEMORY LOGIC MEM [LENGTH-1][NB_BIT] ;

 BUILD {
   ......
 }
 INITIALIZE {
   ......
 }
 SIMULATE {
   ......
 }
}
```

The M module is actually a function defined by the keyword **MODULE**, with arguments (NB_BIT, LENGTH) being the generic parameters of the module and containing predefined sections. After the first bracket, the terminals or ports of the module (DATA_IN, CLOCK, DATA_OUT), as well as the internal memory variables (MEM), are defined. Then three different (and some optional) sections become available:

- the *BUILD section*, executed once at the very beginning of the simulation (i.e., the elaboration), which includes the structural description of the module (instantiation of other modules);

- the *INITIALIZE section*, executed once in the simulation just after the BUILD section, which contains code performing the initialization of the module variables;
- the *SIMULATE section*, which includes the portion of code that is executed during the simulation each time there is an event on one input terminal (the Lsim kernel simulating M models being event driven).

The VHDL source corresponding to the previous M module could be the aggregate of the following design units:

```
library STD_LIB ; use STD_LIB.XBIT_PKG.all ;
entity SHIFT_REGISTER is
 generic ( NB_BIT :      NATURAL ;
           LENGTH :      NATURAL ) ;
 port (  DATA_IN :       in XBIT_VECTOR (NB_BIT-1 downto 0) ;
         CLOCK :         in XBIT ;
         DATA_OUT :      out XBIT_VECTOR (NB_BIT-1 downto 0) ) ;
end SHIFT_REGISTER ;

architecture STRUCTURE of SHIFT_REGISTER is
  type MATRIX is array (0 to LENGTH-2) of XBIT_VECTOR(NB_BIT-1 downto 0) ;
  signal MEM : MATRIX ;
begin
...
end STRUCTURE ;
-- or
architecture BEHAVIOR of SHIFT_REGISTER is
  type MATRIX is array (0 to LENGTH-2) of XBIT_VECTOR(NB_BIT-1 downto 0) ;
begin
  process
    variable MEM : MATRIX ;
      ...
  end process ;
end BEHAVIOR ;
```

The declarative part of the M module, beginning with the keyword **MODULE** and finishing just before the declaration of the MEMORY variables, is equivalent to the VHDL entity, as shown above. The meaning of this equivalence has to be interpreted in the sense that both codes represent the same objects: here, generics and ports.

In VHDL, the port clause corresponds to the definition of the M terminals, and the generic clause to the list of arguments of the M module (which is actually translated into a C function further linked with the simulator kernel). The definition of ports, signals (for VHDL), and terminals (for M) will be discussed in more details in section 8.4.

Since the code included in the three sections BUILD, INITIALIZE, and SIMULATE represents the internal structure and/or the behavior of the M module, it is more or less similar to the VHDL source of a single architecture (two examples of architecture are shown above).

The BUILD section is equivalent to a VHDL structural description (component instantiation statements), whereas the SIMULATE section is equivalent to a VHDL behavioral description (with process statement(s), for example). These VHDL structural and behavioral statements, which do not need to appear in a given order and can be mixed together, are included in a single architecture of the previous entity. The code in the INITIALIZE section is equivalent to the VHDL sequential statements placed before the first wait statement in a process. Detailed examples of structural and behavioral descriptions in both languages will be given in section 8.7.

The declaration of the *MEMORY variables* (internal variables) of the M module corresponds to the declaration of *signals* (first architecture) or to the declaration of *variables in a process* (second architecture). In fact, the second equivalence is more relevant, as is shown in paragraph 8.4.1.

8.2.2. Software Design Unit

In the C programming language, functionally related functions are frequently placed in the same source file (".c"). With each ".c" file is associated a header file (".h") including the declaration of the C functions exported by the ".c" file. To be able to use macros and functions exported by a file "FA.c" from a file "FB.c", the ANSI-C programmer must include the file header "FA.h" of "FA.c" into the current "FB.c" file by using the **#include** directive.

The same mechanism is available in M. As an example, let us suppose that we have defined a set of arithmetical functions operating on bit vectors, interpreting these bit vectors with the two's complement notation. This set of functions is included in a file "arithmetic.c" and has the following aspect in M:

```
int SIGNED_VALUE ( VECT )
        LOGIC_BUS VECT ;
{ ...... }
LOGIC_BUS SIGNED_VECTOR ( INTEG, SIZE )
        int INTEG, SIZE ;
{ ...... }
LOGIC_BUS TWO_COMPLEMENT ( VECT )
        LOGIC_BUS VECT ;
{ ...... }
LOGIC_BUS MULTIPLY ( A, B, SIZEOF_A, SIZEOF_B )
        LOGIC_BUS A, B ; int SIZEOF_A, SIZEOF_B ;
{ ...... }
```

A file containing the definition of a module M using one of these functions may have the following aspect (assuming that a file header "arithmetic.h" corresponding to the previous file "arithmetic.c" is available):

```
#include "arithmetic.h"
MODULE M (...)
{
 IN LOGIC VECT [8] ;
 MEMORY int INTEG = 0 ;
 SIMULATE
 {
      INTEG = SIGNED_VALUE (VECT) ;
      ...
 }
}
```

In VHDL, the corresponding mechanism is based on the use of packages. Let us continue with the previous example. Two design units are defined:

• a package specification arithmetic:

```
package ARITHMETIC is
 function SIGNED_VALUE ( VECT : BIT_VECTOR ) return INTEGER ;
 function SIGNED_VECTOR ( I : INTEGER ; SIZE : NATURAL ) return BIT_VECTOR ;
 function TWO_COMPLEMENT ( VECT : BIT_VECTOR ) return BIT_VECTOR ;
 function MULTIPLY ( A, B : BIT_VECTOR ) return BIT_VECTOR ;
 ......
end ARITHMETIC ;
```

• a package body corresponding to the package specification:

```
package body ARITHMETIC is
 function SIGNED_VALUE ( VECT : BIT_VECTOR ) return INTEGER is
      ...
 end SIGNED_VALUE ;
 function SIGNED_VECTOR ( I : INTEGER ; SIZE : NATURAL ) return BIT_VECTOR is
      ...
 end SIGNED_VECTOR ;
 function TWO_COMPLEMENT ( VECT : BIT_VECTOR ) return BIT_VECTOR is
      ...
 end TWO_COMPLEMENT ;
 function MULTIPLY ( A, B : BIT_VECTOR ) return BIT_VECTOR is
      ...
 end MULTIPLY ;
 ......
end ARITHMETIC ;
```

These two design units may be placed in one or two files, and so can be compiled separately (the compiler checks that the package specification has been analyzed before the corresponding package body). Furthermore, these units may be stored after compilation in a specialized library including only related packages (for example, in a mathematical library, named "math").

In order to use these functions in another design unit, the VHDL designer must make the previous library visible by its current design unit (using the **library** clause) and must use (or not) the **use** clause to have direct access to the function names. Listed below is the example of an architecture using these functions:

```
library MATH ;
use MATH.ARITHMETIC.all ;
architecture A of E is
        signal VECT : BIT_VECTOR (0 to 7) ;
        signal INTEG : INTEGER := 0 ;
begin
        INTEG <= SIGNED_VALUE (VECT) ;
        ...
end A ;
```

8.2.3. What about VHDL Configurations?

One VHDL design unit has not yet been compared: the configuration declaration. The reason is that there is no equivalence of the VHDL concept of configuration is the M language.

In M, modules are instantiated in other modules. In VHDL, components are instantiated in architectures, and these components are linked to couples entity-architecture via a configuration declaration or a configuration specification. The components are actually sockets into which the circuit will be plugged by the configuration. The VHDL instantiation mechanism is detailed in paragraph 8.7.1 (with a configuration specification) and compared with the equivalent M instantiation.

8.2.4. Conclusion

Figure 8.2 sums up the different design units in both languages and establishes the connections between the similar concepts seen in this paragraph.

The main conclusion is that the VHDL language provides more features enabling separate compilation and consistency checking of the design units, whereas M is based on the C notion of files.

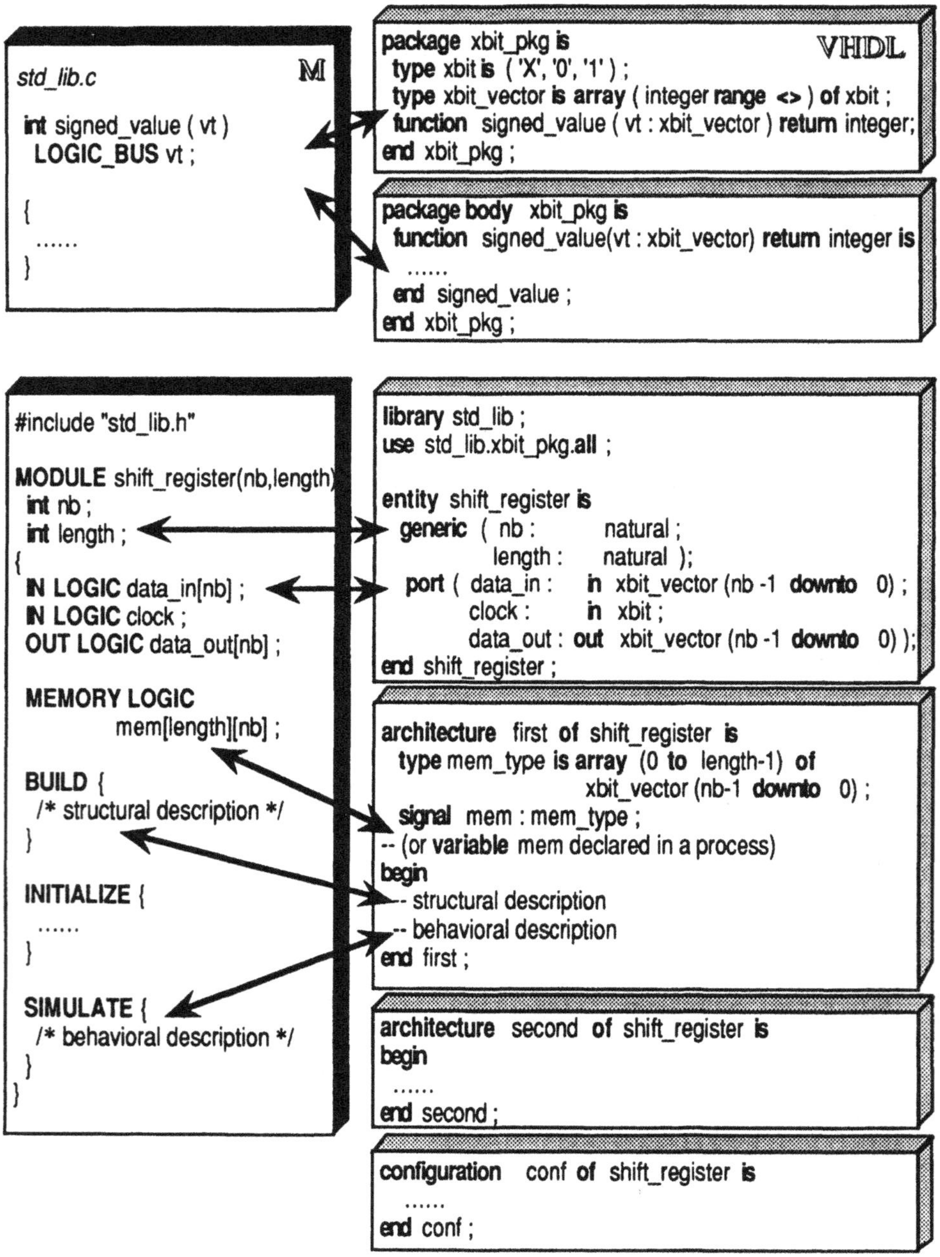

Fig 8.2 Design Units

8.3. SEQUENTIAL AND CONCURRENT DOMAINS

From a practical point of view, we may say that the concurrent world consists of black boxes interconnected by wires (thus being concurrent!), whereas the sequential world includes mainly (sequential) algorithms.

In the concurrent world, black boxes are modeled by the M modules or the VHDL entities, and the wires by interconnected M terminals or VHDL ports and signals.

In the sequential world, M or VHDL variables are used to describe algorithms. The "container" of these algorithms is the M module (without the BUILD section) or the VHDL process. Of course, M terminals or VHDL ports and signals can be accessed (but never created!) from within the sequential world.

The distinction between these objects (signals, variables, etc.) will be detailed in section 8.4.

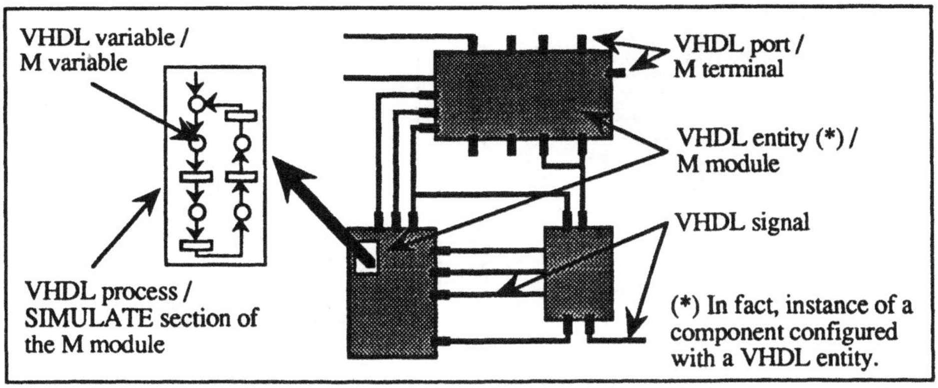

Fig 8.3 Sequential and Concurrent Domains

The VHDL Language Reference Manual defines sequential and concurrent statements. Sequential statements are used to define algorithms for the execution of a subprogram or a process, and are executed in the order in which they appear. Concurrent statements are used to define interconnected blocks and processes that jointly describe the overall behavior or structure of a design, and are executed asynchronously with respect to each other.

On the base of these definitions, we may consider the statements included in the INITIALIZE and SIMULATE sections of an M module as sequential statements: indeed, they are executed in the order in which they appear because they are equivalent to C code. *The domain of an M module description is the sequential domain.* Concurrent behaviors are mainly expressed by the interconnection of blocks (i.e., other M modules) in the BUILD section (containing also "sequential" C code) or in a netlist of the N format.

On the other hand, *the domain of a VHDL architecture is the concurrent domain.* In this domain, the interconnection of blocks as well as the concurrent execution of (sequential) processes can be directly expressed. Processes communicate with each other via signals, variables being only defined inside each (sequential) process.

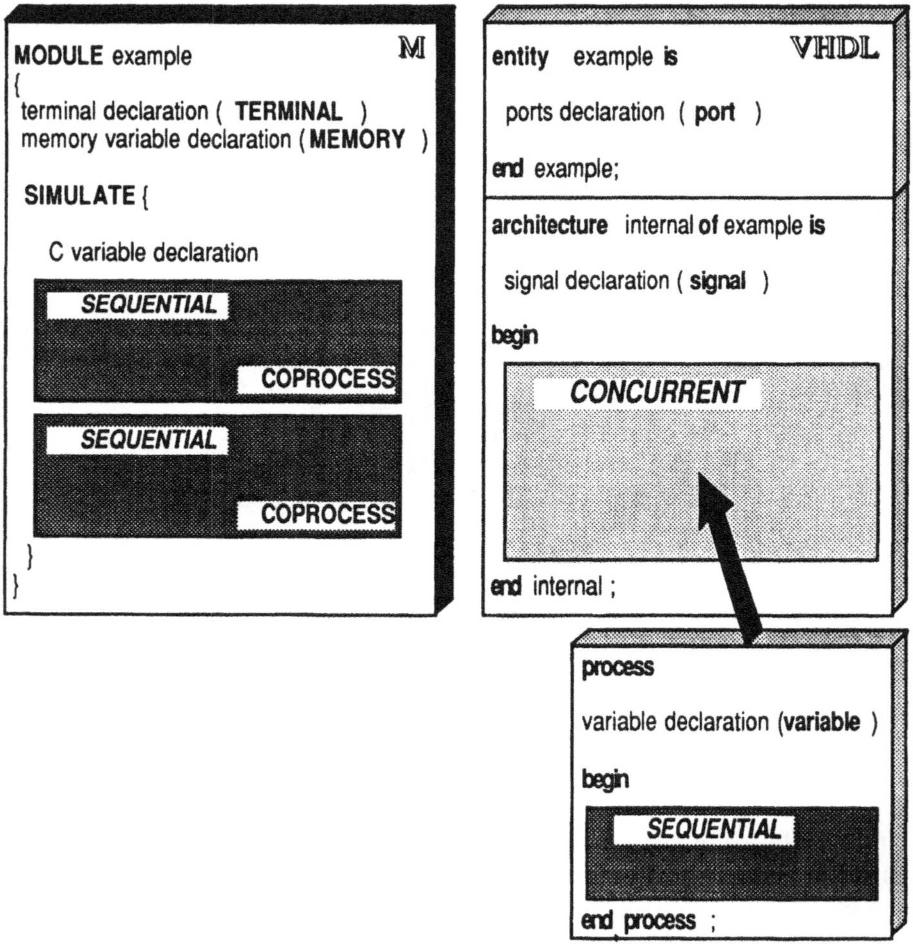

Fig 8.4 Sequential and Concurrent Domains

Within a single M module, concurrent behaviors may be implemented via the M mechanism of *coprocesses.* In fact, coprocesses are the representation of UNIX processes: they execute in an order chosen by the system (possibly in parallel), but the main point is that they can communicate via variables (MEMORY variables, for example), whereas VHDL processes communicate via signals only.

Consequently, the communication between processes in VHDL is totally controlled by the user: if two processes are trying to drive the same *signal* at the same time point, the user resolution function is called, and there is no ambiguity on the result.

On the other hand, if two M (UNIX) processes are trying to have simultaneous access to a common *variable*, the resolution of this conflict is handled by the system, not by a call of a user's resolution function. The user is therefore strongly advised to write a kind of semaphore protocol to prevent such unpredictable results (which are not detectable at the simulation). This user's control can be achieved through wait statements on terminals or memory variables inside coprocesses.

Another way to model concurrent statements within an M module is to write *FORK statements*. According to the M manual, statements within a FORK section are executed in parallel. The same remarks stated just above apply for this concurrency: the FORK statement is derived from the UNIX FORK statement (there is a JOIN statement, too), and is thus not equivalent to a VHDL concurrent block.

The fundamental VHDL mechanism that is not implemented in M is the *delta-delay* mechanism. Each concurrent statement inside an architecture (or block) is executed each time there is an event on one of the signals appearing in the sensitivity list of the equivalent process. As a consequence, a zero-delay assignment can be executed many times at the same simulation time point. The best example is the asynchronous RS flip-flop with zero-delay gates:

```
S1 : Q <= NQ nand NSET ;          -- Two concurrent signal assignments
S2 : NQ <= Q nand NRESET ;        -- statements inside an architecture
```

If NSET falls from '1' to '0' (becomes active), assignment S1 is executed, changing thus the value of Q. So, at the same time point, assignment S2 is executed, changing the value of NQ. Finally, always at the same time point, assignment S1 is re-executed (because of the change of NQ), with no change on the final value of Q. Such a mechanism is doing much more than simply executing two UNIX processes in "parallel".

This distinction between sequential and concurrent domains in M and in VHDL is fundamental. The difference between the sequential domain of M and the concurrent domain of VHDL has a parallel in the different ways both languages handle signal and variable assignment.

A VHDL equivalent to a typical behavioral M module (without a BUILD section and without coprocess or fork statements) is illustrated in the following figure.

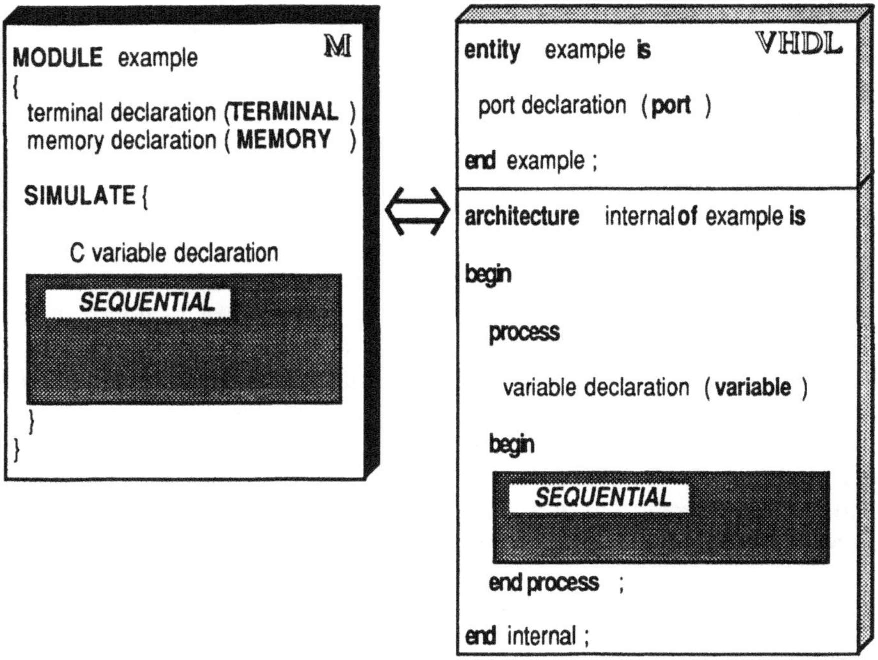

Fig 8.5 VHDL Equivalent of an M Module

8.4. OBJECTS AND TYPES

Three classes of objects are defined in VHDL: signal, variable, and constant. Each object has a given type: this type may be a scalar type (integer type, floating point type, physical type, enumeration type), a composite type (array and record types), an access type (providing access to objects of a given type), or a file type (providing access to objects that contain a sequence of values of a given type).

The M objects are the following: terminal (and bus), memory variable, C variable, and datafile. M also defines some specific types, such as LOGIC, ADEPT, VOLTAGE, NORTHON, and THEVENIN (very IC design oriented), and includes most of the C types. An abstract datatype mechanism allows the user to encapsulate any C type in order to move abstract data through the terminals.

In both languages, some restrictions on an object's type are specified. In VHDL, a constant or a signal cannot be of access or file type. Beyond this exception, variables and signals share the same VHDL types. In M, variables defined inside a module may be of any C type, whereas terminals are defined with a predefined M type or an abstract datatype, which is slightly different from a standard C type (see the mechanism of abstract datatype in paragraph 8.4.3).

	M
TERMINAL (BUS)	LOGIC, ADEPT, VOLTAGE, NORTON,THEVENIN Abstract Datatype
MEMORY	LOGIC, LOGIC_BUS, ADEPT, VOLTAGE, NORTON, THEVENIN, Any standard C type
VARIABLE	LOGIC, LOGIC_BUS, Any standard C type
DATAFILE	ASCII file with specific M syntax

	VHDL
signal (**port**)	Any VHDL type except **access** and **file**
variable	Any VHDL type
constant	Any VHDL type except **access** and **file**
file	file containing objects of any VHDL type, except **access** and **file**

Fig 8.6 Objects and Types

8.4.1. Terminal, Port, and Signal

The M terminal is equivalent to a VHDL port, which is a signal declared in the interface list of an entity or component declaration.

The M memory variable is equivalent to a VHDL variable defined in the declaration region of a process: they are both static variables (keeping their value during all the simulation).

A C variable defined inside an M module is equivalent to a VHDL variable defined in the declaration region of a subprogram: they are both local variables, and they are not saved during the whole simulation.

Finally, the M datafile is equivalent to a VHDL file of STRING (array of characters), which is defined in the TEXTIO package. Other type of VHDL files are not implemented in M.

Among all these equivalences, the one between the M terminal and the VHDL port points out an important difference: the notion of signal, as it is defined in VHDL, is not really implemented in M. Of course, Lsim has access to signals representing the nodes of the circuit being simulated, but the only signals the user is able to access with the M language are the terminals, equivalent to VHDL ports (objects declared with the keyword **port**). M does not implement real internal signals as VHDL does (objects declared with the keyword **signal**).

This is due to the different nature of the languages: an M module is described by a sequential code using variables (with the restriction of the UNIX fork and join statements), whereas a VHDL architecture is described by a set of concurrent statements (using signals). This distinction was described in section 8.3.

Another related difference is the way M terminals and VHDL signals are assigned. A zero-delay assignment of a M terminal or memory variable is totally equivalent to a C variable assignment, i.e., it changes its value at the end of its statement. The syntax is also the same:

```
Variable_object = value ;
Terminal_object = value ;
```

In VHDL, there is an important difference between a zero-delay signal assignment and a variable assignment: the variable changes its value at the end of its assignment statement, whereas the signal changes its values at the next wait statement in the current process. Syntaxes are thus different:

```
Variable_object := value ;  -- in sequential domain
Signal_object <= value ;    -- in sequential or concurrent domain
```

This distinction proves that M objects are internally handled as variables (immediate assignment), whereas VHDL includes a specific signal semantics (driver mechanism) that makes the signals totally different from the variables.

8.4.2. Predefined M Types

In the M language, the LOGIC type is the predefined bit type used for functional/gate levels. Each signal (terminal) of type LOGIC is represented within Lsim by a set of two values:

- the signal state, which has three possible values:
  - **Low** (L),
  - **High** (H),
  - **Unknown** (X).
- the signal strength (from the weakest to the strongest value):
  - **Initial** (strength always 0),
  - **Charged** (strength from 0 to 31),
  - **Driven** (strength from 0 to 31),
  - **Supply** (strength always 31).

The signal strength is internally stored in 12 bits (providing a range from 0 to 4095), but the user only has access to the five leading bits (range from 0 to 31). The conflict resolution is based on the comparison between the strengths of the different signals driving the same node.

In VHDL, the only predefined bit type is the enumerated type BIT:

```
type BIT is ( '0' , '1' ) ;
```

The user may define any other type he prefers to use, such as:

```
type XBIT is ( 'X' , '0' , '1' ) ;
```

A VHDL type equivalent to the M LOGIC type could be

```
subtype STRENGTH_TYPE is INTEGER range 0 to 31 ;
type CLASS_TYPE is (INITIAL, CHARGED, DRIVEN, SUPPLY) ;

type LOGIC is
   record
    STATE :        XBIT ;
    STRENGTH :     STRENGTH_TYPE ;
    CLASS :        CLASS_TYPE ;
  end record ;
```

Then the VHDL user may write a resolution function based on the comparison between the strength and class fields of the LOGIC type. There is no predefined resolution function in VHDL, even for type BIT: such a function has to be written by the user.

The M language has some other predefined types: ADEPT, VOLTAGE, NORTHON, and THEVENIN. These types are used for analog or analog-digital simulations: an ADEPT terminal is recognized by the Lsim simulator as an analog terminal, and thus is handled differently from a LOGIC terminal. This M-Lsim feature allows efficient mixed-mode simulation.

VHDL does not include the features allowing efficient analog or analog-digital simulation.

8.4.3. M Abstract Datatypes

As mentioned in the previous figure, M provides the user with a mechanism to define abstract terminals in order to perform system-level modeling. The definition of a new datatype in the M language proceeds as following:

- Structure definition: the user declares a C structure (*struct*) encapsulating objects of any C standard type. For example, floating point terminals will use

```
typedef struct { double value ; } DOUBLE ;
```

- Arbitration function: the user has to define an arbitration function, which is used by the Lsim simulator when multiple terminals attempt to force a value on a single abstract node. Every abstract node has an arbitration data structure associated with it: this structure has also to be defined by the user. This M arbitration function is totally equivalent to a VHDL resolution function.
- Display functions and others: the user specifies a set of functions used by the Lsim simulator to perform probe creation, abstract terminal display, memory allocation for abstract node creation, and so on.

With VHDL, the user may declare a signal or port of any abstract VHDL type (user-defined type). In case of conflicts, a user-defined resolution function is needed for signals having multiple drivers. But no display functions need to be specified: VHDL is not linked to a particular simulator, and these display functions may be implemented differently by each VHDL simulator. In M, the display functions are in fact interface functions needed by the Lsim simulator: they should not really be considered as part of the M language.

8.5. PREDEFINED OPERATORS AND FUNCTIONS

Many predefined operators are available in M: arithmetic and relational operators, bit-level shift and logic operators, and so forth. Arrays of LOGIC elements (LOGIC_BUS) are handled as unsigned integers, thereby leading to implicit conversions between LOGIC vectors and integers. This set of M operations is a super-set of the standard C operations.

On the other hand, the set of predefined operators in VHDL is very poor! Following are some tables listing the predefined operators in both languages.

8.5.1. Bit Logic and Boolean Operators

In VHDL, six logic operators are predefined on the following types:

```
type BIT is ('0','1') ;
type BOOLEAN is (FALSE, TRUE) ;
```

If bit '0' is taken for boolean FALSE and bit '1' for boolean TRUE, then all these operators have the same functionality.

<table>
<tr><th>Bit Logic Operators</th><th>M</th><th>VHDL</th></tr>
<tr><td>unary bitwise invert</td><td>~</td><td>not</td></tr>
<tr><td>binary bitwise and</td><td>&</td><td>and</td></tr>
<tr><td>binary bitwise nand</td><td>~&</td><td>nand</td></tr>
<tr><td>binary bitwise or</td><td>|</td><td>or</td></tr>
<tr><td>binary bitwise nor</td><td>~|</td><td>nor</td></tr>
<tr><td>binary bitwise xor</td><td>^</td><td>xor</td></tr>
<tr><td>binary bitwise xnor</td><td>~^</td><td>not predefined</td></tr>
<tr><td>unary reduce and</td><td>@&</td><td>not predefined</td></tr>
<tr><td>unary reduce nand</td><td>@~&</td><td>not predefined</td></tr>
<tr><td>unary reduce or</td><td>@|</td><td>not predefined</td></tr>
<tr><td>unary reduce nor</td><td>@~|</td><td>not predefined</td></tr>
<tr><td>unary reduce xor</td><td>@^</td><td>not predefined</td></tr>
<tr><td>unary reduce xnor</td><td>@~^</td><td>not predefined</td></tr>
</table>

<table>
<tr><th>Boolean Logical Operators</th><th>M</th><th>VHDL</th></tr>
<tr><td>logical invert</td><td>!</td><td>not</td></tr>
<tr><td>logical and</td><td>&&</td><td>and</td></tr>
<tr><td>logical nand</td><td>not predefined</td><td>nand</td></tr>
<tr><td>logical or</td><td>||</td><td>or</td></tr>
<tr><td>logical nor</td><td>not predefined</td><td>nor</td></tr>
<tr><td>logical xor</td><td>not predefined</td><td>xor</td></tr>
<tr><td>logical xnor</td><td>not predefined</td><td>not predefined</td></tr>
</table>

In M, many operators are predefined for the LOGIC and LOGIC_BUS values. On the other hand, only the INVERSION, AND, and OR operators are defined for logical operations (as is the case in C), and they do not have the same functionality as those defined on LOGIC values.

Example: **HIGH & UNKNOWN = UNKNOWN** but **HIGH && UNKNOWN** = 1.

The small number of predefined logic operators in VHDL is a good illustration of the philosophy of this language: VHDL will cover a wide range of applications (not only ICs, but also PCBs, high-level specified systems, etc.). Thus, most of the constructs must be user defined.

On the other hand, the M language was designed for a given tool (the Lsim simulator) and for a given application: IC design. This is why the LOGIC type and the associated predefined operators seem to cover all the possible needs.

8.5.2. Relational Operators

Relational Operators	M	VHDL
equal to	==	=
not equal to	!=	/=
less than	<	<
less than or equal to	<=	<=
greater than	>	>
greater than or equal to	>=	>=

There is no difference between the two languages.

8.5.3. Arithmetic Operators

Arithmetic Operators	M	VHDL
addition	+	+
subtraction	-	-
multiplication	*	*
division	/	/
modulus	*not predefined*	**mod**
remainder	%	**rem**
exponentiation	*not predefined*	**
absolute value	*not predefined*	**abs**
postfix and prefix increment	++	*not implemented*
postfix and prefix decrement	--	*not implemented*

The capabilities of both languages are comparable in arithmetic operations.

8.5.4. Miscellaneous

<table>
<tr><th>Miscellaneous Operators</th><th>M</th><th>VHDL</th></tr>
<tr><td>concatenation</td><td><||></td><td>&</td></tr>
<tr><td>left shift operator</td><td><<</td><td>not predefined</td></tr>
<tr><td>right shift operator</td><td>>></td><td>not predefined</td></tr>
<tr><td>conditional operator</td><td>... ? ... : ...</td><td>not predefined</td></tr>
</table>

Assignment Operators	M	VHDL
assign	=	= *or* <=
add assign	+=	*not implemented*
subtract assign	-=	*not implemented*
multiply assign	*=	*not implemented*
divide assign	/=	*not implemented*
remainder assign	%=	*not implemented*
left shift assign	<<=	*not implemented*
right shift assign	>>=	*not implemented*
bitwise and assign	&=	*not implemented*
bitwise or assign	|=	*not implemented*
bitwise xor assign	^=	*not implemented*

The shift and conditional operators, as well as most of the assignment operators, are very specific to the C language: M provides a super-set of them (with bitwise logic assignments), but VHDL has no equivalent features (the assignment operators cannot be implemented in VHDL).

8.5.5. Predefined M Functions

M provides to the user some built-in functions. Among them:

- detection of signal value changes:
  CHANGED, FALL, RISE;
- control of the simulation event list:
  DESENSITIZE, SENSITIZE, RESCHEDULE;
- get terminal values or terminal attribute values:
  GET_CAPACITANCE, GET_CONDUCTANCE,
  GET_DVDT, GET_VOLTAGE;
- set terminal values or terminal attributes values:
  SET_CAPACITANCE, SET_CONDUCTANCE, SET_CURRENT,
  SET_IG, SET_RESISTANCE, SET_VOLTAGE, SET_VR,
  UNSET_VOLTAGE.

The functions detecting signal edges can be easily modeled in VHDL with the predefined attributes 'EVENT and 'LAST_VALUE. For example, on signals of type BIT:

```
function CHANGED(signal S : BIT) return BOOLEAN is
begin
        return S'EVENT ;
end CHANGED ;
```

```
function FALL(signal S : BIT) return BOOLEAN is
begin
        return (S'EVENT and S'LAST_VALUE = '1') ;
end FALL ;

function RISE(signal S : BIT) return BOOLEAN is
begin
        return (S'EVENT and S'LAST_VALUE = '0') ;
end RISE ;
```

The M functions DESENSITIZE and SENSITIZE are provided to the user because each M module is implicitly sensitive on *all* the terminals of mode IN or INOUT. In VHDL, the sensitivity list of a process is built by the wait statement(s) included in that process. Thus, a given port is activating a given process if and only if the name of that port appears in a "**wait on**..." or "**wait until**..." statement of that process. A VHDL process can therefore have different temporary sensitivity lists. Finally, the RESCHEDULE function is close to the VHDL "**wait for**..." statement.

The other functions (SET_CAPACITANCE, GET_CAPACITANCE, etc.) implement a *backannotation mechanism* and are thus very useful for the IC designer. There are no such functions available in VHDL.

As a matter of fact, these functions should be considered as Lsim features more than M features, in the sense that they are more linked to the simulator (and the characteristics of its environment) than to the language itself. It is thus not obvious how to model such functions in VHDL (see paragraph 8.8.3).

8.6. STATEMENTS

8.6.1. Assignments

In both M and VHDL, variable assignments are always immediate:

```
M_variable_name       =   expression ;
VHDL_variable_name    :=  expression ;
```

On the other hand, for the M terminals as well as for the VHDL signals, assignments need a delay specification (that can be null, of course):

```
M_terminal_name {Trise : Tfall } = { expression  @ time1 } { expression  @ time2 } ... ;
VHDL_signal_name  <=  [ transport ]  expression after time_expression1 ,
                                     expression after time_expression2 , ... ;
```

Inertial and transport delays are implemented in both languages.

In the M language, inertial delays are specified on the left side of the assignment (a different delay can be given for each edge of the output bit-terminal), whereas delays on the right side are transport delays (@ TIME). The total delay of the assignment is the sum of inertial and transport delays.

In VHDL, the first time expression (always on the right side of the assignment) is assumed to be the inertial delay of the assignment, unless the keyword **transport** is used. In such a case, delays are considered as transport delays.

8.6.2. Structured Statements

8.6.2.1. Sequential Structured Statements

Most of the constructs available in the C language can be used in M. Similarly, many Ada constructs are present in VHDL and can be used in the sequential domain. Following is a list of all the main structured sequential statements in both languages.

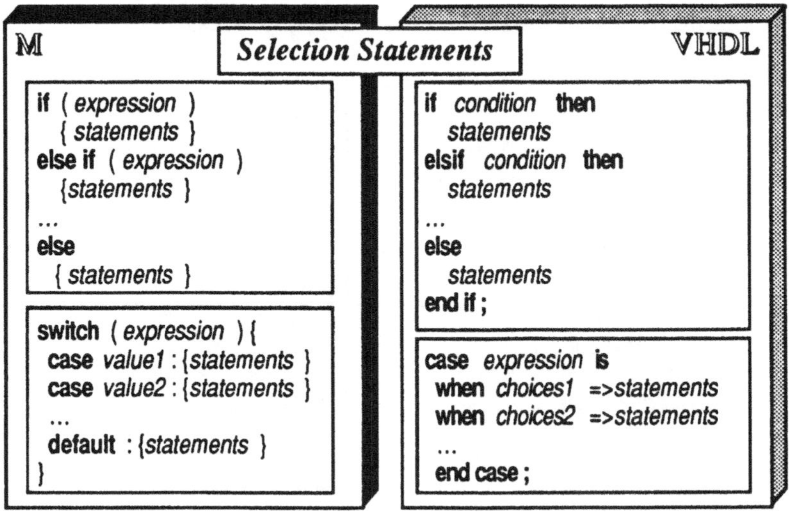

Fig 8.7 Selection Statements

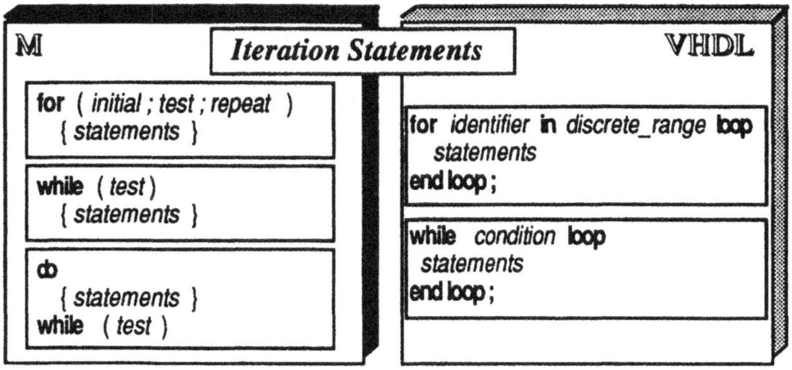

Fig 8.8 Iteration Statements

8.6.2.2. Concurrent Structured Statements

Concurrent structured statements are also available in VHDL. According to the VHDL LRM, each of these statements is totally equivalent to one of the previous sequential structured statements inside a process. Indeed:

VHDL Structured Concurrent Statement	VHDL Equivalent Process Statement

target <= [**transport**]
waveform1 **when** *condition1* **else**
waveform2 **when** *condition2* **else**
...
waveformN ;
	if *condition1* **then**
	target <= [**transport**] *waveform1* ;
	elsif *condition2* **then**
	target <= [**transport**] *waveform2* ;
	else
	target <= [**transport**] *waveformN* ;
	end if ;

with *expression* **select** *target* <= [**transport**] *waveform1* **when** *choices1* , *waveform2* **when** *choices2* , ... *waveformN* **when** *choicesN* ;	**case** *expression* **is** **when** *choices1* => *target* <= [**transport**] *waveform1* ; **when** *choices2* => *target* <= [**transport**] *waveform2* ; ... **when** *choicesN* => *target* <= [**transport**] *waveformN* ; **end case** ;

Thus, if these VHDL concurrent constructs have no equivalent in the M language (because there is no dataflow description in M: see paragraph 8.7.3), they can always be translated into an equivalent sequential code in both languages.

Another structured concurrent statement in VHDL is the generate statement, which is used, for example, to model arrays of components. This generate statement is a kind of loop statement (with the keyword **generate** instead of **loop**), which is executed once at the beginning of the simulation, i.e., during the elaboration phase. In the M language, the standard C iteration statement is used in the BUILD section of the module (this section is only executed once at the beginning of the simulation: see paragraph 8.2.1).

M

```
BUILD {
 int i , j ;
 for (i=1 ; i<= length ; i++) {
 INSTANCE (REG, c[i], NB_BIT) ;
 NET ( c[i].ck , clock ) ;
 for (j=0 ; j< nb_bit ; j++)
    NET ( c[i-1].q[j] , c[i].d[j] ) ;
 }
}
```

VHDL

```
LBL: for I in 1 to LENGTH generate
   C : REG
   generic map  ( NB_BIT )
   port map  ( S( I-1 ), CK, S( I ) ) ;
end generate  LBL ;
```

Fig 8.9 Loop vs. Generate Statement

The generate statement was intentionally used to model structural arrays of components, but it has a more general purpose in VHDL: according to the LRM. (§ 9.7), a generate statement provides a mechanism for iterative or conditional elaboration of a portion of description (including all possible concurrent statements, such as signal assignments or processes). Such a mechanism is not implemented in M, because there is no such behavioral concurrent domain inside the M module.

Another VHDL concurrent statement not implemented in M is the block statement, in particular the guarded block statement.

8.7. DESCRIPTION LEVEL

The purpose of this section is to find equivalent descriptions of a typical design at the different levels allowed by both languages. To do this, we will take the example of a synchronous shift-register, where the number of registers (LENGTH) and the number of bits (NB_BIT) are generic.

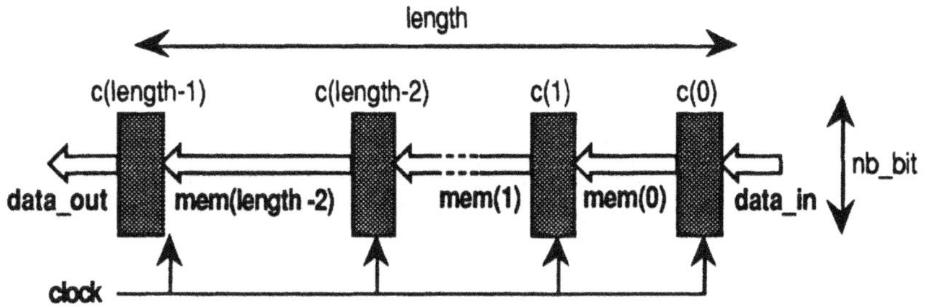

Fig 8.10 Shift Register

8.7.1. Structural Description

Here is a typical structural M model of a shift register:

```
MODULE shift_register (nb_bit, length)
  int nb_bit ;
  int length ;
{
 IN LOGIC data_in [nb_bit] ;
 IN LOGIC clock ;
 OUT LOGIC data_out [nb_bit] ;

 BUILD {
  int i , j ;
  for (i=0 ; i< length ; i++) {
     INSTANCE ( reg_cmos , c[i] , nb_bit) ;
     NET ( c[i].ck , clock ) ;
  }
  for (j=0 ; j< nb_bit ; j++) {
     NET ( c[0].d[j] , data_in [j] ) ;
     for (i=1 ; i< length ; i++)  NET ( c[i-1].q[j] , c[i].d[j] ) ;
     NET ( c[length-1].q[j] , data_out [j] ) ;
  }
 }
}
```

Following is a corresponding VHDL structural description. First, the entity declaration, using a predefined XBIT_PKG package:

```
library STD_LIB ; use STD_LIB.XBIT_PKG.all ;
entity SHIFT_REGISTER is
  generic ( NB_BIT :  NATURAL ;  LENGTH :  NATURAL  ) ;
   port (    DATA_IN :          in  XBIT_VECTOR (NB_BIT-1 downto 0) ;
             CLOCK :            in  XBIT ;
             DATA_OUT :         out  XBIT_VECTOR (NB_BIT-1 downto 0) ) ;
end SHIFT_REGISTER ;
```

Second, the corresponding architecture:

```
architecture STRUCTURE of SHIFT_REGISTER is
 component REGIST
    generic ( N_BIT : NATURAL ) ;
    port ( D :    in   XBIT_VECTOR ( N_BIT-1 downto 0) ;
           CK :   in   XBIT ;
           Q :    out  XBIT_VECTOR ( N_BIT-1 downto 0)  ) ;
 end component ;
 type MEM_TYPE is array (0 to LENGTH-2) of XBIT_VECTOR ( NB_BIT-1 downto 0) ;
 signal MEM : MEM_TYPE ;
 for all : REGIST use  entity  WORK.REG_CMOS ( BEHAV_ARCH )
                       generic map (NB => N_BIT)
                       port map ( DIN => D , CK => CK , DOUT => Q ) ;
begin
  COMPF : REGIST
    generic map ( N_BIT => NB_BIT )
    port map ( D => DATA_IN , CK => CLOCK , Q => MEM( 0 ) ) ;
  COMPI : for I in 1 to LENGTH-2 generate
    COMP : REGIST
     generic map ( N_BIT => NB_BIT )
     port map ( D => MEM( I-1 ) , CK => CLOCK , Q => MEM( I ) ) ;
  end generate COMPI ;
  COMPL : REGIST
    generic map ( N_BIT => NB_BIT )
    port map (D => MEM( LENGTH-2 ) , CK => CLOCK , Q => DATA_OUT) ;
end STRUCTURE ;
```

There is one important distinction between the M description and the VHDL description. In the M module, reg_cmos represents another M module. But in the VHDL architecture, REGIST is a component that is declared in the declarative region of the architecture.

This component has to be linked to an entity (with a corresponding architecture) in a configuration declaration (another design unit) or in a configuration specification, which may be written in the declarative region of the architecture. The example above illustrates the second choice: a configuration specification has been written with the entity aspect REG_CMOS (BEHAV_ARCH).

This configuration between component and entity may seem redundant or verbose for the M modeler. In fact, it may be very useful when the user has different component libraries (one for each technology, for example): the VHDL structural description remains the same; only the configuration changes from one library to another.

Finally, there is a VHDL default configuration mechanism (associating components with entities having the same name) that may provide some convenient shortcuts in such a description.

8.7.2. Behavioral Description

Here is a typical behavioral M model of the same shift register:

```
MODULE shift_register (nb_bit, length)
 int nb_bit ;
 int length ;
{
 IN LOGIC data_in [nb_bit] ;
 IN LOGIC clock ;
 OUT LOGIC data_out [nb_bit] ;

 MEMORY LOGIC mem [length-1] [nb_bit] ;

 INITIALIZE {
  mem = UNKNOWN ;
 }
 SIMULATE {
  int i ;
  if ( RISE (clock) ) {
     data_out { 1 }  = mem [length-2] ;
     for ( i = length-2 ; i > 0 ; i -- )
         mem [ i ] = mem [ i-1 ] ;
     mem [ 0 ] = data_in ;
  }
 }
}
```

One corresponding VHDL description (at a behavioral level) is the following. The entity remains the same as the one written in the previous paragraph, i.e.,

```
library STD_LIB ; use STD_LIB.XBIT_PKG.all ;
entity SHIFT_REGISTER is
  generic (  NB_BIT :   NATURAL ;   LENGTH :   NATURAL  ) ;
  port (    DATA_IN :        in  XBIT_VECTOR(NB_BIT-1 downto 0) ;
            CLOCK :          in  XBIT ;
            DATA_OUT :       out  XBIT_VECTOR(NB_BIT-1 downto 0)  );
end SHIFT_REGISTER ;
```

But another architecture (at a behavioral level) could be

```
architecture BEHAVIOR of SHIFT_REGISTER is
begin
 process
   type MEM_TYPE is array (0 to LENGTH-2) of XBIT_VECTOR (NB_BIT-1 downto 0);
   variable MEM : MEM_TYPE ;
 begin
   wait on CLOCK ;
   if RISE (CLOCK) then
       -- RISE is an user-defined function detecting rising edges on XBIT signals and
       -- declared in package XBIT_PKG (such a function is defined on type BIT in 8.5.5)
        DATA_OUT <= MEM (LENGTH-2) after 1 ns ;
     for I in LENGTH-2 downto 1 loop
        MEM ( I ) := MEM ( I-1) ;
     end loop ;
      MEM ( 0 ) := DATA_IN ;
   end if ;
 end process ;
end BEHAVIOR ;
```

8.7.3. Dataflow Description

In VHDL, we currently make the distinction between three kinds of description levels: structural, behavioral, and dataflow. The two first have been detailed above, and they both have an equivalent description level in the M language. What about the dataflow level?

As explained in section 8.3, there is no "true" concurrent domain in a M module: consequently, it is not surprising that there is no equivalent M feature to the VHDL concurrent dataflow description.

In VHDL, another architecture of the same SHIFT_REGISTER entity could be written at a dataflow description level as follows:

```
architecture DATAFLOW of SHIFT_REGISTER is
 type MEM_TYPE is array (0 to LENGTH-1) of XBIT_VECTOR ( NB_BIT-1 downto 0);
 signal MEM : MEM_TYPE ;
begin
 COMPF : MEM(0) <= DATA_IN after 1 ns when RISE (CLOCK) else MEM(0) ;
 COMPI : for I in 1 to LENGTH-1 generate
         MEM(I) <= MEM(I-1) after 1 ns when RISE (CLOCK) else MEM (I) ;
 end generate COMPI ;
  COMPL : DATA_OUT <= MEM(LENGTH-1) ;
-- DATA_OUT is a out port, so its value cannot be read.
-- Thus, signal MEM(LENGTH-1) is compulsory!

end DATAFLOW ;
```

8.8. TRANSLATING FROM M TO VHDL

In this section, we will focus more deeply on some specific problems that arise when translating M models into VHDL models, such as the choice of a VHDL bit logic type corresponding to the M-Lsim LOGIC type or the backannotation.

8.8.1. Logic Type

As shown in paragraph 8.4.2, the Lsim simulator works on a predefined M LOGIC type. Each LOGIC signal is defined by:

- a state: **Low** (L), **High** (H), or **Unknown** (X).
- a strength (from the weakest to the strongest): **Initial** (strength always 0), **Charged** (strength from 0 to 31), **Driven** (strength from 0 to 31), or **Supply** (strength always 31).

The conflict resolution between two signals driving the same node is based on the comparison between the strengths of both signals.

A VHDL representation of the Lsim LOGIC type is described below:

```
type XBIT is ( 'X' , '0' , '1' ) ;
subtype STRENGTH_TYPE is INTEGER range 0 to 31 ;
type CLASS_TYPE is ( INITIAL, CHARGED, DRIVEN, SUPPLY ) ;
type LOGIC is   record
                    STATE :         XBIT ;
                    STRENGTH :      STRENGTH_TYPE  ;
                    CLASS :         CLASS_TYPE  ;
                end record ;
type LOGIC_BUS is array ( INTEGER range <> ) of LOGIC ;
```

The resolution function on the LOGIC type could be defined as follows:

```
function LOGIC_RESOLVE ( SOURCES : LOGIC_BUS ) return LOGIC is
        variable RESULT : LOGIC ;   -- initialized with ('X',0,INITIAL)
begin
        for i in SOURCES'range loop
            if SOURCES(I).CLASS > RESULT.CLASS then
                RESULT := SOURCES(I) ;
            elsif SOURCES(I).CLASS = RESULT.CLASS then
                if SOURCES(I).STATE = RESULT.STATE then
                    if SOURCES(I).STRENGTH > RESULT.STRENGTH then
                        RESULT.STRENGTH := SOURCES(I).STRENGTH ;
                    end if ;
```

```
                else      -- Conflict
                    if SOURCES(I).STRENGTH > RESULT.STRENGTH then
                        RESULT := SOURCES(I) ;
                    elsif SOURCES(I).STRENGTH = RESULT.STRENGTH then
                        RESULT.STATE := 'X' ;
                    end if ;
                end if ;
            end if ;
        end loop ;
        return RESULT ;
    end LOGIC_RESOLVE ;
```

Although this resolution function has the safe properties of being associative and commutative, it is only a first draft. Indeed, if such a resolved LOGIC signal probably has the "correct" state and class values ("correct" in the sense that these values are the same as the values computed by Lsim with M), its resulting strength may be different.

In our case, the resulting strength is simply the strength of the strongest source (giving the final state result). In most cases, this result may not be very significant: it could seem more realistic to have a resulting strength lower than that of the strongest source, thereby taking into account the weakening due to the conflict.

Another consideration is the internal use by Lsim of a strength stored on 12 bits (0 to 4095), the user having only access to the five most significant bits (0 to 31). Thus, another VHDL LOGIC type could be:

```
type XBIT is ( 'X' , '0' , '1' ) ;
subtype STRENGTH_TYPE is INTEGER range 0 to 31 ;
subtype FULL_STRENGTH_TYPE is INTEGER range 0 to 4095 ;
type CLASS_TYPE is ( INITIAL, CHARGED, DRIVEN, SUPPLY ) ;
type LOGIC is
    record
        STATE :        XBIT ;
        STRENGTH :     FULL_STRENGTH_TYPE ;
        CLASS :        CLASS_TYPE ;
    end record ;
```

Exactly the same resolution function may be written (using the FULL_STRENGTH_TYPE instead of the STRENGTH_TYPE), but a mechanism has to be built in order to prevent the designer from accessing the signal strength on 12 bits: he may only use the five first bits.

In order to do that, two functions are defined:

```
function GET_LOGIC_STRENGTH ( L : LOGIC ) return STRENGTH_TYPE is
begin
    return ( L.STRENGTH / 128 ) ;
end GET_LOGIC_STRENGTH ;
```

```
function SET_LOGIC_STRENGTH (L : LOGIC ; S : STRENGTH_TYPE) return LOGIC is
      VARIABLE RESULT : LOGIC := L ;
begin
      RESULT.STRENGTH := S * 128 ;
      return RESULT ;
end SET_LOGIC_STRENGTH ;
```

The designer is then supposed to modify the strength field of a LOGIC signal only by using the two previous functions. But the user will always be able to access directly the full strength of the signal: no mechanism, such as the private types in Ada, is implemented in VHDL to prevent this.

8.8.2. Timing Information

With the M language (and Lsim), a designer may decide to change the timing information during the simulation. For example, he may start the simulation with zero-delay signal assignments and, at a given time point, change an environment variable (one of *Useri0* to *Useri9*) to simulate with given delays written in the models.

In VHDL, the lack of global variables requires the designer to use a global signal. The following package specification may be defined:

```
package TIMING_PKG is
      type TIMING_OPTIONS is ( ZERO_DELAY, UNIT_DELAY, PHYSIC_DELAY ) ;
      signal TIMING_CHOICE : TIMING_OPTIONS ;
end TIMING_PKG ;
```

Then the designer is allowed to write such a design entity:

```
entity LATCH is
      port (   DATA_IN :      in BIT ;
               CLOCK :        in BIT ;
               DATA_OUT :     out BIT ) ;
end LATCH ;
library TIMING_LIB ; use TIMING_LIB.TIMING_PKG.all ;
architecture ARC of LATCH is
begin
      process
      begin
          wait until CLOCK = '1' ;
          case TIMING_CHOICE is
           when ZERO_DELAY      =>  DATA_OUT <= DATA_IN ;
           when UNIT_DELAY      =>  DATA_OUT <= DATA_IN after 1 ns ;
           when PHYSIC_DELAY    =>  DATA_OUT <= DATA_IN after 1.385 ns ;
          end case ;
      end process ;
end ARC ;
```

During the simulation, the designer may change the timing nature of this simulation by changing the value of the global TIMING_CHOICE signal.

8.8.3. Backannotation

This is one of the most important features in Lsim: the possibility of backannotation within the M models. A typical M description using this mechanism is

```
MODULE NAND2 ( )
{
  IN LOGIC A ;
  IN LOGIC B ;
  OUT LOGIC S ;

  INITIALIZE
  {
                    SET_CAPACITANCE ( A , 0.1 ) ;
                    SET_CAPACITANCE ( B , 0.1 ) ;
                    SET_CAPACITANCE ( S , 0.025 ) ;
  }

  SIMULATE
  {
                    S { 27.3 * GET_CAPACITANCE ( S ) } = A ~& B ;
  }
}
```

The GET_CAPACITANCE and SET_CAPACITANCE functions are Lsim built-in functions that have direct access to the global netlist of the circuit being simulated. Let us build the following network of three NAND2 models:

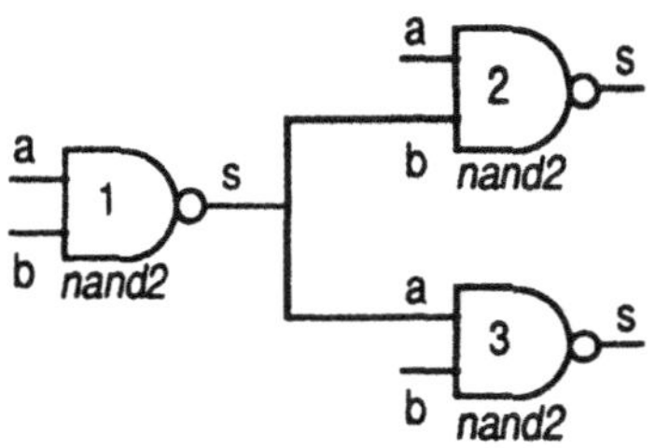

Fig 8.11 Backannotation Example

For the NAND2 model labeled 1 in the previous schema, the GET_CAPACITANCE function retrieves the value 0.1 + 0.1 + 0.025 = 0.225 for terminal S.

Implementing such a mechanism in VHDL is not easy. How can a given instance know the number of ports of other instances connected to one of its ports? How can the capacitance information be attached to each node of the netlist?

The use of an attribute may have been judicious, but the attribute information cannot be transmitted through the hierarchy. Actually, there is one VHDL mechanism that is able to count the number of drivers (of the ports) driving the same node: it is the resolution function. But the resolution function is called every time a port wants to drive a value on a node: it is a dynamic mechanism that repeats many times during simulation, whereas our capacitance process should be processed once at the very beginning of the simulation (after or during the circuit parse). Furthermore, such a mechanism would not have been possible for input ports (which are not driving values outside the component).

The idea is to create specific ports (capacitance ports) that are associated with the signal ports and that use a specific resolution function. Thus, the following package is declared:

```
package CAPACITANCE_PKG is
 type SINGLE_CAPACITANCE is range 0 to 1E9
      units
           fF ;
           pF = 1000 fF ;
           nF = 1000 pF ;
           uF = 1000 nF ;
      end units ;
 type CAPACITANCES is array ( INTEGER range <> ) of SINGLE_CAPACITANCE ;
 function SUM_CAP (SOURCES: CAPACITANCES) return SINGLE_CAPACITANCE ;
 subtype CAPACITANCE is SUM_CAP SINGLE_CAPACITANCE ;
end CAPACITANCE_PKG ;

package body CAPACITANCE_PKG is
 function SUM_CAP (SOURCES: CAPACITANCES) return SINGLE_CAPACITANCE is
      variable RESULT : SINGLE_CAPACITANCE := 0 fF ;
 begin
      for I in SOURCES'range loop
           RESULT := RESULT + SOURCES(I) ;
      end loop;
      return RESULT ;
 end SUM_CAP ;
end CAPACITANCE_PKG ;
```

Then the above NAND2 M model may be described by the following VHDL design units using the CAPACITANCE_PKG package:

```
library STD_LIB ;
use STD_LIB.CAPACITANCE_PKG.all ;
entity NAND2 is
      port (   A :                  in BIT ;
               B :                  in BIT ;
               S :                  out BIT ;
               A_CAPACITANCE :  inout CAPACITANCE ;
               B_CAPACITANCE :  inout CAPACITANCE ;
               S_CAPACITANCE :  inout CAPACITANCE ) ;
end NAND2 ;

architecture ONE of NAND2 is
begin
SET_CAPACITANCE : process
begin
      A_CAPACITANCE <= 0.1 pF ;
      B_CAPACITANCE <= 0.1 pF ;
      S_CAPACITANCE <= 0.025 pF ;
      wait ;     -- process definitively suspended
end process ;
S <= A nand B after ( 27.3 * S_CAPACITANCE / pF * ns ) ;
end ONE ;
```

Such a strategy to implement the capacitances may seem too verbose because of the need for capacitance ports (the total number of signals in the circuit is exactly doubled). But these capacitance signals are only assigned once during the initialization, so their use is not very damaging for simulation efficiency.

Another constraint is the need to duplicate the netlist information: if the designer connects port S of component 1 to port A of component 2, he also needs to connect port S_CAPACITANCE of component 1 to port A_CAPACITANCE of component 2.

This solution may be more usable if the VHDL netlist description is generated from a schematic editor. In that case, the designer may only draw the connections between signal ports, and a program will automatically add in the netlist the capacitance ports connection.

8.9. CONCLUSION

The main equivalences between M and VHDL have been explained in this chapter. They may be seen as guidelines to a translation of M models into VHDL models, but they do not intend to resolve all the problems posed by this task. Typical ways of using VHDL capabilities to implement some Lsim built-in features have been presented.

9. VERILOG AND VHDL

9.1. INTRODUCTION

Verilog is a hardware description language originally designed by Gateway for a proprietary simulation product. This company later merged with Cadence Design Systems, Inc. Since then, Verilog has been known and used as the language of the Cadence digital simulator.

Later faced with the competition of VHDL, Cadence decided in 1990 to transfer Verilog to the public domain, thereby attempting to give a "standard label" to its language. The "Verilog" name is still a registered trademark of Cadence, but the language itself can be supported by any CAD tool.

The use of the Verilog HDL is promoted by Open Verilog International (OVI), which published in October 1991 the first version of the Verilog Hardware Description Language Reference Manual [VER91]. In this version 1.0, the definition of the language is exactly the one provided by Cadence at the time the company transferred its language to OVI. Changes or additions approved by OVI have not been included yet.

Verilog is currently widely used by IC designers (especially those using Cadence tools) and is integrated in many different existing CAD products (simulators, synthesis tools, etc. from Cadence, Synopsys, and others). As we will see in this chapter, Verilog has many predefined features very specific (and necessary) to IC design.

The current state of VHDL is very different. Although VHDL has been an IEEE standard since 1987, its integration in the front-end of existing tools is quite recent, as is its use by designers. Its purpose is to model hardware — not only ICs, but also PCBs and systems. Thus, the set of predefined features is very restricted in VHDL: most of the specific constructs must be defined by the user. Nevertheless, the real use of VHDL as a modeling language is increasing very quickly, and now most CAD vendors do propose a VHDL interface.

Although this comparison is based on both Verilog and VHDL reference manuals, many of the general concepts handled in the beginning of the chapter are taken from the VHDL Language Reference Manual [IEE87]: design units, sequential and concurrent worlds, types, etc. This choice is certainly not a partial choice but comes from the fact that the VHDL standardization process took concepts having a common meaning for a large number of designers.

After these general concepts, the rest of the comparison is, in contrast, based on most of the typical Verilog features: nets and registers, predefined operators, continuous and procedural assignments, user-defined primitives, and so forth. In each case, the Verilog type, object, statement or construct is explained in terms of equivalent VHDL declarations and statements. The reader must be aware that such a methodology necessarily gives an advantage to Verilog: for every one or two lines of a given Verilog construct, several lines of equivalent VHDL code are provided. Of course, explaining all the VHDL constructs in terms of equivalent Verilog models would have given an opposite result. Conclusions about the efficiency and the conciseness of both languages must thus be stated with care and without apriorism.

9.2. DESIGN UNIT

According to the VHDL Language Reference Manual, the design units are the constructs that may be separately analyzed and inserted into a design library. Each VHDL design library can be mapped to a system-dependent directory (via an index file not specified in the language).

A VHDL design unit is either an *entity declaration*, a *configuration declaration*, a *package declaration* (three primary units), an *architecture body*, or a *package body* (two secondary units). From the compilation point of view, only the design unit — and not the file — is relevant.

In Verilog, the fundamental design unit is the *module*, describing a piece of hardware with its interface and its content. From a functional point of view, a Verilog module holds the information found in both the VHDL entity and a corresponding architecture body.

Verilog does not provide separate compilation: on the contrary, all modules related to the same design (and simulation) must be present in the same file. This file is called the *design file*. The main drawback is, of course, a slower compilation cycle, especially when most of the modules of a given file are correct and have already been compiled.

There is another Verilog construct defined at the same lexical level as the module, which is the *user-defined primitive (UDP)*. Verilog primitives are user-defined basic hardware units described with a truth-table syntax (using the Verilog table definition). This style of description will be covered deeply in paragraph 9.8.1, but for the time being, we will handle Verilog primitives as specific modules.

As just mentioned, a Verilog module is equivalent to both a VHDL entity and an architecture body. Thus, each Verilog design unit has only one internal description: the VHDL mechanism of associating several architectures with a single entity describing the interface is not implemented in Verilog.

The instantiation mechanism is also very different in the two languages. In VHDL, components are instantiated in architectures, and these components are linked to couples entity-architecture via a configuration declaration or a configuration specification. Components may be seen as sockets into which circuits are plugged by the configuration.

In Verilog, modules (and primitives) are directly instantiated in other modules. An example of structural description in both languages is given in paragraph 9.7.1. As a consequence of this direct instantiation and of the unicity of the internal description of the module, there is no configuration mechanism in Verilog, and thus no configuration declaration design unit.

However, Verilog provides the *specify block*, which can be present in any module and which allows the specification or overriding of any parameter value (such as a delay or a bus size) of any block in the design hierarchy. Thus, this Verilog parameter specification is functionally more or less equivalent to the generic map of a VHDL configuration.

Following is the skeleton of a typical Verilog module and a corresponding couple entity-architecture in VHDL:

module module_name (list_of_ports) ;	**entity** entity_name **is**
	--> *generic clause :*
// parameter declaration	-- parameter declaration
	--> *port clause :*
// input declaration	-- in port declaration
// output declaration	-- out port declaration
// inout declaration	-- inout port declaration
	end entity_name ;
	architecture architecture_name **of** entity_name **is**
// register declaration	-- signal declaration
// net declaration	-- signal declaration
// task declaration	-- procedure declaration
// function declaration	-- function declaration
	begin
// continuous assignment statement	-- concurrent signal assignment statement
// gate instantiation	-- component instantiation
// primitive instantiation	-- component instantiation
// module instantiation	-- component instantiation
// initial statement	-- process statement
// always statement	-- process statement
endmodule	**end** architecture_name ;

The correspondence between the declarations and statements in both languages will be covered more deeply in the next paragraphs.

Besides the hardware design units, VHDL offers to the designer two software design units: the package declaration and its corresponding package body. The purpose of these two units, analyzed separately from the others, is essentially to group together related functions and procedures (and also types, constants, signals, etc.) that may be used in any other design unit.

Such a facility does not exist in Verilog. In this language, functions and tasks (more or less equivalent to VHDL procedures) used within a module have to be defined *inside* the same module. This is a major limitation of the language: Verilog has most of the powerful constructs of software programming languages such as C or Ada (see section 9.6), but this capacity of description is limited by the fact that all tasks and functions must be defined inside a module (i.e., a *hardware* design unit). So they cannot be shared conveniently by different modules, in the same way that procedures and functions, defined in a package (i.e., a *software* design unit), may be called by different design units in VHDL.

Nevertheless, Verilog provides to the user the ability of using in a given module tasks and functions that have been defined inside another module. This can be achieved when the two previously mentioned modules are instantiated inside the same top module (at some hierarchical levels, which may be different), as illustrated by the following example:

```
module TopModule() ;
        AnotherModule_    AnotherModule (...) ;
        Module2_          Module2 (...) ;
endmodule

module AnotherModule_ (...) ;
        Module1_          Module1 (...) ;
endmodule

module Module1_ (...) ;
        function MyFunction ;
            MyFunctionBody
        endfunction
endmodule

module Module2_ (...) ;
        assign output_name = AnotherModule.Module1.MyFunction ( input_name ) ;
endmodule
```

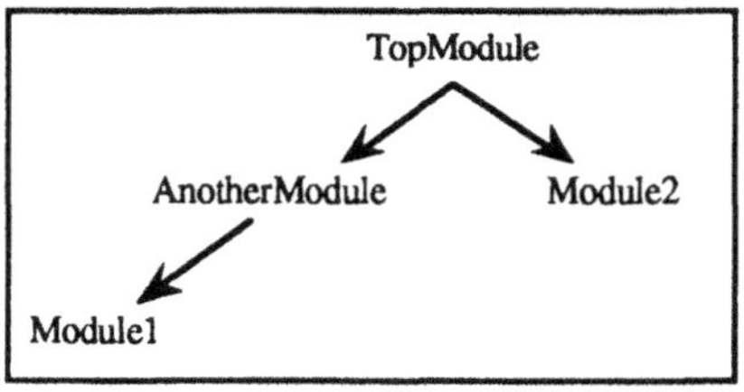

Fig 9.1 Hierarchical Names

Each object defined at some level in a Verilog module may be accessed in another module by using its hierarchical name. Unlike the VHDL objects, which are local to their declarative region (a subprogram defined in an architecture may only be called inside that architecture), Verilog objects are global and may be accessed within the entire hierarchy being simulated.

Of course, this mechanism does not at all have the same possibilities as the use of packages in VHDL. In order to group together related types and subprograms, the Verilog designer should create a *fictitious* module (i.e., a hardware design unit without *real* hardware meaning) containing all these definitions and instantiate this somehow strange module inside the *real* module hierarchy being currently modeled.

The design methodology consisting in writing packages of usual subprograms and storing them by related subjects into different design libraries cannot be practiced with Verilog, which does not specify library management and does not support separate compilation. As stated previously, Verilog supports the opposite concept of *design file*: all modules used in the same simulation must be written in the same file. Actually, packages may become part of the Verilog language at its first revision.

Figure 9.2 sums up the different design units in both languages and establishes the connections between the similar concepts seen in this section.

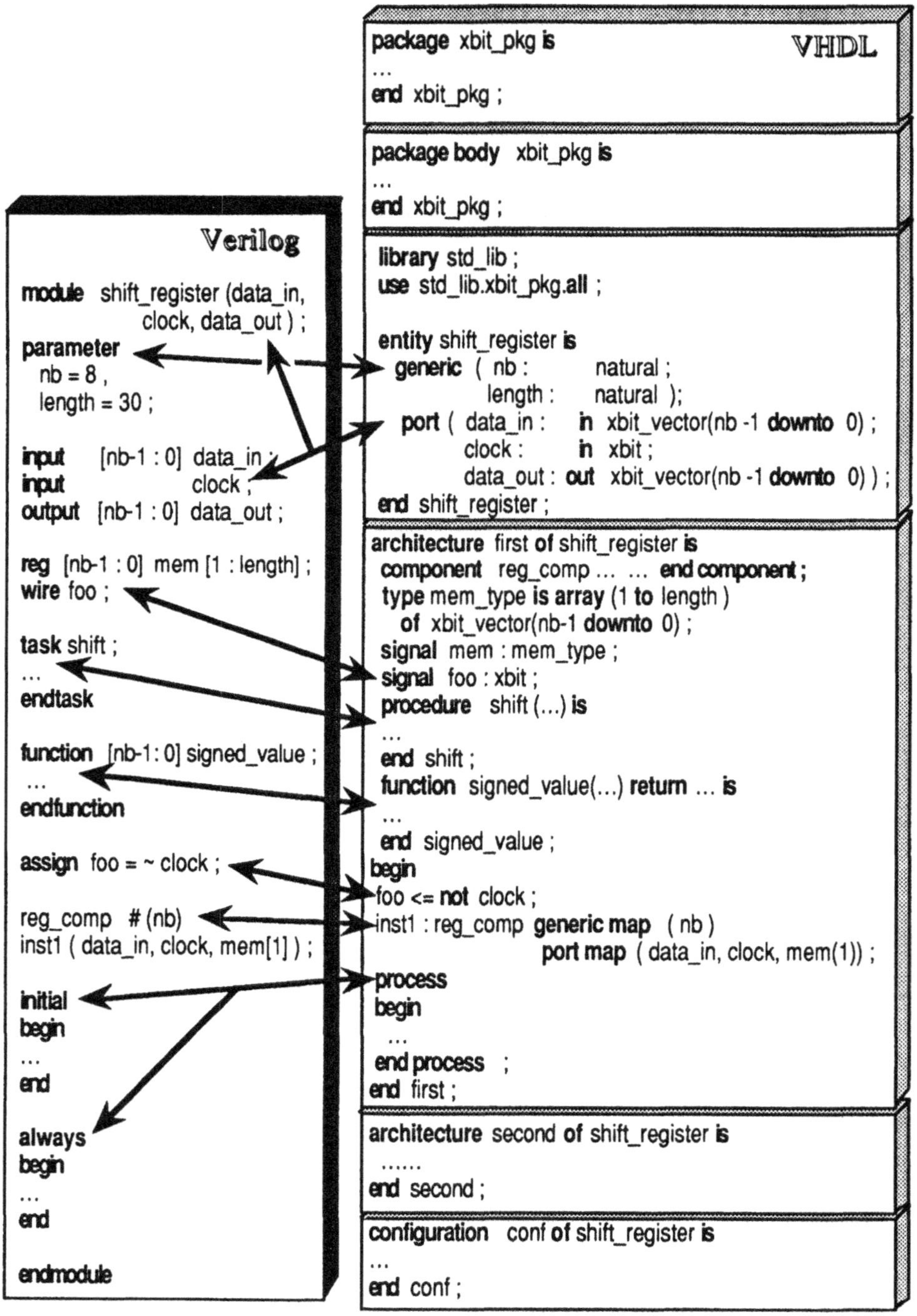

Fig 9.2 Design Units

9.3. SEQUENTIAL AND CONCURRENT DOMAINS

The distinction between sequential and concurrent domains could be roughly stated as follows: the sequential world is mainly intended to implement the algorithmic part of the model, whereas the concurrent world basically represents the interconnection of black boxes, themselves containing algorithms or interconnected black boxes.

Putting these two worlds together gives the designer the ability to describe concurrent processes, i.e., several sequential algorithms executing in parallel and synchronizing between each other.

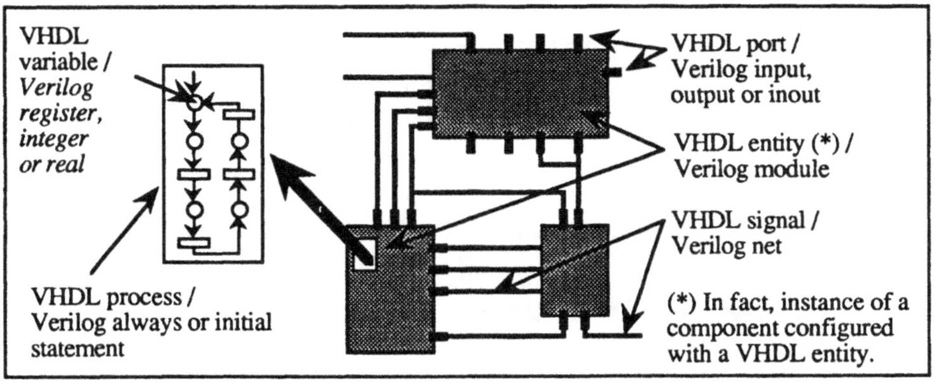

Fig 9.3 Sequential and Concurrent Domains

The implementation of these sequential and concurrent domains in both languages is very close:

- the domain of a Verilog module or a VHDL architecture (or block) is the concurrent domain;
- in both languages, concurrent statements may describe structural descriptions (Verilog module or VHDL component instantiation), dataflow descriptions (Verilog continuous assign statement or VHDL concurrent signal assignment statement), and behavioral descriptions (Verilog always and initial statements or VHDL process statements);
- these equivalent Verilog always (or initial) statements and VHDL process statements encapsulate sequential algorithms.

The next schema (figure 9.4) describes this fundamental separation between the sequential world and the concurrent world, a separation that is implemented in a similar way in both languages.

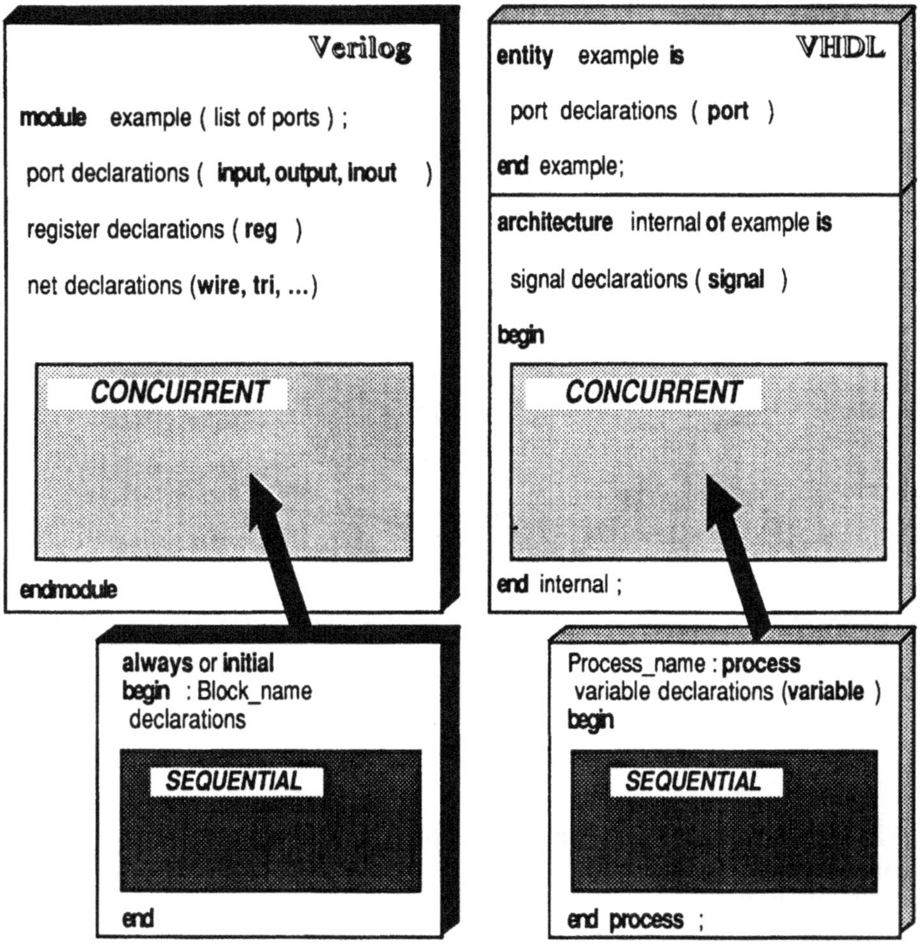

Fig 9.4 Sequential and Concurrent Domains

9.3.1. Sequential Processes

Each VHDL process contains an implicit infinite loop: at the initialization, instructions included in the process are executed in a sequential order until the first wait statement (if found), where the process is suspended. Then, still in this initialization phase or later in a simulation cycle, an external event may reactivate the process, which resumes, and so on. After having executed its last instruction, a process re-executes its first instruction and the following ones, and thus loops during the entire simulation time.

In Verilog, the always statement is executed in the same manner as the VHDL process statement (implicit infinite loop). On the other hand, the initial statement does not have this infinite loop: it is executed once at the initialization and suspends forever after the last instruction is executed. In fact, the initial statement is equivalent to a VHDL process with a *wait forever* (**wait**;) as the last instruction.

The next two figures sum up the equivalences between the Verilog initial or always statement and the VHDL process statement.

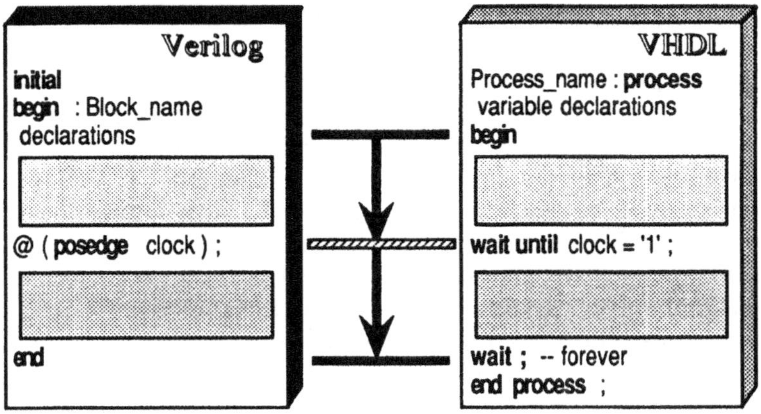

Fig 9.5a Verilog Initial Statement and Corresponding VHDL Process

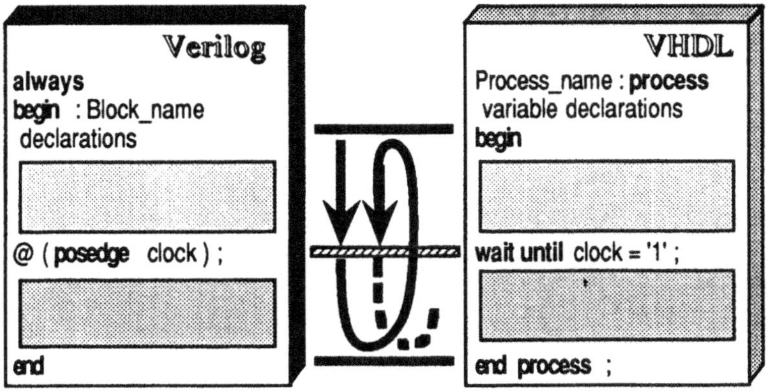

Fig 9.5b Verilog Always Statement and Corresponding VHDL Process

In Verilog, the always and initial statements have the same modeling purpose as the VHDL process statement: to encapsulate sequential algorithms that are executed asynchronously in the concurrent world. Sequential statements available in both languages will be detailed in the next sections.

The way these concurrent processes are executed and synchronized is the semantic basis of any hardware description language. This simulation semantics is much more detailed in the VHDL reference manual than in the Verilog manual: for example, all the VHDL concurrent statements are explained in terms of equivalent process statements. Nevertheless, as just shown, VHDL and Verilog obviously share many implementation rules.

9.3.2. Concurrent Assignments and Delta-Delay Mechanism

A typical construct implemented by both languages that highlights the semantics of the concurrent domain is the concurrent assignment statement. These similar concurrent statements are the continuous assignment statement (in Verilog) and the concurrent signal assignment statement (in VHDL). Although this comparison will be detailed in paragraph 9.6.1, it is interesting at this point to focus on the mechanism implied by these statements.

In both languages, the assignment statement is executed each time any right-hand-side signal value changes. For zero-delay assignments, this can lead to the execution of an assignment several times during the same simulation time point: the VHDL reference manual calls it the *delta-delay* mechanism. The following (classical) figure illustrates this mechanism:

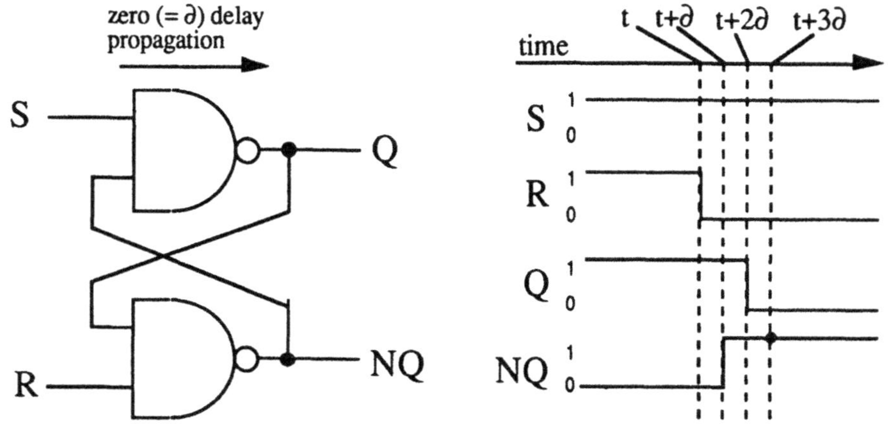

Fig 9.6 Timing in Concurrent Assignments

At the same current time t, the VHDL assignment "NQ <= R **nand** Q" has been executed twice and "Q <= S **nand** NQ" once. The same happens with the equivalent Verilog assignments:

```
architecture DATAFLOW of RS is
-- declarations
begin
        Q <= S nand NQ ;
        NQ <= R nand Q ;
end ;
```

```
module RS (S,R,Q,NQ) ;
//declarations

        assign   Q = S ~& NQ ,
                 NQ <= R ~& Q ;
endmodule
```

The delta delay is compulsory to model causality (and therefore true concurrency) when using zero-delay assignments. Both languages implement this semantics: Verilog and VHDL concurrent assignment statements have similar execution rules.

9.3.3. Verilog Fork and Join Statement

Another Verilog feature modeling a kind of concurrency is the *fork* and *join* statement, which allows the designer to describe concurrent statements inside a sequential block (initial or always). VHDL does not have this capability: the lowest level is always described by sequential statements, and there is no way from the sequential world to go back to the concurrent world.

Nevertheless, the reader must be aware of the fact that the concurrency encapsulated inside a fork and join block is not of the same kind as the concurrency described just above. Let us consider the following example:

```
module DetectConflict (request1, request2, collision) ;
input  request1 ;
input  request2 ;
output      collision ;
always
begin
        fork
        @ ( posedge request1 ) ;
        @ ( posedge request2 ) ;
        join
        if (request1 == 1 && request2 == 1) collision = 1 ;
end
endmodule
```

This description is maybe not the most straightforward way to describe a collision detector, but it is a good illustration of the fork and join statement: the previous module is waiting a rising edge of each requesting signal, appearing in any order, and once these two edges have been detected, a collision signal may be set to 1.

The two wait instructions (beginning with symbol @) correspond to two processes executing in parallel, but not according to the concurrency explained before. The main difference between concurrent statements in a module and statements inside a fork and join block is the following:

- concurrent statements inside a module (such as continuous assignments) are separate processes that may be re-executed *several* times during the same simulation time point (according to the delta-delay mechanism) ;
- statements inside a fork and join block are separate processes that begin when control is passed to the fork and that are executed only *once*, with the join waiting for all these processes to be completed before continuing with the next statement.

No delta-delay mechanism is applied in such concurrency: the fork and join block just gives the designer the possibility of specifying that he or she does not care about the order of execution of some statements that are nevertheless executed only once as in any sequential block.

9.3.4. Subprograms

As shown by figure 9.2, Verilog implements tasks and functions that are more or less equivalent to VHDL procedures and functions respectively.

```
module FOO ;
  ...
  task TSK ;
   ... // task declaration
  endtask
  function FCT ;
   ... // function declaration
  endfunction


  always
    TSK(WIRE1) ; -- task enable

  assign WIRE1 = #10 FCT(INPUT1);
endmodule
```

```
architecture FOO of ENT is
  ...
  procedure TSK (A : BIT) is
  ... -- procedure declaration
  end TSK;
  function FCT(A : BIT) return BIT is
  ... -- function declaration
  end FCT;
begin
  process
  begin
    TSK(WIRE1) ; -- procedure call
  end process ;
  WIRE1 <= FCT(INPUT1) after 10 ns ;
end FOO;
```

Each Verilog and VHDL subprogram encapsulates sequential code. Functions have very similar semantics in both languages: they return a value, so that a function call can be used in an expression; they cannot contain timing control statements (statements with @, #, or **wait** in Verilog, and wait statements in VHDL); and they may include local declarations (temporary variables are created at each function call).

Verilog tasks and VHDL procedures have also many similarities: they encapsulate sequential code, may declare local variables, and do not return value as functions but may include timing control statements. But two main differences exist between tasks and procedures.

First of all, a Verilog task call (*task enable* in the Verilog reference manual) is a *sequential* statement that can be present in initial statements, always statements, and task and functions bodies. Such a Verilog statement is thus equivalent to a VHDL *sequential procedure call,* which can appear in the process statement part, and in the procedure and function bodies. But Verilog does not support concurrent task enable such as the VHDL concurrent procedure call (equivalent to a VHDL process with an implicit wait statement).

Also, variables declared in the declarative part of a VHDL procedure are local and temporary: they are dynamically created at each procedure call and cannot be accessed from outside the procedure (side effects are impossible). In Verilog, it seems that all invocations of the same task may share the local variables of the same task. Actually, the Verilog Language Reference Manual states in paragraph 9.2.4. that "some tools allow multiple copies of a task to execute concurrently, but it does not copy or otherwise preserve the task arguments or local variables; some tools use the same storage for each invocation of the task". Although the Verilog LRM refers to this mechanism as an "implementation specific detail", it would have been much safer to specify a tool-independent behavior concerning these local variable declarations. Consequences on a design are indeed too important and critical to let the implementation specify this mechanism.

Finally, excepting the fork and join statement (which is an additional feature of Verilog having no equivalent in VHDL) and the previous remarks on the Verilog tasks, it is possible to conclude that sequential and concurrent worlds are very similarly defined in both languages.

9.4. OBJECTS AND TYPES

Three classes of objects are defined in VHDL: signal, variable and constant. Each object has a given type: this type may be a scalar type (integer type, floating point type, physical type, enumeration type), a composite type (array and record types), an access type (providing access to objects of a given type), or a file type (providing access to objects that contain a sequence of values of a given type).

On the other hand, Verilog has only a very restrictive set of types, which is very specific to IC modeling. The following figure lists all the objects and their possible types in both languages.

	Verilog
Port	**input, output, inout**
Register	**reg**
Net	**wire, wand, wor, tri, triand, trior, tri0, tri1, trireg, supply0, supply1**
Variable	**integer, real time**

	VHDL
signal (**port**)	Any VHDL type except **access** and **file**
variable	Any VHDL type
constant	Any VHDL type except **access** and **file**
file	file containing objects of any VHDL type, except **access** and **file**

Fig 9.7 Objects and Types

In both languages, *ports* represent the interface elements of the module or the entity that allow connection with the outside world. Basically, they are handled inside the model just as other nets or signals. Each port may be an input, output, or bidirectional port, respectively an **input**, **output**, or **inout** in Verilog and an **in**, **out**, or **inout** port in VHDL.

Inside the module or the architecture, physical wires may be represented in Verilog with two datatypes, *nets* and *registers*, whereas VHDL models them with a single class of objects, the *signals*.

In Verilog, nets are used to model an electrical connection: they do not store a value, but only transmit values that are driven on them by gate outputs or continuous assigns (except the *trireg* net, which models a physical wire as a capacitor). On the other hand, registers hold their value during simulation: acting as memory elements, they may be updated by procedural assignments (inside always or initial blocks).

There is no such distinction in VHDL: signals are both used to connect ports of different component instances and to store values provided by signal assignment statements. In VHDL, each physical wire (signal) acts as a memory element.

Actually, Verilog has clearly distinguished the two fundamental functions of a physical wire:

- the *connecting or routing function*: a physical wire connects together two or more subelements ports ;
- the *memory function*: a physical wire may or may not store any value that has been driven on it by a gate output or a functional assignment.

These two separate properties may lead to some common Verilog descriptions, such as the following:

```
module    example ( regout , wirein ) ;
output    regout ;
input     wirein ;
reg       regout ;
wire      wirein ;  // facultative, wire being the default type...
...
```

Such a declaration could look somewhat strange to a VHDL designer if he or she thinks (erroneously!) that regout has been declared first as an output port and second as a signal (as a physical memory element). In fact, the output direction is a property associated with the connecting function of regout (as a port), whereas the reg datatype is a property associated with its memory function. When just declaring "**input** wirein", the Verilog designer implicitly declares a net (not a register) of type wire (the default type), named wirein and having the input direction.

Actually, this distinction between the connecting and memory functionalities is relevant from the *designer point of view*: nets and registers correspond to different hardware devices that behave according to the above explanation. But from the *simulator point of view*, this distinction may be actually less accurate: each Verilog net is certainly implemented inside the simulation data with a variable representing (or a structure having a field representing) the current value of the net. Indeed, nets are targets of continuous assignments, which are evaluated each time any input changes. Thus, between two input changes, the target net remains at the same value (which needs to be stored). Actually, nets may be computed in a very similar way as registers: the simulator is event driven, there is thus no real *continuous computation*, and nets (like registers) do store a value between two assignment computations.

In Verilog, the general form of a net declaration is the following:

```
Net_Type Charge_Strength    Range   Delay   List_of_Names ;
```

As illustrated by figure 9.7, nets may be of different types:
- **wire** (or **tri**), representing the basic net type
- **wand** (or **triand**), representing a net with a wired-and logic function
- **wor** (or **trior**), representing a net with a wired-or logic function
- **tri0** and **tri1**, representing resistive pull-down and pull-up
- **supply0** and **supply1**, representing power supply
- **trireg**, representing wire as a capacitor (with charge storage)

In the net declaration, the first argument is the net type. The charge strength may be specified for the trireg only, and may have one of three values:

small, **medium**, and **large**. These values are capacitor sizes that correspond to strength levels 1, 2, and 4 (in Verilog, strengths range from 0 (high impedance) to 7 (supply drive)). The optional range specifies the bus width.

Finally, before the list of names, a delay may be specified for the net(s). This delay may consist of up to nine values: a different value may be given for each transition to 0, 1, or Z and for each minimum, typical, and maximum case. Such a delay attached to a given net is added to the delay of each assign statement driving this net, the result being the total delay between any change of the assign statement input and a corresponding change in one of the inputs to which the net is connected (delay specifications and propagations are detailed in 9.6.2.2).

These strength and delay properties of the Verilog net emphasize its physical aspects. No such notion exists in VHDL, where the signal does not have any delay or strength associated with it. In VHDL, all the delays are provided by the assignments, and there are no predefined strengths.

Besides the functional distinction between nets and registers seen previously, these two classes of physical objects, together with the ports, have the same logic parent type, which is based on the classical four-level logic system:

- 0, representing a logic zero (or FALSE)
- 1, representing a logic one (or TRUE)
- X, representing an unknown logic value
- Z, representing a high-impedance state

A VHDL equivalent to the Verilog four-level logic type could be

```
type FOUR_LEVEL_BIT is ('X', '0', '1', 'Z') ;
```

Verilog wire, wand, and wor (tri, triand, and trior) net types may be considered as VHDL *subtypes* of the basic four-level logic bit type: their only particularity consists in the associated logic function used to resolve conflicts (wired-X, wired-and, and wired-or). In a first stage and with some restrictions, tri0 and tri1 may be handled in the same way, their associated resolution function computing the resistive pull state when all tristate buffers are off (i.e., driving high impedance).

In VHDL, the user may define the following package NET_TYPES declaring the FOUR_LEVEL_BIT type, the subtypes corresponding to the Verilog net types and their associated resolution functions:

```
package NET_TYPES is

-- types
type FOUR_LEVEL_BIT is ('X', '0', '1', 'Z') ;
type FOUR_LEVEL_BIT_VECTOR is array (NATURAL range <>) of FOUR_LEVEL_BIT ;
```

```
-- resolution functions
function WIRED_X (S : FOUR_LEVEL_BIT_VECTOR) return FOUR_LEVEL_BIT ;
function WIRED_AND (S : FOUR_LEVEL_BIT_VECTOR) return FOUR_LEVEL_BIT ;
function WIRED_OR (S : FOUR_LEVEL_BIT_VECTOR) return FOUR_LEVEL_BIT ;
function WIRED_TRI0 (S : FOUR_LEVEL_BIT_VECTOR) return FOUR_LEVEL_BIT ;
function WIRED_TRI1 (S : FOUR_LEVEL_BIT_VECTOR) return FOUR_LEVEL_BIT ;

-- resolved subtypes
subtype WIRE is WIRED_X  FOUR_LEVEL_BIT ;
subtype TRI is WIRE ;
subtype WAND is WIRED_AND  FOUR_LEVEL_BIT ;
subtype TRIAND is WAND ;
subtype WOR is WIRED_OR  FOUR_LEVEL_BIT ;
subtype TRIOR is WOR ;
subtype TRI0 is WIRED_TRI0  FOUR_LEVEL_BIT ;
subtype TRI1 is WIRED_TRI1  FOUR_LEVEL_BIT ;
-- Warning: TRI0 and TRI1 subtypes must be used to declare only
-- signals assigned by drivers all located in the same architecture or block.
subtype SUPPLY0 is FOUR_LEVEL_BIT range '0' to '0' ;
subtype SUPPLY1 is FOUR_LEVEL_BIT range '1' to '1' ;

end NET_TYPES ;
```

Most of the Verilog net types have been translated into VHDL subtypes of the basic FOUR_LEVEL_BIT type. Wire, wand, wor, tri0, and tri1 are defined by a subtype indication using a resolution function that implements the appropriate conflict arbitration. Tri, triand, and trior are the same subtypes as wire, wand, and wor, respectively (using a subtype indication without constraint). Supply0 and supply1 use as subtype indication a range constraint that is limited to one element, which is the specified supplied logic level.

As indicated by the comment following their declarations, the tri0 and tri1 VHDL subtypes should be used to declare only signals assigned by drivers that are all located in the same block or architecture (at the same hierarchical level). The reason is that the proposed resolution functions wired_tri0 and wired_tri1 are not associative (see resolution functions bodies below). Hierarchical resolutions would thus lead to unexpected results. For example:

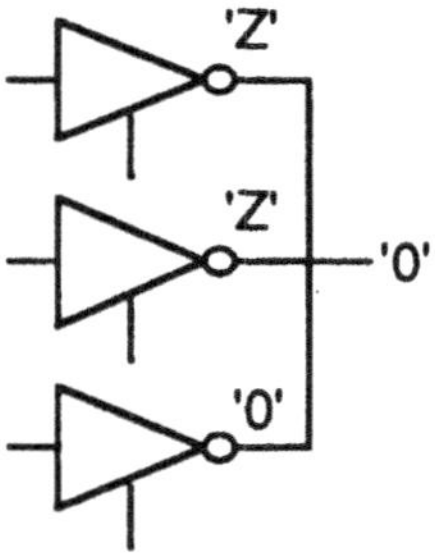

Fig 9.8a Drivers in Same Unit

VHDL flat resolution:
WIRED_TRI1("ZZ0") = '0' => *GOOD*

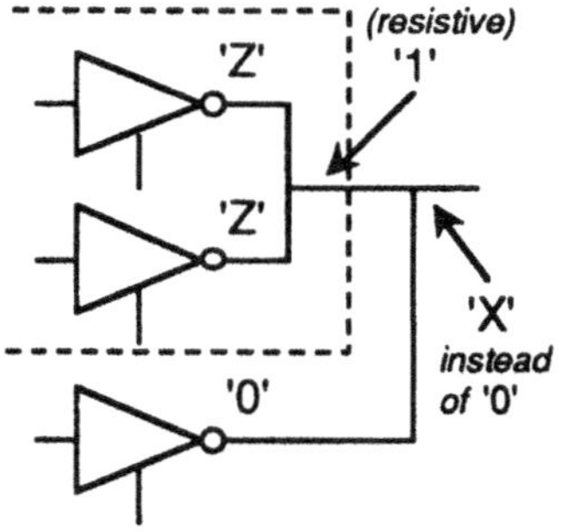

Fig 9.8b Drivers in Different Units

VHDL hierarchical resolution:
WIRED_TRI1("ZZ") = '1'
WIRED_TRI1("10") = 'X' => *WRONG!*

In fact, this tristate modeling would need strength levels in order to differentiate a charged '1' (respectively, '0') from a driven '1' (respectively, '0'). In the example above, the first hierarchical resolution would give a resistive or charged '1', and the conflict of this value with a driven '0' would result in the correct value '0'.

Methodology speaking, it is never recommended to write non-associative resolution functions (see paragraph 7.1.4 in chapter 7, "Tricks and Traps"). Thus, a better way to model the tri1 and tri0 nets would require another bit logic type taking into account strength values. Making the distinction between states and strengths is compulsory for switch-level modeling.

Trireg nets are capacitors (having a charge storage) to which may be attached a charge strength.

Following is the NET_TYPES package body (including resolution functions bodies):

```
package body NET_TYPES is

-- type used to describe two-inputs truth tables
type TABLE is array (FOUR_LEVEL_BIT,FOUR_LEVEL_BIT) of FOUR_LEVEL_BIT;

-- resolution function associated with WIRE and TRI subtypes
function WIRED_X (S : FOUR_LEVEL_BIT_VECTOR) return FOUR_LEVEL_BIT is
        constant RESOLVE : TABLE := (('X','X','X','X'),
                                     ('X','0','X','0'),
                                     ('X','X','1','1'),
                                     ('X','0','1','Z')) ;
        variable RESULT : FOUR_LEVEL_BIT := 'Z' ;
```

```
begin
        for I in S'RANGE loop
            RESULT := RESOLVE(RESULT, S(I)) ;
            exit when RESULT = 'X' ;
        end loop ;
        return RESULT ;
end WIRED_X ;

-- resolution function associated with WAND and TRIAND subtypes
function WIRED_AND (S : FOUR_LEVEL_BIT_VECTOR) return FOUR_LEVEL_BIT is
        constant RESOLVE : TABLE := (('X','0','X','X'),
                                     ('0','0','0','0'),
                                     ('X','0','1','1'),
                                     ('X','0','1','Z')) ;
        variable RESULT : FOUR_LEVEL_BIT := 'Z' ;
begin
        for I in S'RANGE loop
            RESULT := RESOLVE(RESULT, S(I)) ;
            exit when RESULT = '0' ;
        end loop ;
        return RESULT ;
end WIRED_AND ;

-- resolution function associated with WOR and TRIOR subtypes
function WIRED_OR (S : FOUR_LEVEL_BIT_VECTOR) return FOUR_LEVEL_BIT is
        constant RESOLVE : TABLE := (('X','X','1','X'),
                                     ('X','0','1','0'),
                                     ('1','1','1','1'),
                                     ('X','0','1','Z')) ;
        variable RESULT : FOUR_LEVEL_BIT := 'Z' ;
begin
        for I in S'RANGE loop
            RESULT := RESOLVE(RESULT, S(I)) ;
            exit when RESULT = '1' ;
        end loop ;
        return RESULT ;
end WIRED_OR ;

-- resolution function associated with TRI0 subtype (warning: not associative)
function WIRED_TRI0 (S : FOUR_LEVEL_BIT_VECTOR) return FOUR_LEVEL_BIT is
        variable RESULT : FOUR_LEVEL_BIT := 'Z' ;
begin
        RESULT := WIRED_X (S) ;
        if RESULT = 'Z' then  RESULT := '0' ; end if ;
        return RESULT ;
end WIRED_TRI0 ;

-- resolution function associated with TRI1 subtype (warning: not associative)
function WIRED_TRI1 (S : FOUR_LEVEL_BIT_VECTOR) return FOUR_LEVEL_BIT is
        variable RESULT : FOUR_LEVEL_BIT := 'Z' ;
begin
```

```
        RESULT := WIRED_X (S) ;
        if RESULT = 'Z' then  RESULT := '1' ; end if ;
        return RESULT ;
    end WIRED_TRI1 ;

    end NET_TYPES ;
```

As a conclusion for this section, it is important to notice that Verilog has predefined all the basic types commonly used to model nets in integrated circuit design, these nets having low-level physical attributes (delay, strength, etc.). But on the other hand, Verilog does not provide the designer with the possibility of defining nets (or ports or registers) of abstract datatypes, such as the user-defined enumerated or composite types in VHDL.

In Verilog, the only type of values propagated on wires is the four-level logic bit type. There is an implicit conversion of bit vectors into unsigned integers, thereby allowing direct arithmetic operations on buses.

This difference between the two languages — Verilog nets being only of the four-level bit type, and VHDL signals being of any user-defined abstract type — is directly linked to their original philosophy. As already stated in the introduction, Verilog was designed as a hardware description language for ICs in a proprietary simulation product, whereas VHDL is a standard intended to be widely used in many applications, even for high-level system modeling. In the former case, manipulating signals of the bit type only could seem sufficient, but a need for a higher level of abstraction seems justified for the latter case.

Finally, as listed in the previous table, the designer may also define internal variables of type **integer**, **real** or **time**. These variables are handled as registers (class **reg**) by the simulator: they are created once during the initialization step, they exist throughout the simulation (like VHDL variables declared in a process), and they can be used in most of the places that registers are used. The size of integer and real variables is implementation dependent, but the size of a time variable is always 64 bits. In VHDL, the size of variables and of signals of predefined type integer, real, or time is implementation dependent.

Here a fundamental difference between both language semantics must be pointed out (this difference underlies figure 9.7). VHDL provided three *classes* of objects (constants, variables, and signals) that can be of integer, real, physical, enumeration, array, or record *type* (only variables can be of access or file type). Each VHDL object thus has *two* characteristics: a class and a type. On the contrary, any Verilog object is characterized by *one* qualifier only (type or class): net, register, integer, real, or time. In the first two cases, the only possible values for the object are X, 0, 1, and Z (or an array of theses values).

Examples: a Verilog integer is more or less equivalent to a VHDL variable declared in a process (or a VHDL signal) of type integer; a Verilog register is more or less equivalent to a VHDL signal of one of the previously enumeration subtypes of type FOUR_LEVEL_BIT. Here again appears a more general terminology, reflecting the more general concepts of VHDL as compared to those of Verilog.

9.5. PREDEFINED OPERATORS, FUNCTIONS, AND GATES

Following this given philosophy, many predefined bit-level operators are available in Verilog that are very useful in IC design. VHDL provides only a very restricted set of predefined operators. The next paragraphs list all these predefined operators in both languages.

As will be seen at the end of this section, Verilog also provides predefined logic gates. In contrast, VHDL is certainly the only hardware description language without any predefined logic gate.

9.5.1. Bit Logic and Boolean Operators

<table>
<tr><th>Bit Logic Operators</th><th>Verilog</th><th>VHDL</th></tr>
<tr><td>unary bitwise invert</td><td>~</td><td>not</td></tr>
<tr><td>binary bitwise and</td><td>&</td><td>and</td></tr>
<tr><td>binary bitwise nand</td><td>~&</td><td>nand</td></tr>
<tr><td>binary bitwise or</td><td>|</td><td>or</td></tr>
<tr><td>binary bitwise nor</td><td>~|</td><td>nor</td></tr>
<tr><td>binary bitwise xor</td><td>^</td><td>xor</td></tr>
<tr><td>binary bitwise xnor</td><td>~^ or ^~</td><td>not predefined</td></tr>
<tr><td>unary reduce and</td><td>&</td><td>not predefined</td></tr>
<tr><td>unary reduce nand</td><td>~&</td><td>not predefined</td></tr>
<tr><td>unary reduce or</td><td>|</td><td>not predefined</td></tr>
<tr><td>unary reduce nor</td><td>~|</td><td>not predefined</td></tr>
<tr><td>unary reduce xor</td><td>^</td><td>not predefined</td></tr>
<tr><td>unary reduce xnor</td><td>~^ or ^~</td><td>not predefined</td></tr>
<tr><td>Verilog logical equality</td><td>==</td><td>not predefined</td></tr>
<tr><td>Verilog logical inequality</td><td>!=</td><td>not predefined</td></tr>
</table>

The Verilog logical equality and inequality are comparison operators that may produce as a result the unknown value (X). For example, "A == 1" evaluates to TRUE if A is equal to 1, to FALSE if A is equal to 0, and to unknown (X) if A is either X or Z. In the last case, if such an expression is used in an if statement, the unknown result will be interpreted as FALSE. The notion of logical equality is also described in paragraph 7.1.6 ("Tricks and Traps").

As shown by the previous table, Verilog has many more predefined bit logic operators than VHDL. The main difference is the definition of Verilog *reduce* operators, i.e., unary operators producing the single bit resulting from the logical operation on all the bits of the input (bus). *Bitwise* operators are the classical logical operators having two operands. As both families of operators use the same notation, they are distinguished by the syntax (reduce operators are unary, whereas bitwise operators are binary).

Boolean Logical Operators	Verilog	VHDL
logical invert	!	**not**
logical and	&&	**and**
logical nand	*not predefined*	**nand**
logical or	||	**or**
logical nor	*not predefined*	**nor**
logical xor	*not predefined*	**xor**
logical xnor	*not predefined*	*not predefined*

VHDL uses the same overloaded operators for bit and boolean operands, whereas Verilog uses the standard C operators as boolean operators.

9.5.2. Relational Operators

Relational Operators	Verilog	VHDL
equality *(Verilog case equality)*	===	=
inequality *(Verilog case inequality)*	!==	/=
less than	<	<
less than or equal to	<=	<=
greater than	>	>
greater than or equal to	>=	>=

The Verilog case equality and inequality are totally equivalent to the VHDL equality and inequality operators: these operators perform a lexical comparison, and there is no logical interpretation as for the Verilog logical equality and inequality. With the same example introduced in the previous paragraph, "A=== 1" evaluates to TRUE if A is equal to 1 and directly to FALSE if A is either 0, X, or Z.

A better illustration of the difference between the logical equality (==) and the case equality (===) in Verilog is the following:

- X === X evaluates to TRUE
- X == X evaluates to X, thus to FALSE!

The name of the Verilog relational operators comes from the fact that the Verilog case statement (see 9.6.3.1) uses this lexical comparison (===).

Conditional statements in Verilog may use either the logical equality (==) or the case equality (===), depending on designer's needs. VHDL conditional statements may only use lexical comparisons.

9.5.3. Arithmetic Operators

Arithmetic Operators	Verilog	VHDL
addition	+	+
subtraction	-	-
multiplication	*	*
division	/	/
modulus	*not predefined*	**mod**
remainder	%	**rem**
exponentiation	*not predefined*	**
absolute value	*not predefined*	**abs**

Compared to Verilog, VHDL additionally provides modulus and remainder operators (the distinction is only relevant for the negative first operand), as well as the exponentiation and the absolute value.

9.5.4. Miscellaneous

Miscellaneous Operators	M	VHDL
concatenation	{ , }	&
left shift operator	<<	*not predefined*
right shift operator	>>	*not predefined*
conditional operator	... ? ... : ...	*not predefined*

Except for concatenation, these Verilog operators, directly derived from the corresponding C operators, have no equivalent in VHDL.

9.5.5. Predefined Verilog Functions

The Verilog reference manual describes all the system tasks implemented by a simulator. Some of them are listed below:

• Simulation results display tasks:

```
$write("Bit coding of decimal %d is equal to: %h", SigD, SigH);
$display("Bit coding of decimal %d is equal to: %h", SigD, SigH);
$monitor("Bit coding of decimal %d is equal to: %h", SigD, SigH);
$strobe("Bit coding of decimal %d is equal to: %h", SigD, SigH);
```

These tasks print on standard output (screen) the specified string with the provided arguments. They are derived from the C printing procedures (similar format control specification).

$write and $display print whenever the statement including them is executed, the only difference between them being the addition of a newline character by the $display function. The $monitor command is executed whenever there is a change in one of the specified arguments. The $strobe command has the same function as $monitor, except that it prints values before the simulation cycle begins.

• Simulation results print tasks:

```
$fwrite(FileID,"Bit coding of decimal %d is equal to: %h", SigD, SigH);
$fdisplay(FileID,"Bit coding of decimal %d is equal to: %h", SigD, SigH);
$fmonitor(FileID,"Bit coding of decimal %d is equal to: %h", SigD, SigH);
$fstrobe(FileID,"Bit coding of decimal %d is equal to: %h", SigD, SigH);
```

These tasks are equivalent to the previous tasks, except that the results are written in a file. File open and close functions are provided.

• Timing check tasks:

```
$setup(), $hold(), $setuphold(), $period(), $width(), $skew(), $recovery()
```

These tasks implement timing verification procedures usually met in synchronous design (constraints between data and clock, constraints on clock, etc.).

• Function returning the simulation time:

```
$time      //returns current time as a 64-bit value
```

• Function generating a random number:

```
$random    //may be called with an inout parameter to control the random output
```

• Functions stopping the simulation:

```
$stop      //to halt the simulator
$finish    //to quit the simulator
```

The Verilog display and print functions seem very restrictive compared to the VHDL predefined TEXTIO package, which contains declarations of types and subprograms that support formatted ASCII I/O operations. Verilog has a lack of read functions that are very convenient — for example, when specifying a ROM content or importing stimuli from a file.

Timing verification procedures are not predefined in VHDL, but can be easily implemented with passive processes or passive procedures using assert statements. A user-defined package containing such timing checks is very convenient and is commonly available in any VHDL IC design environment.

VHDL has the predefined function NOW, which returns the current simulation time, but does not provide a function for random number generation or simulation control.

9.5.6. Predefined Verilog Gate-Level Primitives

Verilog provides a set of 26 predefined gate-level primitives:

- 12 logic gates, i.e., **and, nand, or, nor, xor, xnor, not, buf** (non-inverting buffer), **bufif0, bufif1** (tristate buffers), **notif0, notif1** (tristate inverters);
- 14 switch level gates, i.e., **nmos, pmos** (non-resistive unidirectional mos transistors), **rnmos, rpmos** (resistive unidirectional mos transitors), **cmos** (non-resistive unidirectional transmission gate), **rcmos** (resistive unidirectional transmission gate), **tran, tranif0, tranif1** (non-resistive bidirectional transmission gates), **rtran, rtranif0, rtranif1** (resistive bidirectional transmission gates), **pullup, pulldown** (gates driving pull strength values).

These gates may be instantiated in any user-defined module with the following syntax:

```
Primitive_Gate_Type    Drive_Strength         Delay       List_of_Instances ;
```

For example:

```
nand                   (strong0, strong1 )    # (2,3,2)   nand1(out1, in1, in2) ,
                                                          nand2(out2, in3, in4, in5) ;
```

Any gate may be instantiated with delay specifications: these specifications may contain up to nine delay values, i.e., rising (to 1), falling (to 0), and turn-off (to Z) delays for each minimum, typical, and maximum case. Delay specifications are explained in 9.6.2.2.

The 12 logic gates plus the pullup and pulldown gates support driving strength specifications. A different strength may be associated with each logic value 0 and 1 according to four levels (among a total number of eight levels): weak, pull, strong, and supply (strength levels 3, 5, 6, 7).

VHDL does not provide any predefined gate. Furthermore, there is *neither a predefined entity nor a predefined component in VHDL.*

Delays are implemented in VHDL with the predefined physical type TIME, but strengths are not predefined. While the Verilog designer uses an implicit resolution function based on the previously mentioned strengths, the VHDL designer needs to define its own strength type and the resolution function associated with it. After that, entities modeling the Verilog basic gates may be written in VHDL.

Nevertheless, many VHDL packages defining a multi-value logic system are now available: some of them are provided with some particular VHDL simulators (they usually implement the native logic type of the simulator); some others are defined by the VHDL standardization process (such as the STD_LOGIC_1164 package defining a nine-value logic type). Typical VHDL logic systems are presented in chapter 4 ("Structuring the Environment").

9.6. STATEMENTS

9.6.1. Signal Assignments

This paragraph compares the signal assignment statements between both languages. The focus is on the behavior of these statements, in particular when and how they are executed, and on the kind of logic they may specifically model (if any). Delay and event control specifications will be detailed in the two next paragraphs.

In VHDL, signal assignments are either concurrent or sequential statements. A concurrent signal assignment statement is executed whenever one of the signals in the right-hand-side waveform changes its value. On the other hand, a sequential signal assignment statement is executed when control passes to it, as a classical sequential statement within a process or a subprogram.

The Verilog language provides three kinds of assignments: the continuous assignment, the procedural assignment, and the procedural continuous (or quasi-continuous) assignment. The last of these is an additional feature of Verilog that is somewhat particular and not as widely used as the continuous and the procedural assignments.

9.6.1.1. The Verilog Continuous Assignment

The continuous assignment is a module statement executed each time any input of the right-hand-side expression changes. It drives values to any kind of *nets*, but only to them and not to *register, integer, real,* or *time* variables. For example:

```
wire WIRE_S ;
...
assign #10 WIRE_S = ANY_FUNCTION(WIRE_A, WIRE_B) ;
```

Such a continuous assignment is equivalent to the VHDL concurrent signal assignment statement:

```
signal WIRE_S : A_TYPE ;
...
WIRE_S <= ANY_FUNCTION(WIRE_A, WIRE_B) after 10 ns ;
```

The VHDL concurrent assignment statement is executed each time any signal value on the right-hand-side of the statement changes, having thus the same semantics as the Verilog statement.

But there are some differences between these two statements. First of all, as shown above, the Verilog continuous assignment statement is equivalent to a VHDL concurrent assignment statement with an inertial delay (glitches of less than the specified delay are not propagated). Unlike VHDL, the Verilog language does not implement the transport delay (in VHDL, the keyword **transport** after the symbol "<=" indicates that all input changes are propagated through the assignment). *All Verilog delays are inertial delays.*

Furthermore, the right-hand side of the assignment statement may be more complex in VHDL: the designer may assign *waveforms*, i.e., specify different values associated with different timing steps in the same assignment statement. Verilog does not provide this possibility: all assignments are composed of a single pair value/delay.

The two Verilog and VHDL statements have another major difference. In VHDL, no distinction is made between the kinds of signals being assigned either by a concurrent or a sequential statement. In Verilog, as already stated in section 9.4, an important difference is made between *nets* (assigned by continuous statements) and *registers* (assigned by procedural statements: see below).

Moreover, any concurrent signal assignment statement in VHDL is totally equivalent to a process containing a corresponding sequential signal assignment statement:

```
                                         Label : process
                                         begin
Label : Target <= Waveform ;      <=>        Target <= Waveform ;
                                             wait on List_of_signals_in_waveform ;
                                         end process ;
```

Thus, both VHDL concurrent and sequential assignments are able to model the same kind of logic. Even if the dataflow style (i.e., concurrent assignments) is more adapted for combinational logic, there is no modeling specificity of one type of statement as compared to the other.

On the other hand, Verilog continuous and procedural assignments are intended to model different kinds of logic. Indeed, the continuous assignment statement provides a means of describing combinational logic at a description level that cannot be reached by the procedural assignment statement, which is more often used to model sequential logic with event control statements (see 9.6.1.2 below and paragraph 9.6.3).

The complete syntax of the continuous assignment statement is

```
assign    Drive_Strength       Delay    List_of_Assignments ;
```

For example:

```
assign    (strong0, strong1)   # 10     sum = a ^ b ^ cin ,
                                        cout = (a & b) | ((a | b) & cin) ;
```

Any continuous assignment statement contains optional strength and delay specifications followed by a *list* of assignments. Strengths and delays have the same characteristics as those of the primitive gate instantiations (see paragraph 9.5.6), except that the possible strength levels are **weak**, **pull**, and **strong**, but not **supply** (strength levels 3, 5, 6).

In contrast, a procedural assignment cannot be specified with driving strengths. Then, as will be seen in paragraph 9.6.2, delays are much more complex in continuous assignment than in procedural assignment: if both assignments provide minimum, typical, and maximum delays, only the continuous assignment may be specified with a different delay for each output transition.

This proves that the Verilog continuous assignment statement is much more adapted than the procedural statement to model combinational logic. Any continuous assignment cannot have an equivalent procedural assignment, as is the case in VHDL between concurrent and sequential signal assignments. Furthermore, if the syntaxes of continuous assignment statements and primitive gate instantiations (paragraph 9.5.6) are compared,

```
assign                     Drive_Strength   Delay   List_of_Assignments ;
// continuous assignment statement

Primitive_Gate_Type        Drive_Strength   Delay   List_of_Instances ;
// primitive gate instantiation
```

it becomes obvious that these two statements are very close in their modeling purpose (remember that delays and strengths are similar in both statements): the continuous assignment statement models combinational logic without its implementation, whereas the gate instantiation models the implemented combinational logic.

No such difference of modeling features (different delays, strengths, etc.) between continuous and procedural assignments appears in VHDL between concurrent and sequential signal assignments.

9.6.1.2. The Verilog Procedural Assignment

The procedural assignment is a statement present in initial and always statements, and in tasks and functions (no delay or wait or event control is allowed in Verilog functions). It is executed when control is transferred to it, as with any other sequential statement. It can assign values to *register, integer, real,* or *time* variables, but not to *nets*. For example:

```
reg     WIRE_S ;
always
        @(posedge CLOCK)  WIRE_S =  #10  ANY_FUNCTION(WIRE_A, WIRE_B) ;
```

Such a procedural assignment is equivalent to a VHDL sequential signal assignment statement found in a process or in a procedure (signal assignment is not allowed in a function body):

```
signal WIRE_S : A_TYPE ;
...
process
begin
        wait until CLOCK = '1' ;
        WIRE_S <= ANY_FUNCTION(WIRE_A, WIRE_B) after 10 ns ;
        wait for 10 ns ;
        -- wait statement added to match the blocking assignment (see below)
end process ;
```

As for the concurrent assignment, only VHDL allows writing a waveform on the right side of a sequential signal assignment. In Verilog, successive simple assignments have to be written to model one VHDL assignment with waveform. But the designer must be aware of the different methods used by both languages to schedule and generate events (see paragraph 9.6.2).

The previous Verilog example used one of the two available forms of procedural assignments, namely, the *blocking procedural assignment* (using the symbol"="). The other kind of procedural assignment provided by Verilog is the *non-blocking procedural assignment* (using the symbol "<=" instead of "="). The next Verilog always statement includes a non-blocking assignment:

```
reg     WIRE_S ;
always
        @(posedge CLOCK)  WIRE_S <=  #10  ANY_FUNCTION(WIRE_A, WIRE_B) ;
```

and is equivalent to the following VHDL process statement:

```
signal WIRE_S : A_TYPE ;
...
process
begin
        wait until CLOCK = '1' ;
        WIRE_S <= transport ANY_FUNCTION(WIRE_A, WIRE_B) after 10 ns ;
        -- no other wait statement, but a transport delay is specified (see below)
end process ;
```

The blocking procedural assignment is the Verilog original, the introduction of the second one being more recent (the Verilog language reference manual refers to the latter as the *new* non-blocking assignment, and most of the Verilog books do not even mention it). This non-blocking procedural assignment is actually much closer to the VHDL sequential signal assignment statement than the first one.

Indeed, there is a major difference between the Verilog blocking procedural assignment and the VHDL sequential assignment in the way signal values are updated by both statements. The execution of signal assignments (even *zero-delay* assignments) inside a VHDL process always generates *projected* events, and these scheduled events may be *cancelled* by future events that have been generated in the meantime (i.e., before first events are effectively generated). For example:

```
process
begin
    A <= '1' after 10 ns ;    -- first event (change to '1') scheduled at time 10 ns ;
    A <= '0' after 5 ns ;     -- second event (change to '0') scheduled at time 5 ns,
    wait ;                    -- cancelling the previous projected event at 10 ns!
end process ;
```

Considering that signal A is of type BIT (thus initialized at '0'), the simulation of this process will show signal A always remaining at the initial value '0'. Actually, the first assignment will have no effect: the projected first event has been removed from the event queue of the simulator by the second projected event (which will occur in this case). The semantics of the VHDL timing model is characterized by this *preemption* mechanism [AUG91].

In Verilog, although it looks very similar, the following Initial statement is not at all equivalent to the previous VHDL process:

```
initial
begin
    #10 A = 1 ;   // or equivalent statement : A = #10  1 ;
    #5  A = 0 ;   // or equivalent statement : A = #5   0 ;
end
```

The simulation of this Verilog model will generate the two events (change to 1 at time 10 ns, change to 0 at time 15 ns). Actually, the previous Verilog statement is totally equivalent to the following VHDL process:

```
process
begin
        A <= '1' after 10 ns ;    -- first event (change to '1') scheduled at time 10 ns ;
        wait for 10 ns ;          -- first event occurs
        A <= '0' after 5 ns ;     -- second event (change to '0') scheduled at 10+5 = 15 ns ;
        wait for 5 ns ;           -- second event occurs
        wait ;
end process ;
```

Verilog events generated by blocking procedural assignments always occur. The execution of a Verilog blocking procedural assignment with a delay ΔT will always make the simulation time advance for ΔT so that the projected assignment takes effect (if no conflict is encountered). The execution of the following statements in the initial or always statement is thus suspended (*blocked*) during this delay ΔT.

In VHDL, simulation time is *stopped* during the execution of all the sequential statements (including even delayed signal assignments) that are comprised between two wait statements. The VHDL simulation time only advances at the execution of a wait statement. This can lead to cancellations of projected events.

On the other hand, simulation time *advances* after the execution of each Verilog blocking procedural assignment statement, so that such scheduled events are always effective. Figure 9.9 illustrates this difference.

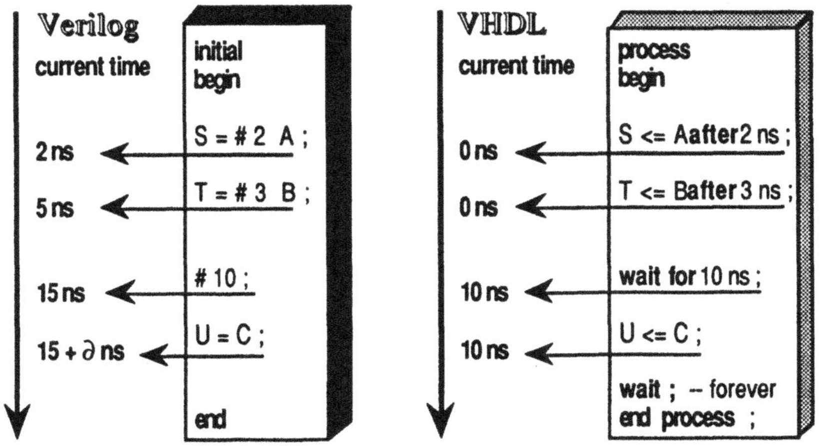

S becomes equal to A at	t = 2 ns	- - - - - - - - - - - - - - -	t = 2 ns
T becomes equal to B at	t = 5 ns	- - - - - - - - - - - - - - -	t = 3 ns
Process is suspended from	t = 5 to 15 ns	- - - - - - - - - - - - -	t = 0 to 10 ns
U becomes equal to C at	t = 15 ns + ∂	- - - - - - - - - - - - -	t = 10 ns + ∂

(∂ is a Delta Delay)

Fig 9.9 Timing with Verilog Blocking Assignments

To correspond to the Verilog ones, the VHDL sequential assignment statements must be followed by a wait statement for a duration equal to the delay specified in the assignment. It is referred to here as *the* delay of the assignment because the syntax of Verilog assignments does not include waveforms (with several pairs time/value).

Concerning the way the time advances (or not) during its execution, the Verilog non-blocking procedural assignment is very similar to the VHDL sequential assignment statement: like the VHDL statement, it does not suspend the execution of the process. Figure 9.10 shows this similarity.

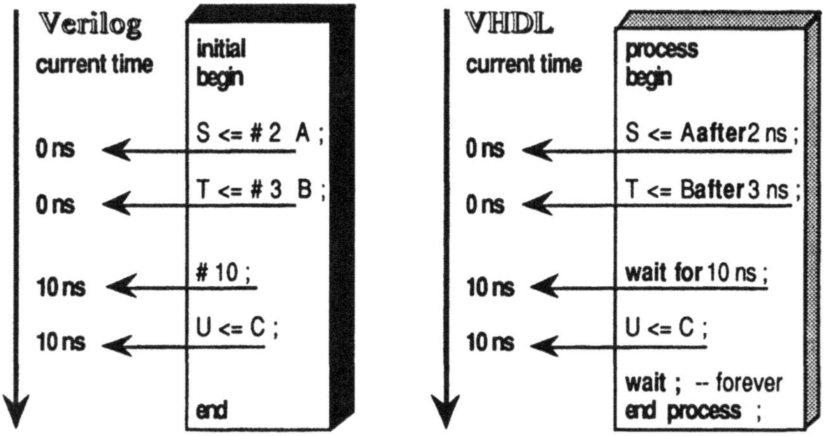

S becomes equal to A at t = 2 ns - - - - - - - - - - - - - - t = 2 ns
T becomes equal to B at t = 3 ns - - - - - - - - - - - - - - t = 3 ns
Process is suspended from t = 0 to 10 ns - - - - - - - - - - - - t = 0 to 10 ns
U becomes equal to C at t = 10 ns + ∂ - - - - - - - - - - - - t = 10 ns + ∂
(∂ is a Delta Delay)

Fig 9.10 Timing with Verilog Non-Blocking Assignments

There is still an important difference between both statements. In Verilog, a non-blocking assignment can be executed without cancelling a previous non-blocking assignment to the same register. For example, the execution of

```
initial
begin
        REG_S <= #5    1;
        REG_S <= #10   0;
end
```

will generate on register REG_S a change to 1 at time 5 ns and a change to 0 at time 10 ns. On the other hand, in the simulation of the VHDL process

```
process
begin
        REG_S <= '1' after 5 ns ;
        REG_S <= '0' after 10 ns ;
        wait ;
end process ;
```

the execution of the first statement schedules a change to '1' in 5 ns, but the execution of the second statement cancels the first projected event and schedules a transaction to '0' in 10 ns: the delay of the second assignment has been handled as an inertial delay, thereby removing all events scheduled in less than 10 ns. In fact, the previous Verilog initial statement is equivalent to

```
process
begin
        REG_S <= transport '1' after 5 ns ;
        REG_S <= transport '0' after 10 ns ;
        wait ;
end process ;
```

In Verilog, *only the delays associated with continuous assignments are handled as inertial delays*. In contrast, delays associated with procedural assignments are handled as transport delays. For blocking procedural assignments, this distinction has no real meaning because such statements always suspend the execution of the process for the amount of time equal to the delay. But for non-blocking procedural assignments, this difference is fundamental. According to the examples provided by the Verilog language reference manual, it is clear that *delays specified in non-blocking procedural assignments are handled as transport delays.*

9.6.1.3. The Procedural Continuous Assignment

The procedural continuous assignments, or quasi-continuous assignments, are an additional feature of Verilog having no real correspondence in VHDL. Two kinds of statements — the **assign** and **deasssign** procedural statements — allow continuous assignments to be placed onto registers for controlled periods of time. For example:

```
module REGISTER_ELEMENT (RESET, CLOCK, D, Q) ;
input     RESET, CLOCK, D ;
output    Q ;
reg       Q ;

always @(posedge CLOCK) Q = #2 D ;  // procedural assignment.

always @(RESET)
     if (! RESET)
          #2 assign Q = 0;  // procedural continuous assignment or assign statement ;
     else                   // it overrides the previous procedural assignment.
          #2 deassign Q ;   // procedural deassign statement ; Q keeps the last value
                            // assigned by the procedural continuous assignment ;
                            // future procedural assignments are enabled.
endmodule
```

This procedural continuous assignment may drive registers for certain specified periods of time. It is a procedural statement and therefore only executed when control passes to it, and it is included in always or initial statements, tasks, or functions.

As illustrated in the previous example of a memory element with reset, the procedural continuous assignment overrides any procedural assignment to registers: when RESET is equal to 0, the register Q is continuously held to 0 and is not affected by the procedural assignment synchronized by the positive edge of CLOCK. The deassign statement ends this procedural continuous assignment and allows the procedural assignment to be effective.

VHDL does not provide an assignment having such a priority over others assignments: all the assignments are equal, any conflict being resolved by the mechanism of resolution function.

Actually, a kind of priority can be found in VHDL with the selected signal assignment, which allows, for example, the following model of the register:

```
Q <=  '0' after 2 ns when RESET = '0' else
      D after 2 ns when not CLOCK'STABLE and CLOCK = '1' else
      Q ;
```

The value effectively assigned to Q is the value of the waveform associated with the first true condition, thus giving a priority to the reset signal. The reader familiar with VHDL may notice that this statement is correct only if Q has been declared as an inout port (its value is read in the statement). Otherwise, another statement (and another local signal!) can be used as follows:

```
INTQ <=   '0' after 2 ns when RESET = '0' else
          D after 2 ns when not CLOCK'STABLE and CLOCK = '1' else
          INTQ ;
Q <= INTQ ;
```

These considerations about some Verilog and *equivalent* VHDL descriptions call for some clarifications. Indeed, the previous VHDL models are *equivalent* to the above procedural continuous assignment statement in the sense that they represent the same hardware device. On the other hand, they are very *different* if the semantics of simulation is taken into account.

With this simulator point of view, the compilation of the Verilog description will generate at least two concurrent processes (and probably the same number of drivers or sources) for signal Q: one (or two?) for the procedural continuous assignment, and another one for the procedural assignment. The previous VHDL description will generate for certain only one process (and one driver) for signal Q.

Thus, if the Verilog description generates multiple drivers for the same signal, the **deassign** statement can be seen as a VHDL **disconnect** statement: when this Verilog statement is executed, it will disconnect the driver associated with the procedural continuous assignment, allowing the remaining driver (from the procedural statement) to take effect. Furthermore, such a Verilog register would be modeled in VHDL as a **guarded** signal of the **register** kind: after its disconnection, the signal value remains the last assigned value.

This notion of driver disconnection is strongly linked to the simulator implementation. In VHDL, the characteristics of the implied simulator are provided by the VHDL language reference manual. Apart from the fact that the simulator is event driven, this reference manual also describes the initialization, elaboration, and simulation cycles of a generic VHDL simulator.

In particular, the mechanism of drivers is very explicit: creation, contribution to a signal value, conflict resolution, disconnection, guarded blocks and assignments, two kinds of guarded signals (**bus** and **register**) that differ by the value taken after disconnection, and so on. Rules are provided that precisely give the number of drivers created for each signal (one driver per signal per process!), the meaning of all statements (such as the disconnection) being clearly defined in this kernel of simulator implementation.

On the other hand, equivalent semantics information is missing in the Verilog language reference manual (in the current version 1.0). Some behaviors of the simulator are directly implied by the language itself (event-driven simulator, delta-delay mechanism, etc.). The mechanisms of drivers and their possible disconnection are almost implied by the procedural continuous assignment statements. But since this language does not include waveforms (see 9.6.1.1), the semantics of drivers as defined for VHDL (future scheduled signal values) seems less justified for Verilog.

This chapter will not describe further the simulator kernels implied by both languages, but will remain at the level of a comparative study between the Verilog and VHDL language constructs provided to the user, i.e., the *designer*. And the first priority of the designer is to know not how his simulator is built, but rather what he himself is able to build with his simulator.

9.6.2. Delay Specifications

As already illustrated in the previous paragraphs, both languages provide a way to specify delays in assignments. Actually, a Verilog designer is able to specify delays in procedural and continuous assignments, inside always and initial statements, in primitive gate instantiations, and for nets. The VHDL designer may only specify delays in sequential and concurrent signal assignments and inside processes. The comparison is thus divided according to the different Verilog statements.

9.6.2.1. Delays in the Verilog Procedural Assignment

In Verilog, the delay specification of a procedural assignment uses the character # followed by an integer representing a number of time units (the time unit is specified by the *'timescale* compiler directive; this chapter will always use the nanosecond as unit). Two possible syntaxes are available to specify a delay in a procedural assignment. For blocking procedural assignments:

TARGET = #10 VALUE ;	*equivalent to*	TARGET <= VALUE **after** 10 ns ; **wait for** 10 ns ;
#10 TARGET = VALUE ;	*equivalent to*	**wait for** 10 ns ; TARGET <= VALUE ; **wait for** 0 ns ;

For non-blocking procedural assignments:

TARGET <= #10 VALUE ;	*equivalent to*	TARGET <= **transport** VALUE **after** 10 ns ;
#10 TARGET <= VALUE ;	*equivalent to*	**wait for** 10 ns ; TARGET <= VALUE ;

The difference between Verilog blocking and non-blocking procedural assignments has been explained in paragraph 9.6.1. During the execution of the blocking assignment, the initial or always statement is suspended for a period of time equal to the specified delay, so that the right-hand-side value of the blocking assignment is assigned to the target signal. To be equivalent to the Verilog statement, a "**wait for...**" statement has to be added in the VHDL process. The nature of the delay (inertial, transport) specified in the VHDL assignment has thus no real importance.

Verilog non-blocking assignments are more similar to the VHDL sequential signal assignments: both statements only schedule future events but do not suspend the execution of the process. But, as detailed in 9.6.1.2, a delay specified in the right-hand side of a non-blocking assignment is equivalent to a VHDL **transport** delay: if a first non-blocking assignment with a delay T1 is followed by a second non-blocking assignment to the same register with a delay T2 (T2>T1), then each value of both statements is successively assigned to the register.

For both blocking and non-blocking assignments, two possible syntaxes have been presented above. The difference between both delay specifications is the date when the signal values that are present in the right-hand side (in VALUE) of the assignment are read. In the former case, these values are read immediately at the execution of the statement: this statement therefore corresponds to the VHDL sequential assignment statement. In the latter case, values are read after the specified delay (and can thus be different from the former ones): this form therefore corresponds to a VHDL wait statement for a period equal to the same delay, followed by a sequential assignment statement with zero delay.

This example recalls the way delays are specified in VHDL: the delay is a value of the predefined physical type TIME (10 ns, 25 fs, etc.) placed after the keyword **after** in a waveform (S <= '1' **after** 5 ns) or after the keyword **for** in a wait statement (**wait for** 10 ns).

In Verilog, a delay control may be a part of a procedural statement (as illustrated above) or even a procedural statement itself:

```
#10 ;          equivalent to          wait for 10 ns ;
```

The Verilog delay specification in a procedural assignment may also have three values instead of a single one. These three values are the minimum delay, the typical delay and the maximum delay. The syntax is

```
TARGET = #(Min_Delay : Typ_Delay : Max_delay)  VALUE ;
#(Min_Delay : Typ_Delay : Max_delay)  TARGET = VALUE ;
```

VHDL does not provide a syntax with minimum, typical, and maximum delays. Nevertheless, it is possible to define the following package:

```
package TIMING is
        function DELAY ( DMIN, DTYP, DMAX : TIME) return TIME ;
        -- minimum, typical and maximum delay specification
end TIMING ;

package body TIMING is
        type TIMING_CHOICE is (MIN, TYP, MAX ) ;
        constant CHOICE : TIMING_CHOICE := TYP ;
        -- constant specifying the timing choice ; as it is declared in the package body,
        -- only this package body needs to be recompiled after any change of timing option
```

```
        function DELAY ( DMIN, DTYP, DMAX : TIME) return TIME is
        begin
            case CHOICE is
             when MIN  =>   return DMIN ;
             when TYP =>    return DTYP ;
             when MAX =>    return DMAX ;
            end case ;
        end DELAY ;
end TIMING ;
```

Assert statements can be added to check that DMIN < DTYP < DMAX. Once compiled, this package may be used by any design unit to implement minimum, typical, and maximum delays. For example:

```
use WORK.TIMING.all ;
architecture ANY_ARC of ANY_ENT is
        signal S : BIT ;
begin
        S <= '1' after DELAY(2 ns, 5 ns, 8 ns) ;
end ;
```

Of course, once the constant CHOICE is set, any design unit will have the same kind of delays (i.e., minimum, typical, or maximum). Such an implementation does not allow the mixing of minimum and maximum delays, which could have been more realistic.

9.6.2.2. Other Delays in Verilog

As seen in paragraph 9.5.6, Verilog provides a set of 26 predefined primitive gates (and, or, nand, nor, xor, and so forth) that may be instantiated with delay specifications. The syntax is

```
Primitive_Gate_Type        Drive_Strength    Delay    List_of_Instances ;
```

Similarly, continuous assignment statements may be specified with delay specifications (see 9.6.1.1):

```
assign              Drive_Strength    Delay    List_of_Assignments ;
```

Finally, the general form of a net declaration is (see section 9.4)

```
Net_Type       Charge_Strength       Range    Delay    List_of_Names ;
```

VHDL provides two kinds of delays: the inertial delay and the transport delay (introduced by the keyword **transport** in a signal assignment). The inertial delay is the implicit delay used for combinational logic modeling (removing glitches), whereas the transport delay is intended to model propagation delays (on transmission lines, for example).

In Verilog, delays specified in continuous assignment, primitive gate instantiation, and net declaration are *inertial delays*, suitable for combinational logic, but not appropriate for transmission lines or similar devices. But the inertial delay used in these three module items is more complex than the procedural assignment delay. Indeed, a different delay may be specified for the transition to 1, the transition to 0, and the transition to Z. Examples are listed below:

```
nand   #5           inst1(o1, i1, i2),
                    inst2(o2, i3, i4) ;
nor    #(3, 5)      inst3(o3, i5, i6, i7) ;
notif0 #(4, 6, 3)   inst4 (o4, i8, en) ;
```

When only one value is given, this value will be used for all transitions (to 0, 1, and Z).

When two values are given, the first one is the rising delay (transition to 1) and the second one the falling delay (transition to 0): the delay associated with the transition to Z is the minimum of both provided delays.

Finally, when three values are given, the first one is the rising delay, the second one the falling delay, and the third one the turn-off delay (transition to Z).

The following table sums up the different delay specifications:

Transition	Name	#d1	#(d1,d2)	#(d1,d2,d3)
0, X or Z to 1	rising	d1	d1	d1
1, X or Z to 0	falling	d1	d2	d2
0, 1 or X to Z	turn-off	d1	min(d1,d2)	d3
0, 1 or Z to X	-	d1	min(d1,d2)	min(d1,d2,d3)

These delays may be combined with minimum, typical, and maximum delays. Thus, the most complex delay syntax is the following:

```
#(   Rising_Min_Delay : Rising_Typ_Delay : Rising_Max_delay ,
     Falling_Min_Delay : Falling_Typ_Delay : Falling_Max_delay ,
     Turnoff_Min_Delay  : Turnoff_Typ_Delay  : Turnoff_Max_delay )
```

Such a delay syntax may be used in primitive gate instantiation, net declaration, and continuous assignment statements, but not in procedural assignment statements which have the syntax described above.

To specify different delays depending on the output transition, the VHDL designer may use the concurrent selected signal assignment. If this designer uses the package NET_TYPES described in section 9.4, he may write, for example:

```
use WORK.NET_TYPES.all ;
entity TRISTATE_BUFFER is
        generic ( RISING_DELAY, FALLING_DELAY, TURNOFF_DELAY : TIME ) ;
        port ( INPUT, ENABLE : in WIRE ;  OUTPUT : out WIRE ) ;
end ;
architecture EXAMPLE of TRISTATE_BUFFER is
begin
        OUTPUT  <=      '1' after RISING_DELAY when INPUT = '1' and ENABLE = '1' else
                        '0' after FALLING_DELAY when INPUT = '0' and ENABLE = '1' else
                        'Z' after TURNOFF_DELAY when ENABLE = '0' else
                        'X' ;
end ;
```

Using the package TIMING of 9.6.2.1, the VHDL designer may also specify minimum, typical, and maximum delays:

```
use WORK.TIMING.all ;
use WORK.NET_TYPES.all ;
entity TRISTATE_BUFFER is
        generic (  MIN_RISING, TYP_RISING, MAX_RISING,
                   MIN_FALLING, TYP_FALLING, MAX_FALLING,
                   MIN_TURNOFF, TYP_TURNOFF, MAX_TURNOFF : TIME ) ;
        port ( INPUT, ENABLE : in WIRE ;  OUTPUT : out WIRE ) ;
end ;
architecture EXAMPLE of TRISTATE_BUFFER is
begin
        OUTPUT  <=
            '1' after DELAY(MIN_RISING, TYP_RISING, MAX_RISING)
                when INPUT = '1' and ENABLE = '1' else
            '0' after DELAY(MIN_FALLING, TYP_FALLING, MAX_FALLING)
                when INPUT = '0' and ENABLE = '1' else
            'Z' after DELAY(MIN_TURNOFF, TYP_TURNOFF, MAX_TURNOFF)
                when ENABLE = '0' else
            'X' ;
end ;
```

In VHDL, a waveform has to be written for each output value in order to specify different rising, falling, and turnoff delays. Such VHDL syntax is thus not as compact as the Verilog syntax, which is more adapted to combinational logic delay modeling.

9.6.3. Event Control and Wait Statements

Verilog provides event control and wait statements in order to synchronize and/or suspend the execution of processes. These statements may be used inside any initial or always statement and inside tasks, but not in functions that include neither event control nor timing statements. For the same purpose, VHDL provides wait statements that may be written in processes and procedures, but not in functions (furthermore, VHDL functions may not include any signal assignment).

In Verilog, an event control statement is a procedural statement waiting for a *change* in a value. The execution of a process *always* suspends at an event control statement, and resumes when a change occurs.

Here are three typical examples of event controls inside an assignment statement:

```
@ (clock)  q = d ;              // waiting any edge on clock
@ (posedge clock)  q = d ;      // waiting a rising edge on clock
@ (negedge clock)  q = d ;      // waiting a falling edge on clock
```

An event control statement models *edge-sensitive* logic. The first statement is waiting for any change of clock, the second one is waiting for a positive edge (i.e., a transition from 0 to 1, 0 to X, X to 1), and the third one for a negative edge (i.e., a transition from 1 to 0, 1 to X, X to 0).

The event control statement may be sensitive to more than one event by using the keyword **or** :

```
@ (posedge clock1 or posedge clock2)  q = d ;
```

Finally, an event control statement may be sensitive to a *named event*. The named event is an abstract object declared as another module object (input, output, inout, reg, wire, etc.):

```
event  the_event ;   // any name chosen by the user
```

Any process (initial or always statement) may trigger this event by using the following statement:

```
—> the_event ; // process triggers the named event the_event
```

Any other process (initial or always statement) may be sensitive to the previous named event by using the following statement:

```
@ the_event ;   // process suspended here until the named event the_event triggers
```

Verilog also provides a wait statement that can be used in initial or always statements to suspend the equivalent process if a given condition is not verified. This procedural statement uses the keyword **wait** followed by an expression:

```
wait (clock) ;    // suspends the process if (and only if) clock is not equal to 1
```

If the expression associated with a wait statement evaluates to true (1), then the execution of the process proceeds without any interruption. On the other hand, if the expression evaluates to false (0), then the process suspends and will resume as soon as the expression changes to become true. The wait statement thus models *level-sensitive logic*.

As an illustration of the difference between event control and wait statements, models of a register and a latch are given below:

Verilog edge-sensitive register	VHDL edge-sensitive register
always @ (**posedge** CLOCK) Q = D ; // event control statement	-- **process** **process** -- **begin** **begin** -- **wait on** CLOCK ; **wait until** CLOCK = '1' ; -- **if** CLOCK = '1' **then** Q <= D ; -- Q <= D ; **end process** ; -- **end if** ; -- **end process** ;
Verilog level-sensitive latch	VHDL level-sensitive latch

always
begin
wait (CLOCK) ;
// wait statement, suspends the
// process if clock is not equal to 1
Q = D ;
@ (D) ; //waiting for a change of D
end
	process
	begin
	if not (CLOCK = '1') **then** -- suspends the
	wait until CLOCK = '1' ; -- process if clock
	end if ; -- is not equal to '1'
	Q <= D ;
	wait on D ; -- waiting for a change of D
	end process ;

As shown by the previous table, VHDL provides only the sequential wait statement to synchronize processes. Following are typical examples of wait statements that may be found within VHDL processes or procedures:

```
wait on SIGNAL1, SIGNAL2 ;
-- process suspended until SIGNAL1 or SIGNAL2 changes its value
wait on SIGNAL1, SIGNAL2 until SIGNAL3 = '1' ;
-- process suspended until SIGNAL1 or SIGNAL2 changes its value and SIGNAL3 is '1'
wait until CLOCK = '1' ;
-- process suspended until CLOCK changes its value and CLOCK is equal to '1'
-- statement equivalent to: wait on CLOCK until CLOCK = '1' ;
```

The VHDL wait statement is *edge-sensitive*, i.e., is sensitive to a change in a signal value. It thus corresponds to the Verilog event control statement (using @). There is no VHDL primitive corresponding to the Verilog wait statement (level-sensitive).

Nevertheless, templates of edge-sensitive and level-sensitive wait structures may be written in both languages:

• Verilog and VHDL edge-sensitive waits:

```
@ (CLOCK) ;              equivalent to      wait on CLOCK ;
@ (posedge CLOCK) ;      equivalent to      wait until CLOCK = '1' ;
@ (negedge CLOCK) ;      equivalent to      wait until CLOCK = '0' ;
```

• Verilog and VHDL level-sensitive waits:

```
wait (CONDITION) ;          equivalent to        if not CONDITION then
                                                     wait until CONDITION ;
                                                 end if ;
```

VHDL does not provide abstract objects such as the Verilog named event. Practically, an event may be implemented in VHDL by using a BOOLEAN (or BIT) signal and by making processes sensitive to that signal (with a *wait on* statement). For example:

```
event THE_EVENT ;     may be translated into    signal THE_EVENT : BOOLEAN ;
—> THE_EVENT ;        may be translated into    TRIGGER(THE_EVENT) ;
@THE_EVENT ;          may be translated into    wait on THE_EVENT ;
```

where TRIGGER is a user-defined procedure like the following:

```
procedure TRIGGER ( signal EVENT : inout BOOLEAN ) is
begin
      EVENT <= not EVENT ;
end;
```

Finally, as with the delay control (see 9.6.2.1), an event control of a procedural assignment statement may be specified in two ways. For blocking procedural assignments:

```
TARGET = @THE_EVENT  VALUE ;     equivalent to     TEMP_VAR := VALUE ;
                                                   wait on THE_EVENT ;
                                                   TARGET <= TEMP_VAR ;
                                                   wait for 0 ns ;

@THE_EVENT  TARGET = VALUE ;     equivalent to     wait on THE_EVENT ;
                                                   TARGET <= VALUE ;
                                                   wait for 0 ns ;
```

The difference between both statements is, of course, the date when signal values in the right side of the assignment are read. Equivalent VHDL sequential statements are given above: the statement "**wait for** 0 ns" is an artifact to make the VHDL models totally similar to the Verilog ones (detailed in 9.6.2.1 for delay specifications).

According to the Verilog reference manual, non-blocking procedural assignments may also be event controlled. For example, the Verilog designer may write the following Verilog always statement:

```
always
begin
      TARGET1 <= @THE_EVENT1  VAL1 ;
      TARGET2 <= @THE_EVENT2  VAL2 ;
      ...
end
```

In such a description, the execution of the first statement schedules the assignment of value VAL1 to TARGET1 at the next event THE_EVENT1. Then the second statement is executed, which schedules the assignment of value VAL2 to TARGET2 at the next event THE_EVENT2. Then other statements may be executed, and so on. Actually, this is a kind of concurrency, similar to the concurrency implemented by the fork and join statements: two events (here THE_EVENT1 and THE_EVENT2) may be expected independently.

But what is the exact semantics of such a description? What and where is the sensitivity list of the process equivalent to this always statement? As already stated in a previous paragraph, there is definitely a lack of unambiguous semantic specifications in the Verilog reference manual.

Finally, an event control in Verilog may be specified as part of a procedural statement or as a procedural statement itself (see examples above). In VHDL, an event control is always a single sequential wait statement.

9.6.4. Structured Statements

Structured statements may be divided into two families: sequential and concurrent structured statements.

9.6.4.1. Sequential Structured Statements

Below are tables listing the equivalences between sequential structured statements in both languages. Verilog and VHDL are very similar from this point of view: they both provides high-level programming language features, directly derived from C and Ada, respectively.

- **Selection statements**

These statements include *if* and *case* statements:

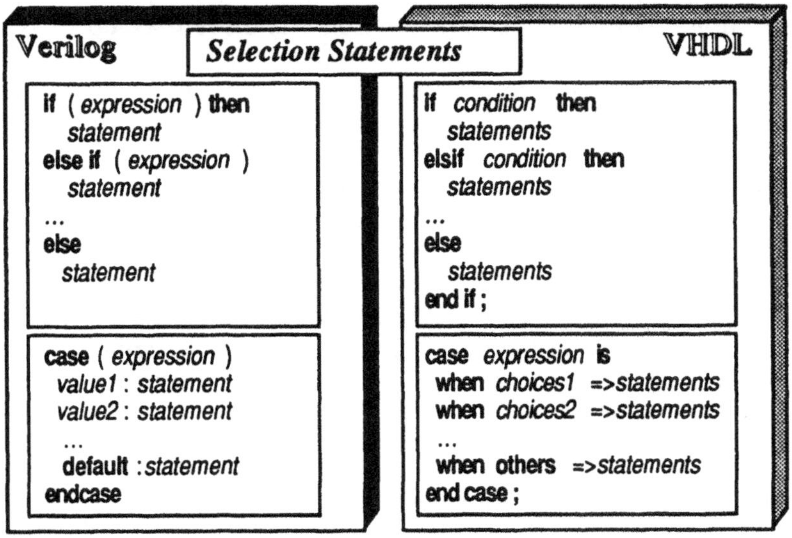

Fig 9.11 Selection Statements

There is a slight difference in the conditional statement (**if**...) between both languages. In VHDL, the condition in the if statement is an expression having a result of the predefined type BOOLEAN (either TRUE or FALSE).

In Verilog, the keyword **if** is followed by an expression that is interpreted as FALSE if this expression is equal to zero (0), unknown (X), or high impedance (Z), and as TRUE for any other value (1 or other integer values for vectors). Particular care must be taken with the logical comparison (==), which may produce a result that is either unknown or high impedance (see paragraph 9.5.1): in that case, the result of the expression is interpreted as FALSE by the if statement.

On the other hand, the case statement uses the same kind of comparison in both languages: the case expression and the different possible values are compared with the case equality (===) in Verilog and the lexical equality (=) in VHDL, leading thus to explicit comparisons with unknown (X) and high impedance (Z) states.

Nevertheless, there is still a difference between case statements in both languages. In VHDL, the case expression must be of a discrete subtype or of a one-dimensional character array subtype, and each value of the subtype must be represented *once and only once* in the set of choices. Any case expression corresponds thus to exactly one choice and to the statement(s) associated with this unique choice.

There is no such rule in Verilog. Indeed, the values of the different choices are evaluated in the order of the description, and if there is a choice corresponding to the case expression, then the statement(s) associated with this matching choice is (are) executed. But some choices may be missing in the different values, thereby specifying *don't-care* choices (for which nothing is executed by the case statement). Such a description (with *don't-care* choices) is very relevant for specifications processed by synthesis tools.

Verilog provides two other selection statements that have no equivalent in VHDL: the *casez* and the *casex* statements. These two statements have the same semantics as the case statement, except that they permit handling Z and X as *don't-care* values.

This approach is again very relevant for a specification description (which may be used by synthesis tools), where all the possible device configurations (especially with X and Z values) may not be explicitly handled by the designer (as is the case for a simulation).

Let us consider the following Verilog functions:

```
function [2:0] Mask1 ;
input [2:0] BusValue ;
casez ( BusValue )
   3'b111 : Mask1 = 3'b111 ;
   3'b110 : Mask1 = 3'b110 ;
   3'b100 : Mask1 = 3'b100 ;
   default : Mask1 = 3'b000 ;
endcase
endfunction
```

```
function [2:0] Mask2 ;
input [2:0] BusValue ;
casez ( BusValue )
   3'b??1 : Mask2 = 3'b111 ;
   3'b?10 : Mask2 = 3'b110 ;
   3'b100 : Mask2 = 3'b100 ;
   default : Mask2 = 3'b000 ;
endcase
endfunction
```

```
function [2:0] Mask3 ;
input [2:0] BusValue ;
casex ( BusValue )
   3'b??1 : Mask3 = 3'b111 ;
   3'b?10 : Mask3 = 3'b110 ;
   3'b100 : Mask3 = 3'b100 ;
   default : Mask3 = 3'b000 ;
endcase
endfunction
```

The casez statement allows Z values to be treated as *don't-care* values, whereas the casex statement allows both Z and X to be treated as *don't-care*. The question mark (?) can also be used in the different choices to specify *don't-care* values.

Thus, the previous functions will, respectively, return the results given below:

```
Mask1(3'b111) =3'b111
Mask1(3'bz10) = 3'b110
Mask1(3'bx10) = 3'b000
Mask1(3'bx00) = 3'b000
Mask1(3'bzzz) = 3'b111 (*)
Mask1(3'bxxx) = 3'b000
```

```
Mask2(3'b111) =3'b111
Mask2(3'bz10) = 3'b110
Mask2(3'bx10) = 3'b110
Mask2(3'bx00) = 3'b000
Mask2(3'bzzz) = 3'b111 (*)
Mask2(3'bxxx) = 3'b000
```

```
Mask3(3'b111) =3'b111
Mask3(3'bz10) = 3'b110
Mask3(3'bx10) = 3'b110
Mask3(3'bx00) = 3'b100
Mask3(3'bzzz) = 3'b111 (*)
Mask3(3'bxxx) = 3'b111 (*)
```

The values followed by an asterisk (*) are hypothetical because they result from the statement associated with the first matching choice in the casez or casex statement. Indeed, in these cases, two or more choices match the controlling expression (3'bzzz matches *all* the choices in the casez and casex statements), and it is not clear whether only the statement of the first matching choice is executed or all the statements of all the matching choices are executed. This is a lack of the Verilog language reference manual.

Nevertheless, the use of such statements (in particular with the question mark to indicate *don't-care* values) is very convenient to describe truth-table-oriented behaviors.

Alas, the translation of such statements in VHDL is impossible with the case statement. Indeed, the equality implicitly used by the VHDL case statement cannot be redefined (i.e., overloaded): any case expression is compared to values of the same type, and it is thus difficult to handle *don't-care* values that actually represent *sets of possible values* (particularly with *don't-care* bit arrays, as in the above example).

On the other hand, expressions included in if statements may use any function call, such as overloaded equality operators. The designer can thus easily handle comparisons with *don't-care* values by using VHDL if statements. This approach is detailed in paragraph 9.8.1 for the gate-level primitives that use the Verilog table definition. Verilog primitives may have *don't-care* choices too, and the methodology used for the VHDL implementation of the Verilog tables can be adapted to the casez and casex statements.

• Iteration statements

These statements include incremental, conditional, and infinite loops.

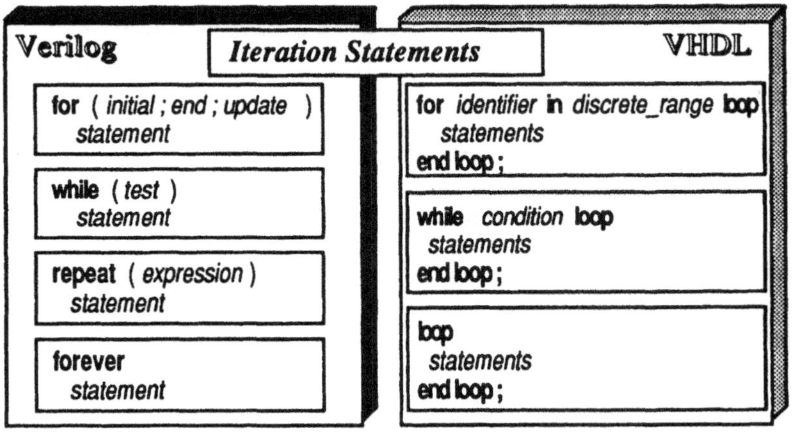

Fig 9.12 Iteration Statements

Identical iteration statements may be found in both languages. The Verilog repeat statement can be easily translated into a VHDL for...loop statement using a decremental index.

9.6.4.2. Concurrent Structured Statements

Concurrent structured statements are available in VHDL. According to the VHDL LRM, each of these statements is totally equivalent to one of the previous sequential structured statements inside a process. Indeed:

VHDL Structured Concurrent Statement	VHDL Equivalent Process Statement

target <= [**transport**]
waveform1 **when** *condition1* **else**
waveform2 **when** *condition2* **else**
...
waveformN ;
	if *condition1* **then**
	target <= [**transport**] *waveform1* ;
	elsif *condition2* **then**
	target <= [**transport**] *waveform2* ;
	else
	target <= [**transport**] *waveformN* ;
	end if ;

with *expression* **select** *target* <= [**transport**] *waveform1* **when** *choices1* , *waveform2* **when** *choices2* , ... *waveformN* **when** *choicesN* ;	**case** *expression* **is** **when** *choices1* => *target* <= [**transport**] *waveform1* ; **when** *choices2* => *target* <= [**transport**] *waveform2* ; ... **when** *choicesN* => *target* <= [**transport**] *waveformN* ; **end case** ;

Thus, if these VHDL concurrent constructs have no equivalent in Verilog, they can always be translated into an equivalent VHDL sequential code, and then into a Verilog sequential code.

VHDL provides another concurrent structured statement that is very convenient to generate concurrent statements such as component instantiations. This statement is the generate statement. The syntax is the following:

```
LABEL : for I in 1 to 8 generate
        INSTANCE : COMPONENT_NAME port map (SIG(I), SIG(I+1), SIG(I-1), CLOCK);
end generate ;
```

Verilog does not provide such a macro-generation scheme. This construct is very powerful in VHDL: structural descriptions of datapath operators are considerably simplified with this statement. It is almost compulsory when the datapath width depends on a generic parameter, as illustrated below:

```
LABEL : for I in 1 to N generate   -- N is a generic parameter of the corresponding entity
        INSTANCE : COMPONENT_NAME port map (SIG(I), SIG(I+1), SIG(I-1), CLOCK);
end generate ;
```

The VHDL generate statement also provides the designer with the possibility of conditionally generating one or more concurrent statements, such as describing conditional instantiation :

```
LABEL : if N mod 2 = 1 generate   -- N is a generic parameter of the corresponding entity
      INSTANCE : COMPONENT_NAME port map (SIG(0), SIG(1), INPUT, CLOCK);
end generate ;
```

Such parameterized structural descriptions (arrays of components, their number being a generic parameter, conditional instances, etc.) are very useful when describing operators generated by synthesis tools. Typical examples are provided by datapath generators: an N-bit adder may be composed of even and odd bit cells, and the most significant bit cell may depend on the parity of N.

The description of this operator should have the same parameters as the corresponding layout generator and reflect the structure based on technology-dependant library cells: in other words, this description should be *generic and structural*. Such a model can be achieved in VHDL, but not in Verilog.

Verilog and VHDL have both powerful *software* structured statements (used to describe algorithms), but Verilog has a lack of *hardware* structured statements compared to VHDL.

9.7. DESCRIPTION LEVEL

The purpose of this section is to give examples of equivalent descriptions at the different levels allowed by both languages. The typical design chosen for this purpose is a one-bit full adder.

9.7.1. Structural Description

A possible structure of the full adder is described below.

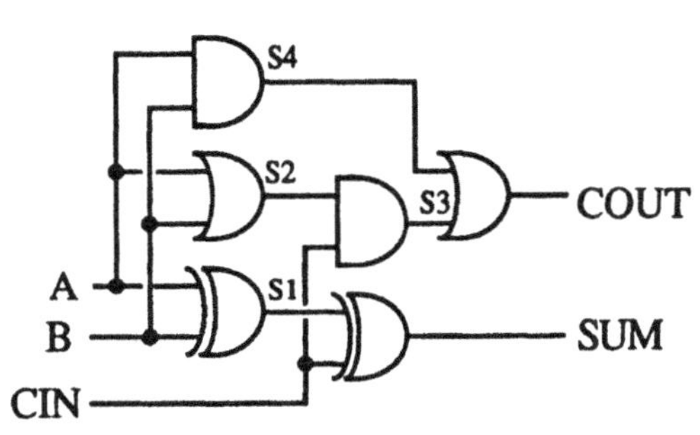

Fig 9.13 Structural Description of a One-Bit Full Adder

The Verilog structural model corresponding to this schema is the following module:

```
module FULL_ADDER (A, B, CIN, SUM, COUT) ;
parameter UNIT_DELAY = 1 ;
input       A, B, CIN ;
output      SUM, COUT ;
wire        S1, S2, S3, S4 ;
xor   #UNIT_DELAY   c1 ( A, B, S1 ),
                    c2 ( S1, CIN, SUM ) ;
or    #UNIT_DELAY   c3 ( A, B, S2 ),
                    c4 ( S3, S4, COUT ) ;
and   #UNIT_DELAY   c5 ( A, B, S4 ),
                    c6 ( S1, S2, S3 ) ;
endmodule
```

The corresponding VHDL entity and architecture could be:

```
entity FULL_ADDER is
      generic ( UNIT_DELAY :   TIME := 1 ns ) ;
      port (    A, B, CIN  :   in BIT ;
                SUM, COUT  :   out BIT ) ;
end FULL_ADDER ;

use WORK.BASIC_PACKAGE.all ;
architecture STRUCTURE of FULL_ADDER is
      signal S1, S2, S3, S4 : BIT ;
begin
      C1 : XOR_GATE  generic map ( UNIT_DELAY ) port map ( A, B, S1 ) ;
      C2 : XOR_GATE  generic map ( UNIT_DELAY ) port map ( S1, CIN, SUM ) ;
      C3 : OR_GATE   generic map ( UNIT_DELAY ) port map ( A, B, S2 ) ;
      C4 : OR_GATE   generic map ( UNIT_DELAY ) port map ( S3, S4, COUT ) ;
      C5 : AND_GATE  generic map ( UNIT_DELAY ) port map ( A, B, S4 ) ;
      C6 : AND_GATE  generic map ( UNIT_DELAY ) port map ( S1, S2, S3 ) ;
end STRUCTURE ;
```

We intentionally wrote the previous VHDL model so that it looks very similar to the Verilog module. Instantiation mechanisms have some similarities in both languages: for each instance, there is a way to specify the generic parameter(s) and the association between ports and signals.

But several fundamental differences have to be pointed out between these two models. First, as explained in section 9.2, Verilog modules are directly instantiated into other modules, whereas VHDL entities are plugged into components (components are configured with couples entity-architecture), which are in turn instantiated into architectures. In the previous model, XOR_GATE is thus the name of a component (and not an entity). Then, considering the given example:

- In the Verilog module, the and, or, and xor gates are *primitive gates* of the language. The designer does not have to write any model for the internal structure of these basic gates, because their truth tables are implicitly known by the Verilog simulator. Delays may be specified at the instantiation of these gates.
- In VHDL, there is no primitive gate! And, or, and xor are indeed VHDL keywords, but they are *primitive functions* on types boolean and bit. Thus, components instantiated in the previous architecture were named AND_GATE, OR_GATE, and XOR_GATE and declared in a package titled BASIC_PACKAGE:

```
package BASIC_PACKAGE is
    component AND_GATE
        generic ( DELAY : TIME ) ;
        port ( A, B : in BIT ; S : out BIT ) ;
    end component ;
    component OR_GATE
        generic ( DELAY : TIME ) ;
        port ( A, B : in BIT ; S : out BIT ) ;
    end component ;
    component XOR_GATE
        generic ( DELAY : TIME ) ;
        port ( A, B : in BIT ; S : out BIT ) ;
    end component ;
end BASIC_PACKAGE ;
```

These components have to be linked before simulation to couples entity-architecture into a configuration declaration. For example, the AND_GATE component could be configured with the following design units:

```
entity AND_ENTITY is
    generic ( DELAY : TIME ) ;
    port ( A, B : in BIT ; S : out BIT ) ;
end ;
architecture DATAFLOW of AND_ENTITY is
begin
    S <= A and B after DELAY ;
end ;
```

This example highlights a VHDL characteristic, namely, that every design ends (at the lower hierarchical level) in a behavioral or dataflow description, such as the concurrent assignment above (using the functional primitive **and** in VHDL).

Furthermore, delay information in VHDL is always attached to dataflow or behavioral descriptions (here in the concurrent assignment of the DATAFLOW architectures), never to structural description, as is the case with the Verilog primitive gates.

Another important difference between both models is the type of wires or signals. In Verilog, all the *nets always have one of the predefined four-level logic values* (classical X, Z, 0, and 1 values). On the other hand, VHDL *signals may be of any user-defined type*. In the previous VHDL model, signals are of the predefined type BIT, i.e., a two-level logic type ('0' and '1').

We could have defined VHDL signals of any type, such as

```
type FOUR_LEVEL_BIT is ('X' , '0', '1', 'Z') ;  -- see section 9.4
```

But then we would have had to redefine all the logical operators (and, or, xor, etc.) for the new type. In fact, in case of unidirectional logic without any conflict, it is often possible to stick with the predefined BIT type. The choice of a relevant multi-value logic system in VHDL is detailed in chapter 4 ("Structuring the Environment").

9.7.2. Dataflow Description

A Verilog dataflow description of the one-bit full adder could be

```
module FULL_ADDER (A, B, CIN, SUM, COUT) ;
parameter UNIT_DELAY = 1 ;
input     A, B, CIN ;
output    SUM, COUT ;
wire      S1, S2, S3, S4 ;
assign    #UNIT_DELAY   S1 = A ^ B ;
assign    #UNIT_DELAY   S2 = A | B ;
assign    #UNIT_DELAY   S3 = S1 & S2 ;
assign    #UNIT_DELAY   S4 = A & B ;
assign    #UNIT_DELAY   SUM = S1 ^ CIN ;
assign    #UNIT_DELAY   COUT = S3 | S4 ;
endmodule
```

The previous model corresponds exactly to the structural description given above. Another dataflow description is

```
module FULL_ADDER (A, B, CIN, SUM, COUT) ;
parameter UNIT_DELAY = 1 ;
input     A, B, CIN ;
output    SUM, COUT ;
assign    #UNIT_DELAY   SUM = A ^ B ^ CIN ;
assign    #UNIT_DELAY   COUT = (A & B) | (B & CIN) | (A & CIN) ;
// Warning : delays are not intended to be consistent with those of the previous model!
endmodule
```

An equivalent VHDL description for both Verilog models consists of the following design units:

```
entity FULL_ADDER is
        generic ( UNIT_DELAY :    TIME := 1 ns ) ;
        port (    A, B, CIN   :   in BIT ;
                  SUM, COUT :     out BIT ) ;
end FULL_ADDER ;

architecture DATAFLOW1 of FULL_ADDER is
        signal S1, S2, S3, S4 : BIT ;
begin
        S1 <= A xor B after UNIT_DELAY ;
        S2 <= A or B after UNIT_DELAY ;
        S4 <= A and B after UNIT_DELAY ;
        S3 <= S1 and S2 after UNIT_DELAY ;
        SUM <= S1 xor CIN after UNIT_DELAY ;
        COUT <= S3 or S4 after UNIT_DELAY ;
end DATAFLOW1 ;

architecture DATAFLOW2 of FULL_ADDER is
begin
        SUM <= A xor B xor CIN after UNIT_DELAY ;
        COUT <= (A and B) or (A and CIN) or (B and CIN) after UNIT_DELAY ;
        -- Warning: delays are not intended to be consistent with those of the previous model!
end DATAFLOW2 ;
```

There is no fundamental difference between Verilog and VHDL dataflow descriptions. The Verilog continuous assignment and the VHDL concurrent assignment statement have the same functionalities and behave according to an identical scheme: every assignment statement is executed each time any signal in the right-hand-side expression changes its value (see 9.6.1.1). The concurrent world is well implemented in both languages.

9.7.3. Behavioral Description

A Verilog behavioral description of the one-bit full adder is

```
module FULL_ADDER (A, B, CIN, SUM, COUT) ;
parameter UNIT_DELAY = 1 ;
input      A, B, CIN ;
output     SUM, COUT ;
reg        SUM, COUT ;
always
        @ (A or B or CIN)      #UNIT_DELAY   { COUT, SUM } = A + B + CIN ;
endmodule
```

An equivalent VHDL model is given below:

```
entity FULL_ADDER is
        generic ( UNIT_DELAY :    TIME := 1 ns ) ;
        port (    A, B, CIN     :    in BIT ;
                  SUM, COUT :    out BIT ) ;
end FULL_ADDER ;

use WORK.BASIC_FUNCTIONS.all ;
architecture BEHAVIOR of FULL_ADDER is
begin
        process
        begin
            wait on A, B, CIN ;
            (COUT, SUM) <= VEC(VAL(A) + VAL(B) + VAL(CIN), 2) after UNIT_DELAY;
        end process ;
end BEHAVIOR;
```

The Verilog behavioral model and the corresponding VHDL model look similar, but are indeed very different.

In Verilog, the addition is predefined on bits and arrays of bits, the last mentioned items being handled as unsigned integers. In VHDL, arrays of bits (type bit_vector) are only *arrays of characters* ('0' and '1') having associated logical operators but no meaning as numeric values. Thus, there is no implicit adding operator on bit vectors, and consequently all these conversion and arithmetic functions have to be user defined.

As an illustration only, the previous VHDL architecture uses a package BASIC_FUNCTIONS that could be written as follows:

```
package BASIC_FUNCTIONS is
  function VAL( B : BIT ) return NATURAL ;
  -- conversion from bit to natural
  function VAL( V : BIT_VECTOR ) return NATURAL ;
  -- conversion from bit array to natural
  function VEC( N : NATURAL ; L : POSITIVE ) return BIT_VECTOR ;
  -- conversion from natural to bit array of a given length L
end BASIC_FUNCTIONS ;

package body BASIC_FUNCTIONS is
  function VAL( B : BIT ) return NATURAL is
  begin
        return BIT'POS(B) ;
  end VAL ;
  function VAL( V : BIT_VECTOR ) return NATURAL is
        variable TEMP : NATURAL ;
  begin
        for I in V'range loop
            TEMP := 2 * TEMP + VAL(V(I)) ;
        end loop;
        return TEMP ;
  end VAL ;
```

```
  function VEC( N : NATURAL ; L : POSITIVE ) return BIT_VECTOR is
        variable TEMP : NATURAL := N ;
        variable VECT : BIT_VECTOR( L downto 1 ) ;
 begin
        for I in 1 to L loop
              VECT(I) := BIT'VAL(TEMP mod 2) ;
              TEMP := TEMP / 2 ;
        end loop;
        assert TEMP = 0 report "Error in VEC: Vector length too small!" severity error;
        return VECT ;
  end VEC ;
end BASIC_FUNCTIONS ;
```

This package has been provided just to illustrate the different ways both languages handle bit vector operations. Actually, a more complete arithmetic package (including bit vector addition) is presented in chapter 4 ("Structuring the Environment").

This typical design points out again the fundamental difference between both languages: Verilog is very IC oriented, and many (necessary) types and functions are predefined or implicit, whereas VHDL is designed for a wide range of applications (IC, PCB, System) and the set of predefined functions is very restrictive. Nevertheless, more and more VHDL utility packages are available in customer environments: these packages implement general routines such as arithmetic and mathematical functions or are specific to a given application (signal processing, system-level design, and so forth).

The philosophy of VHDL is also to be a strongly typed language, leading thus to compulsory explicit user-defined type conversions (between bit vectors and integers, for example).

9.8. TRANSLATING FROM VERILOG TO VHDL

9.8.1. Verilog User-Defined Primitives

Verilog provides a method for describing user-defined primitives (UDPs). This method uses a truth table syntax to define combinational blocks of logic, as well as level-sensitive and edge-sensitive sequential devices. The *primitive* is a kind of basic module written in a very compact, convenient, and readable way and allowing efficient (quick) simulation.

9.8.1.1. Combinational Primitives

Here is an example of a simple combinational block:

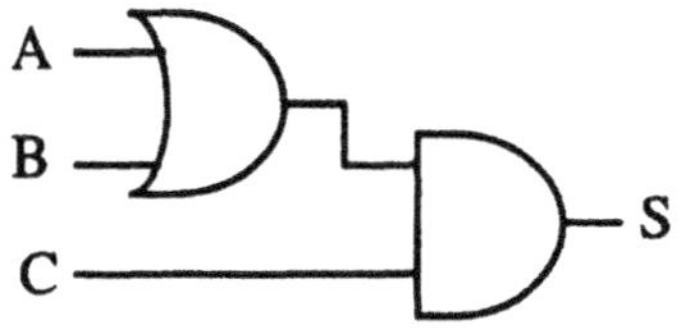

Fig 9.14 S = (A | B) & C

To gain in simulation efficiency, the Verilog designer may describe this block with the following primitive:

```
primitive ANDOR (S, A, B, C) ;
output      S ;
input       A, B, C ;
table
//    A B C  :  S
      0 0 0  :  0  ;
      0 0 1  :  0  ;
      0 1 0  :  0  ;
      0 1 1  :  1  ;
      1 0 0  :  0  ;
      1 0 1  :  1  ;
      1 1 0  :  0  ;
      1 1 1  :  1  ;
      0 0 X  :  0  ;
      X 0 0  :  0  ;
      X 1 0  :  0  ;
      0 X 0  :  0  ;
      1 X 0  :  0  ;
      X X 0  :  0  ;
      X 1 1  :  1  ;
      1 X 1  :  1  ;
endtable
endprimitive
```

A Verilog user-defined primitive must have exactly one output port but any number of input ports, all of the four-level bit type (i.e., 0, 1, X, or Z). The output port must be the first of the list, and only the values 0, 1, and X may be specified on both input and output ports in the table (the Z value on an input is treated as an X).

The only item allowed in a primitive is a table definition. The previous example shows one form of such a table. The table lists the output value corresponding to different sets of input values (inputs are separated from the output by a colon). Any set of input values not appearing in the table leads to the default output value X.

In VHDL, the most straightforward statement that can be used to describe truth tables is the concurrent selected signal assignment or the sequential case statement (both statements are equivalent). Thus, a VHDL design unit corresponding to the previous Verilog primitive may be

```
use WORK.PRIMITIVE_PKG.all ;
entity ANDOR is
        port ( S : out XBIT ; A, B, C : in XBIT ) ;
end ANDOR ;
-- with selected signal assignment
architecture TABLE1 of ANDOR is
        subtype XBIT3 is
            XBIT_VECTOR(1 to 3) ;
begin
        with XBIT3'(A & B & C) select
         S <=    '0' when "000",
                 '0' when "001",
                 '0' when "010",
                 '1' when "011",
                 '0' when "100",
                 '1' when "101",
                 '0' when "110",
                 '1' when "111",
                 '0' when "00X",
                 '0' when "X00",
                 '0' when "X10",
                 '0' when "0X0",
                 '0' when "1X0",
                 '0' when "XX0",
                 '1' when "X11",
                 '1' when "1X1",
                 'X' when others ;

end TABLE1 ;
```

```
-- with equivalent case statement
architecture TABLE2 of ANDOR is
begin
   process
      subtype XBIT3 is
         XBIT_VECTOR(1 to 3) ;
   begin
      case XBIT3'(A & B & C) is
            when "000" =>      S <= '0' ;
            when "001" =>      S <= '0' ;
            when "010" =>      S <= '0' ;
            when "011" =>      S <= '1' ;
            when "100" =>      S <= '0' ;
            when "101" =>      S <= '1' ;
            when "110" =>      S <= '0' ;
            when "111" =>      S <= '1' ;
            when "00X" =>      S <= '0' ;
            when "X00" =>      S <= '0' ;
            when "X10" =>      S <= '0' ;
            when "0X0" =>      S <= '0' ;
            when "1X0" =>      S <= '0' ;
            when "XX0" =>      S <= '0' ;
            when "X11" =>      S <= '1' ;
            when "1X1" =>      S <= '1' ;
            when others =>     S <= 'X' ;
      end case ;
      wait on A, B, C ;
   end process ;
end TABLE2 ;
```

PRIMITIVE_PKG is a package that exports the used bit types:

```
package PRIMITIVE_PKG is
        type ZXBIT is ('X', '0', '1', 'Z') ; -- signal type
        subtype XBIT is ZXBIT range 'X' to '1' ; -- port subtype
        type ZXBIT_VECTOR is array ( NATURAL range <> ) of ZXBIT ;
        type XBIT_VECTOR is array ( NATURAL range <> ) of XBIT ;
        -- ZXBIT and XBIT array types
        function ZTOX ( VALUE : ZXBIT ) return XBIT ;
        -- conversion function
end PRIMITIVE_PKG ;
```

```
package body PRIMITIVE_PKG is
        function ZTOX ( VALUE : ZXBIT ) return XBIT is
        begin
                case VALUE is
                 when 'Z' =>          return 'X' ;
                 when others =>    return VALUE ;
                end case ;
        end ZTOX ;
end PRIMITIVE_PKG ;
```

The used XBIT subtype does not include 'Z'. The Verilog implicit conversion of the high-impedance input state into the unknown state may be implemented in VHDL via the use of the conversion function ZTOX in the port map of any component configuration using the previous entity:

```
use WORK.PRIMITIVE_PKG.all ;
architecture ANY_ARCHITECTURE of ANY_ENTITY is
        component ANDOR_COMPONENT
                port ( CS : out ZXBIT ; CA, CB, CC : in ZXBIT ) ;
        end component ;
        for all : ANDOR_COMPONENT use entity WORK.ANDOR(TABLE1)
                port map ( S => CS, A => ZTOX(CA), B => ZTOX (CB), C => ZTOX (CC) ) ;
        -- configuration specification using conversion function in port map
        signal SS, SA, SB, SC : ZXBIT ;
begin
        AN_INSTANCE : ANDOR_COMPONENT port map (SS, SA, SB, SC) ;
end ;
```

The implicit Verilog conversion has been translated into an explicit VHDL conversion. Similarly, the VHDL select or case statement needs an explicit default choice "**when others**" (see 9.6.3.1), whereas Verilog tables provide an implicit default value. This value assigned to the output for any unspecified set of inputs is the unknown value (explicit 'X' in VHDL, implicit X in Verilog). Besides that, the two previous VHDL architectures look very similar to the original Verilog primitive.

Verilog provides also very useful shortcuts for describing truth tables. For example, the previous ANDOR module can be written in a more compact way:

```
primitive ANDOR (S, A, B, C) ;
output      S ;
input   A, B, C ;
table  / /   A B C    :    S
             ? 1 1    :    1    ;
             1 ? 1    :    1    ;
             0 0 ?    :    0    ;
             ? ? 0    :    0    ;
endtable
endprimitive
```

Indeed, as for the casez and casex statements, the question mark (?) may be used in the table to specify a *don't-care* value. Basically, the question mark indicates an iteration of 0, 1, and X. This shorthand notation is very convenient and readable. Furthermore, it looks very natural and straightforward for the designer, who is often used to thinking with *don't-care* values. VHDL does not implement such a predefined *don't-care* notion.

Similarly, the character B (for Bit) may be used in the table input field to indicate a *don't-care* bit value that is an iteration of 0 and 1.

In Verilog, a *don't-care* value (?) actually represents a set of three possible values (X, 0 and 1), and a *don't-care* bit value (B) a set of two possible values (0 and 1). Thus, each line containing N question mark(s) and M character(s) B is equivalent to (3 ** N)(2 ** M) expanded lines:

```
            1X0:0                          X0X0:0
1?0:0 <=>   100:0                          X0X1:0
            110:0          X0?B:0 <=>      X000:0
                                           X001:0
10B:1 <=>   100:1                          X010:0
            101:1                          X011:0
```

In order to implement these *don't-care* notions in VHDL, the PRIMITIVE_PKG package has been rewritten:

```
package PRIMITIVE_PKG is
    type PRIMITIVE_BIT is ('?', 'B', 'X', '0', '1', 'Z') ;
    -- original bit type including don't-care elements
    subtype ZXBIT is PRIMITIVE_BIT range 'X' to 'Z' ; -- signal subtype
    subtype XBIT is PRIMITIVE_BIT range 'X' to '1' ; -- port subtype
    type ZXBIT_VECTOR is array ( NATURAL range <> ) of ZXBIT ;
    type XBIT_VECTOR is array (NATURAL range <>) of XBIT ;
    -- ZXBIT and XBIT array types
    function ZTOX ( VALUE : ZXBIT ) return XBIT ;
    -- conversion function
    subtype DXBIT is PRIMITIVE_BIT range '?' to '1' ;
    -- possible values of the table
    type DXBIT_VECTOR is array (NATURAL range <>) of DXBIT ;
    -- DXBIT array type
    function EQ(V : XBIT ; W : DXBIT) return BOOLEAN ;
    function EQ(V : XBIT_VECTOR ; W : DXBIT_VECTOR) return BOOLEAN ;
    -- equivalence functions used to write truth tables
end PRIMITIVE_PKG ;
```

The base type has now six elements, including the character literals that represent *don't-care*, i.e., question mark '?' and character 'B'. The previous type ZXBIT is now a subtype of this base type, but any other subtype (e.g., XBIT) and function previously defined remain the same.

In addition, a new subtype has been added: the DXBIT subtype containing all the characters used in a Verilog table, i.e., '?', 'B', 'X', '0', '1'. Together with this subtype, functions have been defined comparing XBIT and DXBIT values, as well as vectors of these values.

Indeed, the idea was to define an equivalence function such that:

XBIT_VECTOR	0 1 0	*is equivalent to*	DXBIT_VECTOR	? 1 0
XBIT_VECTOR	1 1 0	*is equivalent to*	DXBIT_VECTOR	? 1 0
XBIT_VECTOR	0 0 0	*is equivalent to*	DXBIT_VECTOR	0 B 0

...

Of course, this function is not transitive and does not need to be commutative (input values of type XBIT or XBIT_VECTOR are compared to a set of DXBIT or DXBIT_VECTOR values in the table, and not the contrary...). Two overloaded EQ functions have been defined that correspond to these specifications:

```
package body PRIMITIVE_PKG is
        function ZTOX ( VALUE : ZXBIT ) return XBIT is
        -- same body as before
        end ZTOX ;
        function EQ(V : XBIT ; W : DXBIT) return BOOLEAN is
        begin
            if W = V or W = '?' or (W = 'B' and V /= 'X') then
                return TRUE ;
            else
                return FALSE ;
            end if ;
        end EQ ;

        function EQ(V : XBIT_VECTOR ; W : DXBIT_VECTOR) return BOOLEAN is
            variable WW : DXBIT_VECTOR(V'RANGE) ;
        begin
            if V'LENGTH /= W'LENGTH then
                return FALSE ;
            else
                WW := W ;
                for I in V'RANGE loop
                    if not EQ(V(I),WW(I)) then
                        return FALSE ;
                    end if ;
                end loop ;
                return TRUE ;
            end if ;
        end EQ ;
end PRIMITIVE_PKG;
```

The reader may check that functions EQ verify:

```
EQ( "010", "?10" ) = EQ( "110", "?10" ) = EQ( "X10", "?10" ) = TRUE
EQ( "000", "?10" ) = EQ( "0X", "0B") = FALSE
-- XBIT_VECTOR is the left argument, DXBIT_VECTOR is the right argument
EQ( '0', '?' ) = EQ( '0', 'B' ) = TRUE
EQ( '0', '1' ) = FALSE
-- XBIT is the left argument, DXBIT is the right argument
```

With this new release of PRIMITIVE_PKG, the entity ANDOR may be associated with another architecture that corresponds to the previous Verilog primitive (using shorthand notations):

```
architecture TABLE3 of ANDOR is
begin
        S <=    '1' when EQ(A & B & C, "?11" ) else
                '1' when EQ(A & B & C, "1?1" ) else
                '0' when EQ(A & B & C, "00?" ) else
                '0' when EQ(A & B & C, "??0" ) else
                'X' ;
end TABLE3 ;
```

Instead of defining a function named EQ, we could have declared the following overloaded equality operators in the same PRIMITIVE_PKG package:

```
function "="(V : XBIT ; W : DXBIT) return BOOLEAN ;
function "="(V : XBIT_VECTOR ; W : DXBIT_VECTOR) return BOOLEAN ;
-- with the same bodies as EQ functions
```

Unfortunately, the use of such operators does not help much in readability. Indeed, the rewriting of the previous architecture TABLE3 with "=" instead of EQ, i.e.,

```
-- Warning: this VHDL architecture is not correct!
architecture WRONG of ANDOR is
begin
        S <=    '1' when A & B & C = "?11" else ... ;
end WRONG ;
```

will inevitably lead to a compilation error message looking more or less like the following:

```
$$ Error: ambiguity not resolved in line ...     S <=     '1' when A & B & C = "?11"
                                                                              ^
Two possible meanings of "=" are:
"=" is the implicit equality operator on arrays of elements of type PRIMITIVE_BIT
"=" is the overloaded equality operator declared in package PRIMITIVE_PKG
```

One way to solve this problem would have been to qualify both XBIT_VECTOR and DXBIT_VECTOR arrays to specify to the compiler the exact type of the operands:

```
S <=    '1' when XBIT_VECTOR'(A & B & C) = DXBIT_VECTOR'("?11") else ...
```

but the result would have been much less compact than the previous architecture using functions EQ.

Beside that, functions EQ or overloaded equality operators have in common the other major drawback that they cannot be used within the case statement (or the equivalent concurrent selected signal assignment).

Indeed, such a statement only uses the equality operator implicitly declared at the corresponding type declaration, and overloading this operator has no effect on the case statement execution. The if statement (or the equivalent concurrent conditional assignment statement) must therefore be chosen instead of the case statement, and that makes the VHDL translation of the Verilog primitive table less compact than it could be.

9.8.1.2. Sequential Primitives

Verilog user-defined primitives may be used to describe sequential devices. These devices have an internal state that is modeled by a register (*reg*) that directly drives the primitive output. Compared to the combinational primitive syntax, a new column has been added in the table that indicates the current state of the device. Each line of the primitive table thus consists of a set of input values followed by the current state value (surrounded by colons) and the output value (which is also the next state value).

Sequential primitives may be level-sensitive or edge-sensitive. In the first case, the output value depends only on input levels. In the last case, edges may be specified on one (and only one) input value. Mixed level- and edge-sensitive primitives may also be specified (with a single table).

• Level-Sensitive Primitives

The following is an example of a level-sensitive latch:

```
primitive LATCH (Q, CLOCK, D) ;
output      Q ;     // the output Q is also a register because the output
reg         Q ;     // port value is the next internal state value.
input   CLOCK, D ;
table   //  CLOCK  D   :   STATE   :  Q
              1    1   :     ?     :  1 ;
              1    0   :     ?     :  0 ;
              0    ?   :     ?     :  - ;
endtable
endprimitive
```

The minus sign ('-') in the output column specifies that there is no change in the output value, i.e., the next state will be the same as the current state. The described latch remains in the same state when clock is low and transmits the value of the input to the output when clock is high. Otherwise (i.e., if clock is high and input is unknown, or if clock is unknown), the output value (and the next state value) will be unknown.

Such Verilog primitives may be translated into the following VHDL design unit:

```
use WORK.PRIMITIVE_PKG.all ;
entity LATCH is
      port ( Q : out XBIT ; CLOCK, D : in XBIT ) ;
end LATCH ;

architecture TABLE of LATCH is
      signal STATE : XBIT ;
begin
      STATE <=  '1'      when EQ(CLOCK & D & STATE, "11?" ) else
                '0'      when EQ(CLOCK & D & STATE, "10?" ) else
                STATE    when EQ(CLOCK & D & STATE, "0??" ) else
                'X' ;
      Q <= STATE ;
end TABLE ;
```

It is useless to redefine the type PRIMITIVE_BIT in order to include the character literal '-' with a "no change" meaning, because it is impossible in VHDL to assign to a signal a value (or something) meaning "no change".

One way to achieve this translation is thus to declare an internal signal (STATE) representing the internal state of the device, and to assign to this signal its current value when there is no change. The output port Q is continuously driven by the STATE signal.

Thus, it has not been necessary to change anything in the package PRIMITIVE_PKG, originally designed for combinational primitives, in order to model level-sensitive sequential primitives. The main difference in the translation methodology has been the addition of an internal signal representing the current state of the primitive.

Another strategy could be to model the level-sensitive sequential primitive with a single process and to store the internal state in a variable declared in this process. This process would first compute the next internal state according to the input values and then assign this state value to the output.

Following is another architecture of the previous latch:

```
architecture TABLE2 of LATCH is
begin
    process
        variable STATE : XBIT ;
    begin
        if EQ(CLOCK & D & STATE, "11?" )    then    STATE := '1' ;
        elsif EQ(CLOCK & D & STATE, "10?" ) then    STATE := '0' ;
        elsif EQ(CLOCK & D & STATE, "0??" ) then    null ; -- no change
        else                                        STATE := 'X' ;
        end if ;
        Q <= STATE ;
        wait on CLOCK, D ;
    end process ;
end TABLE2 ;
```

This architecture is more verbose than the previous one, but it contains only one process whereas the first one was composed of two processes, and uses a variable instead of a signal to store the state of the device (STATE). This model is thus more efficient and uses less memory space.

• **Edge-Sensitive Primitives**

Here is an example of an edge-sensitive flip-flop:

```
primitive FLIPFLOP (Q, CLOCK, D) ;
output    Q ;     // the output Q is also a register because the output
reg       Q ;     // port value is also the next internal state value.
input  CLOCK, D ;
table  //  CLOCK  D    : STATE : Q
           (01)   1    :   ?   : 1  ;
           (01)   0    :   ?   : 0  ;
           (0X)   1    :   1   : 1  ;
           (0X)   0    :   0   : 0  ;
           (?0)   ?    :   ?   : -  ;
            ?    (??)  :   ?   : -  ;
endtable
endprimitive
```

In the previous model, the first column includes terms in parentheses, which specify edges: (01) is a rising edge, (0X) is an edge from 0 to X, (?0) is an edge from 0, 1 or X to 0, (??) means "any change".

The previous flip-flop transmits the input value to the output when there is a rising edge of CLOCK (two first lines). It remains in the same state when CLOCK is going from 0 to X and the input value is the same as the state (third and fourth lines reducing the pessimism of the model), when there is a falling edge (10 or X0) on CLOCK (fifth line), or when input D changes while CLOCK is steady (last line). It goes into the unknown state otherwise.

One way to translate such Verilog primitive into VHDL is to add the following function EDGE in the package PRIMITIVE_PKG:

```
package PRIMITIVE_PKG is
        -- same declarations as before
        function EDGE(signal CK : XBIT ; V : DXBIT_VECTOR(0 to 1)) return BOOLEAN ;
end PRIMITIVE_PKG ;
```

This function detects the edge on the clock signal and returns the boolean TRUE if this edge corresponds to the specified one. The other declarations are the same as in the previous release of PRIMITIVE_PKG (see 9.8.1.1).

The body of this new function EDGE has been added in the corresponding package body:

```
package body PRIMITIVE_PKG is
        -- same bodies as before
        function EDGE(signal CK : XBIT; V : DXBIT_VECTOR(0 to 1)) return BOOLEAN is
        begin
            if CK'EVENT and EQ(CK'LAST_VALUE, V(0)) and EQ(CK, V(1)) then
                return TRUE ;
            else
                return FALSE ;
            end if ;
        end EDGE ;
end PRIMITIVE_PKG ;
```

Using this release of the package PRIMITIVE_PKG, the Verilog edge-sensitive primitive may be translated into the following VHDL design unit (two possible architectures are provided, depending on whether the internal state is modeled by a signal or a process variable):

```
use WORK.PRIMITIVE_PKG.all ;
entity FLIPFLOP is
        port ( Q : out XBIT ; CLOCK, D : in XBIT ) ;
end FLIPFLOP ;

architecture TABLE of FLIPFLOP is
        signal STATE : XBIT ;
begin
        STATE <=    '1'     when EDGE(CLOCK, "01") and EQ(D & STATE, "1?" ) else
                    '0'     when EDGE(CLOCK, "01") and EQ(D & STATE, "0?" ) else
                    '1'     when EDGE(CLOCK, "0X") and EQ(D & STATE, "11" ) else
                    '0'     when EDGE(CLOCK, "0X") and EQ(D & STATE, "00" ) else
                    STATE   when EDGE(CLOCK, "?0") and EQ(D & STATE, "??" ) else
                    STATE   when not EDGE(CLOCK, "??") and EDGE(D, "??") else
                    'X' ;
        Q <= STATE ;
end TABLE ;
```

```
architecture TABLE2 of FLIPFLOP is
begin
    process
        variable STATE : XBIT ;
    begin
        if EDGE(CLOCK, "01") and EQ(D & STATE, "1?" )      then    STATE := '1' ;
        elsif EDGE(CLOCK, "01") and EQ(D & STATE, "0?" )   then    STATE := '0' ;
        elsif EDGE(CLOCK, "0X") and EQ(D & STATE, "11" )   then    STATE := '1' ;
        elsif EDGE(CLOCK, "0X") and EQ(D & STATE, "00" )   then    STATE := '0' ;
        elsif EDGE(CLOCK, "?0") and EQ(D & STATE, "??" )   then    null ; -- no change
        elsif not EDGE(CLOCK, "??") and EDGE(D, "??")      then    null ; -- no change
        else                                                       STATE := 'X' ;
        end if ;
        Q <= STATE ;
        wait on CLOCK, D ;
    end process ;
end TABLE2 ;
```

Previous models highlight the limits of our methodology: even if the VHDL syntax is very readable and allows specification of the behavior of a device at exactly the same level of abstraction as that of a Verilog primitive table (using the same notions of *don't-care* values, levels, and edges), the obtained description becomes less and less compact with the repetition of function (EDGE, EQ) and signal names at each line.

Furthermore, Verilog provides other shorthand notations to specify edges. These notations use the symbols listed in the table below:

Table symbol	Interpretation of this symbol
0	Bit Logic 0
1	Bit Logic 1
X or x	Unknown Value
?	*Don't-care*, i.e., iteration of 0, 1 ,and X
B or b	*Don't-care* bit value, i.e., iteration of 0 and 1
-	No change (output of sequential device only)
(VW)	Change from V to W (values in 0, 1, X, B, or ?)
R or r	Rising edge, i.e., (01)
F or f	Falling edge, i.e., (10)
P or p	Positive edge, i.e., iteration of (01), (0X), and (X1)
N or n	Negative edge, i.e., iteration of (10) , (1X), and (X0)
*	Any change, i.e., (??)

Including these symbols in the package PRIMITIVE_PKG would significantly complicate the table descriptions if the same VHDL templates are used.

The next paragraph will describe another approach, which provides the VHDL designer with the same simplicity and compactness found in the original Verilog primitive tables.

9.8.1.3. Writing a Verilog Primitive Analyzer in VHDL

This paragraph will show that VHDL can be used as a programming language, as well as Pascal or C. Of course, it is often not necessary to use such capabilities when describing hardware devices. But in some cases — for example, for library migrations or mixed-mode simulations (of VHDL and Verilog models) — it could be very useful to read in a VHDL environment models written in other languages. The algorithmic power of VHDL is very convenient for such needs.

The approach treated here is based on the use of VHDL functions that analyse ASCII files containing Verilog user-defined primitives (UDPs). The designer will thus be able to include inside his or her VHDL environment primitives that are described with the Verilog syntax.

To be able to read these Verilog files, the VHDL designer will use a library named VLIB that contains four design units (detailed below):

- the entity PRIMITIVE, which has two generic parameters (the name of the Verilog file and a propagation delay applied to any output change) and two ports (an output port of type XBIT and an unconstrained input port of type XBIT_VECTOR);
- its associated architecture VERILOG, which consists of a single process that analyzes the Verilog file during initialization and uses the acquired data to simulate the primitive behavior;
- the package VERILOG_PRIMITIVE, which defines the same types and functions as the previous package PRIMITIVE_PKG, plus other functions used to analyze the Verilog file;
- its associated package body VERILOG_PRIMITIVE.

The designer proceeds as follows. First, he writes or reuses an ASCII file containing a single Verilog primitive (using the Verilog table syntax explained above), — for example, the previous primitive ANDOR:

```
// File andor.ver
primitive ANDOR (S, A, B, C);
        output   S ;
        input    A, B, C ;
        table    //  A B C       S
                     ? 1 1   :   1   ;
                     1 ? 1   :   1   ;
                     0 0 ?   :   0   ;
                     ? ? 0   :   0   ;
        endtable
endprimitive
```

Then, the designer may write in the appropriate architecture:

- the declaration of a component having the same interface (ports) as the Verilog primitive;
- the configuration specification of this component with the design entity PRIMITIVE(VERILOG), the name of the previous Verilog file (andor.ver) being used in the generic map;
- the instantiation of this component.

```
library VLIB ; use VLIB.VERILOG_PRIMITIVE.all;
entity ENT is end ;
architecture ARC of ENT is
        component ANDOR
             port (S : out XBIT; A, B, C : in XBIT);
        end component;
        -- component having the same interface as the Verilog primitive ANDOR
        for all : ANDOR use entity VLIB.PRIMITIVE(VERILOG)
             generic map ("andor.ver", 2 ns)
             port map (OUTPUT=> S, INPUT(1)=>A, INPUT(2)=>B, INPUT(3)=>C) ;
        -- configuration specification with the entity PRIMITIVE and its architecture VERILOG,
        -- one of the generic parameters being associated with the Verilog file name (andor.ver).
        signal S,A,B,C : XBIT ;
begin
        I : ANDOR port map(S, A, B, C);
        -- component instantiation
        C <= '0' after 5 ns, '1' after 15 ns, '0' after 45 ns;
        B <= '0' after 10 ns, '1' after 20 ns, 'X' after 35 ns;
        A <= '0' after 25 ns, '1' after 30 ns, '0' after 40 ns;
end;
```

And that is all. The simulation of the previous design unit on the designer's favorite VHDL simulator gives the following results:

```
# Loading \vhdl\std.standard
# Loading \vhdl\std.textio
# Loading \serge\essai\vlib.verilog_primitive
# Loading \serge\essai\work.ent(arc)
# Loading \serge\essai\vlib.primitive(verilog)
# ** Note: ----------------------------------------
#    Time: 0 ns  Iteration: 0
# ** Note: The Verilog primitive is COMBINATIONAL.
#    Time: 0 ns  Iteration: 0
# ** Note: Number of inputs : 3
#    Time: 0 ns  Iteration: 0
# ** Note: Number of lines  : 4
#    Time: 0 ns  Iteration: 0
# ** Note: ----------------------------------------
#    Time: 0 ns  Iteration: 0
list a b c s
```

```
run
 ns a b c s
  0 X X X X
  5 X X 0 X
  7 X X 0 0
 10 X 0 0 0
 15 X 0 1 0
 17 X 0 1 X
 20 X 1 1 X
 22 X 1 1 1
 25 0 1 1 1
 30 1 1 1 1
 35 1 X 1 1
 40 0 X 1 1
 42 0 X 1 X
 45 0 X 0 X
 47 0 X 0 0
```

During the initialization step (Time = 0), the Verilog file analysis prints some note messages: the primitive is combinational (and not sequential), the number of inputs is three, and four lines have been found in the table. This information is provided for debugging. The reader may check the consistency of the simulation results with the Verilog table definition.

Of course, the VHDL code included in the design units of the library VLIB is much more complicated. But once compiled, these design units become part of the resources usable by many designers.

The package VERILOG_PRIMITIVE is listed below:

```
use STD.TEXTIO.all ;
package VERILOG_PRIMITIVE is
type PRIMITIVE_BIT is ('?', 'B', 'X', '0', '1', 'Z') ;
-- original bit level type including don't-care elements
subtype XBIT is PRIMITIVE_BIT range 'X' to '1' ;
type XBIT_VECTOR is array (NATURAL range <>) of XBIT ;
-- port subtype and array type
subtype ZXBIT is PRIMITIVE_BIT range 'X' to 'Z';
type ZXBIT_VECTOR is array ( NATURAL range <> ) of ZXBIT ;
-- four values subtype
function ZTOX ( VALUE : ZXBIT ) return XBIT ;
-- conversion function between ZXBIT and XBIT ;
subtype DXBIT is PRIMITIVE_BIT range '?' to '1' ;
type DXBIT_VECTOR is array (NATURAL range <>) of DXBIT ;
-- possible level values of a Verilog primitive table
type TABLE_INPUT is
 record
      IS_EDGE :   BOOLEAN ;
      LEVEL   :   DXBIT ;
      EDGE    :   DXBIT_VECTOR(1 to 2) ;
 end record ;
```

```
-- Verilog primitive table input (level or edge)
type TABLE_INPUT_VECTOR is array (NATURAL range <>) of TABLE_INPUT ;
type TABLE_INPUT_PTR is access TABLE_INPUT_VECTOR ;
subtype TABLE_STATE is DXBIT ;
-- Verilog primitive table state
type TABLE_OUTPUT is
 record
      NO_CHANGE :   BOOLEAN ;   -- TRUE if output specified with '-'
      LEVEL     :   XBIT ;      -- new output value (if NO_CHANGE = FALSE)
 end record ;
-- Verilog primitive table output
type TABLE_LINE is
 record
      HAS_EDGE  :   BOOLEAN ;            -- TRUE if edge specified at this line
      INPUTS    :   TABLE_INPUT_PTR ;    -- input values
      STATE     :   TABLE_STATE ;        -- state value
      OUTPUT    :   TABLE_OUTPUT ;       -- output value
 end record ;
-- line of the Verilog primitive table
type TABLE_MATRIX is array(NATURAL range <>) of TABLE_LINE ;
-- Verilog primitive table
function EQ(V : XBIT ; W : DXBIT) return BOOLEAN ;
function EQ(V : XBIT_VECTOR ; W : DXBIT_VECTOR) return BOOLEAN ;
-- equivalence functions comparing level values
-- same functions as in the previous paragraph
function EQ(   V, LAST_V     :   XBIT_VECTOR ;
               W             :   TABLE_INPUT_VECTOR ;
               W_HAS_EDGE    :   BOOLEAN                 ) return BOOLEAN ;
-- equivalence function comparing lines of mixed level- and edge-values.
-- Note: LAST_V has been introduced because V(I) is not a static name,
-- thus V(I)'EVENT and V(I)'LAST_VALUE are illegal.
function NEXT_VALUE(  OUTPUT               :   TABLE_OUTPUT ;
                      CURRENT_STATE        :   XBIT          ) return XBIT ;
-- function returning the next output and/or state value (handles '-')
type PARAMETERS is
 record
      PRIMITIVE_IS_SEQUENTIAL   :   BOOLEAN ;
      NUMBER_OF_INPUTS          :   POSITIVE ;
      NUMBER_OF_TABLE_LINES     :   POSITIVE ;
 end record ;
-- parameters returned by function GET_PARAMETERS
function GET_PARAMETERS(FILENAME : STRING) return PARAMETERS ;
-- returns parameters after a first quick analysis of the Verilog file
function TABLE_DEFINITION(FILENAME : STRING) return TABLE_MATRIX ;
-- returns the table definition of the Verilog file
procedure DISPLAY_TABLE(      FILENAME         :   STRING ;
                              variable TABLE   :   in TABLE_MATRIX ;
                              IS_SEQUENTIAL    :   BOOLEAN             ) ;
-- debug procedure used to write the table content into a file
end VERILOG_PRIMITIVE ;
```

The package VERILOG_PRIMITIVE includes almost all the types and functions (except EDGE) previously declared in the package PRIMITIVE_PKG. Beside that, it declares:

- the types used to model the Verilog primitive table (TABLE_INPUT, TABLE_STATE, TABLE_OUTPUT, TABLE_LINE, TABLE_MATRIX);
- another equivalence function (EQ) comparing real inputs with a table input line (including levels and edges);
- a type PARAMETERS, which contains the preliminary information that is extracted from the Verilog file (if the primitive is sequential or combinational, number of inputs and number of table lines);
- functions that are used to analyze the Verilog file and extract parameters from it (GET_PARAMETERS, TABLE_DEFINITION) or to help debug (DISPLAY_TABLE).

The corresponding package body, including other subprogram definitions, is given below:

```
package body VERILOG_PRIMITIVE is
-- conversion function
function ZTOX ( VALUE : ZXBIT ) return XBIT is
begin
        case VALUE is
         when 'Z' =>         return 'X' ;
         when others =>   return VALUE ;
        end case ;
end ZTOX ;
-- equivalence functions comparing level values
function EQ(V : XBIT ; W : DXBIT) return BOOLEAN is
begin
        if W = V or W = '?' or (W = 'B' and V /= 'X') then
             return TRUE ;
        end if ;
        return FALSE ;
end EQ ;
function EQ(V : XBIT_VECTOR ; W : DXBIT_VECTOR) return BOOLEAN is
        variable WW : DXBIT_VECTOR(V'RANGE) ;
begin
        if V'LENGTH /= W'LENGTH then
             return FALSE ;
        else
             WW := W ;
             for I in V'RANGE loop
                  if not EQ(V(I),WW(I)) then return FALSE ; end if ;
             end loop ;
             return TRUE ;
        end if ;
end EQ ;
```

```
function POS(V: XBIT_VECTOR(1 to 2); W: DXBIT_VECTOR(1 to 2)) return BOOLEAN is
-- return TRUE if positive edge (01, 0X or X1) between V(1) and V(2) and code "11" in W
begin
        if W = "11" and (V = "01" or V = "0X" or V = "X1") then
            return TRUE ;
        else
            return FALSE ;
        end if ;
end ;
function NEG(V: XBIT_VECTOR(1 to 2); W: DXBIT_VECTOR(1 to 2)) return BOOLEAN is
-- return TRUE if negative edge (10, 1X or X0) between V(1) and V(2) and code "00" in W
begin
        if W = "00" and (V = "10" or V = "1X" or V = "X0") then
            return TRUE ;
        else
            return FALSE ;
        end if ;
end ;

function EQ(    V,  LAST_V      :   XBIT_VECTOR ;
                W               :   TABLE_INPUT_VECTOR ;
                W_HAS_EDGE      :   BOOLEAN                ) return BOOLEAN is
-- equivalence function comparing lines of mixed level- and edge-values.
-- LAST_V has been introduced because V(I) is not a static name,
-- thus V(I)'EVENT and V(I)'LAST_VALUE are illegal.
        variable VI_EDGE : XBIT_VECTOR(1 to 2) ;
begin
        assert V'LENGTH = LAST_V'LENGTH and V'LENGTH = W'LENGTH
            and V'LEFT = LAST_V'LEFT and V'LEFT = W'LEFT
            and V'RIGHT = LAST_V'RIGHT and V'RIGHT = W'RIGHT
        report "Internal Bug in EQ"
        severity FAILURE;
        for I in V'RANGE loop
            if W(I).IS_EDGE then  -- edge expected
                VI_EDGE := LAST_V(I) & V(I) ;
                if V(I) = LAST_V(I)  -- no edge
                or not (   EQ(VI_EDGE, W(I).EDGE) or
                           POS(VI_EDGE, W(I).EDGE) or
                           NEG(VI_EDGE, W(I).EDGE) ) then
                    return FALSE ;
                end if ;
            else  -- stable input expected
                if (W_HAS_EDGE and (V(I) /= LAST_V(I)) ) -- edge
                or not EQ( V(I), W(I).LEVEL ) then
                    return FALSE ;
                end if ;
            end if ;
        end loop ;
        return TRUE ;
end EQ ;
```

```
function NEXT_VALUE( OUTPUT           :  TABLE_OUTPUT ;
                     CURRENT_STATE    :  XBIT          ) return XBIT is
-- function returning the next state value (handles '-')
begin
    if OUTPUT.NO_CHANGE then
        return CURRENT_STATE ;
    else
        return OUTPUT.LEVEL ;
    end if ;
end NEXT_VALUE ;

procedure CHECK(CONDITION : in BOOLEAN ; MESSAGE : in STRING) is
begin
    assert CONDITION report MESSAGE severity FAILURE ;
end CHECK;

procedure PRINT(MESSAGE : in STRING) is
begin
    assert FALSE report MESSAGE severity NOTE ;
end PRINT;

procedure GET_NEXT_CHAR(  variable F   :  in TEXT ;
                          L            :  inout LINE ;
                          CHAR         :  out CHARACTER    ) is
    variable C,C2 : CHARACTER ;
    variable L2 : LINE ;
    variable OK : BOOLEAN ;
begin
    while not ENDFILE(F) and L'LENGTH = 0 loop
        READLINE(F,L) ;
    end loop ;
    CHECK(not ENDFILE(F),"GET_NEXT_CHAR : Unexpected end of Verilog file");
    READ(L,C,OK) ;
    CHECK(OK,"GET_NEXT_CHAR : Unexpected error during Verilog file processing.") ;
    if C >= '!' and C <= '~' then        -- blanks and escape characters are removed
        if C = '/' then                  -- discard line of comments
            L2 := new STRING'(L.all) ;
            READ(L2,C2,OK) ;
            DEALLOCATE(L2) ;
            case C2 is
             when '/' => READLINE(F,L) ;                  -- Verilog comment:
                         GET_NEXT_CHAR(F,L,CHAR) ;        -- continue on next line
             when '*' => CHECK(FALSE,"Program not handling C comments (/* .. */)"
                              & ": use comments starting with // instead!") ;
             when others => CHAR := C ;
            end case ;
        else
            CHAR := C ;
        end if ;
    else
```

```
            GET_NEXT_CHAR(F,L,CHAR) ;
        end if ;
end GET_NEXT_CHAR ;

function STRING_EQ( S : STRING ; CAP_S : STRING) return BOOLEAN is
-- not case sensitive string comparison
-- CAP_S is the reference string supposed to be in CAPITAL letters
        variable CAP_STR : STRING(S'RANGE) ;
begin
        if S'LENGTH /= CAP_S'LENGTH then
            return FALSE ;
        else
            CAP_STR := CAP_S ;
            for I in S'RANGE loop
                if S(I) /= CAP_STR(I) and
                S(I) /= CHARACTER'VAL(CHARACTER'POS(CAP_STR(I)) + 32)
                then
                    return FALSE ;
                end if ;
            end loop ;
            return TRUE ;
        end if ;
end STRING_EQ ;

function IMAGE(N : NATURAL) return STRING is
-- converts NATURAL into STRING (for error messages)
        variable L : LINE ;
begin
        WRITE(L,N) ;
        return L.all ;
end IMAGE ;

type TABLE_TOKEN is (   T_OTHER,       -- unexpected token
                        T_LEVEL,       -- level, i.e ?, X, 0, 1,... (DXBIT values)
                        T_EDGE,        -- edge, i.e (01), R, (??), *, P, ...
                        T_MINUS,       -- no state change (-)
                        T_COLON,       -- ':' separating input, state and output
                        T_SEMICOL ) ;  -- ';' ending each line
-- enumeration type used to specify tables values

procedure SEARCH_TABLE( variable VERILOG_FILE   :   in TEXT ;
                        variable CURRENT_LINE   :   inout LINE    ) is
-- searches the line BEGINNING with the keyword TABLE
        variable KEYWORD : STRING(1 to 5) ;
        variable C : CHARACTER := ' ';
begin
        while not ENDFILE(VERILOG_FILE) loop
            READLINE(VERILOG_FILE,CURRENT_LINE) ;
            GET_NEXT_CHAR(VERILOG_FILE, CURRENT_LINE, C) ;
            if C = 'T' or C = 't' then  -- looking for keyword TABLE
                KEYWORD(1) := C ;
```

```
                READ(CURRENT_LINE, KEYWORD(2 to 5));
                if STRING_EQ(KEYWORD, "TABLE") then
                    exit ;
                end if ;
            end if ;
        end loop ;
        CHECK(not ENDFILE(VERILOG_FILE),
            "SEARCH_TABLE: Keyword TABLE not found at the BEGINNING of a line...") ;
        while not ENDFILE(VERILOG_FILE) and CURRENT_LINE'LENGTH = 0 loop
            READLINE(VERILOG_FILE,CURRENT_LINE) ;
        end loop ;
        -- CURRENT_LINE is now addressing the first entry of the table
        CHECK(not ENDFILE(VERILOG_FILE), "SEARCH_TABLE: Unexpected end of file...") ;
end SEARCH_TABLE ;

-- legal Verilog characters in primitive tables are:
-- '?', 'B', 'b', 'X', 'x', '0', '1', '-', '(', ')', 'R', 'r', 'F', 'f', 'P', 'p', 'N', 'n', '*', ':', ';'

procedure CHECK_IF_LEVEL(   CHAR   :   in  CHARACTER ;
                            OK     :   out BOOLEAN ;
                            LEVEL  :   out DXBIT          ) is
begin
        case CHAR is
         when '?'       => OK := TRUE ; LEVEL := '?' ;
         when 'B' | 'b' => OK := TRUE ; LEVEL := 'B' ;
         when 'X' | 'x' => OK := TRUE ; LEVEL := 'X' ;
         when '0'       => OK := TRUE ; LEVEL := '0' ;
         when '1'       => OK := TRUE ; LEVEL := '1' ;
         when others => OK := FALSE ;
        end case ;
end CHECK_IF_LEVEL ;

procedure NEXT_TABLE_VALUE (
                        variable VERILOG_FILE    :   in TEXT ;
                            CURRENT_LINE         :   inout LINE ;
                            IS_SEQUENTIAL        :   in BOOLEAN ;
                            TOKEN                :   out TABLE_TOKEN;
                            LEVEL                :   out DXBIT ;
                            EDGE                 :   out DXBIT_VECTOR(1 to 2) ) is
-- retrieves the next table value
        variable CHAR : CHARACTER ;
        variable OK : BOOLEAN ;
begin
        CHECK(not ENDFILE(VERILOG_FILE),"Next_Table_Value: unexpected end of file");
        GET_NEXT_CHAR(VERILOG_FILE, CURRENT_LINE, CHAR) ;
        CHECK_IF_LEVEL(CHAR, OK, LEVEL) ;
        if OK then
            TOKEN := T_LEVEL ;
        elsif CHAR = ':' then
            TOKEN := T_COLON ;
        elsif CHAR = ';' then
```

```
            TOKEN := T_SEMICOL ;
        elsif not IS_SEQUENTIAL then
            CHECK(FALSE, "Illegal character used in Verilog combinational table definition!");
        elsif CHAR = '-' then
            TOKEN := T_MINUS ;
        elsif CHAR = '(' then
            GET_NEXT_CHAR(VERILOG_FILE, CURRENT_LINE, CHAR) ;
            CHECK_IF_LEVEL(CHAR, OK, EDGE(1)) ;
            CHECK(OK,"Unexpected level value in edge specification") ;
            GET_NEXT_CHAR(VERILOG_FILE, CURRENT_LINE, CHAR) ;
            CHECK_IF_LEVEL(CHAR, OK, EDGE(2)) ;
            CHECK(OK,"Unexpected level value in edge specification") ;
            GET_NEXT_CHAR(VERILOG_FILE, CURRENT_LINE, CHAR) ;
            CHECK(CHAR = ')',"Missing ) in edge specification") ;
            TOKEN := T_EDGE ;
        elsif CHAR = 'R' or CHAR = 'r' then
            EDGE := "01" ;
            TOKEN := T_EDGE ;
        elsif CHAR = 'F' or CHAR = 'f' then
            EDGE := "10" ;
            TOKEN := T_EDGE ;
        elsif CHAR = '*' then
            EDGE := "??" ;
            TOKEN := T_EDGE ;
        elsif CHAR = 'P' or CHAR = 'p' then
            EDGE := "11" ; -- code interpreted by function POS
            TOKEN := T_EDGE ;
        elsif CHAR = 'N' or CHAR = 'n' then
            EDGE := "00" ; -- code interpreted by function NEG
            TOKEN := T_EDGE ;
        elsif CHAR = ')' then
            CHECK(FALSE,"Illegal edge specification: unexpected ')'") ;
        else
            CHECK(FALSE ,"Illegal character used in Verilog sequential table definition!") ;
        end if ;
end NEXT_TABLE_VALUE ;

function GET_PARAMETERS(FILENAME : STRING) return PARAMETERS is
        variable P : PARAMETERS ;
        file F : TEXT is in FILENAME ;
        variable C : CHARACTER ;
        variable L, L2 : LINE ;
        variable NUMBER : NATURAL ;
        variable TOKEN : TABLE_TOKEN ;
        variable LEVEL : DXBIT ;
        variable EDGE : DXBIT_VECTOR(1 to 2) ;
        variable KEYWORD : STRING(1 to 8) ;
begin
        SEARCH_TABLE(F, L) ;
        -- count number of inputs
        while L'LENGTH > 0 loop
```

```
            NEXT_TABLE_VALUE(F, L, TRUE, TOKEN, LEVEL, EDGE) ;
            case TOKEN is
             when T_LEVEL | T_EDGE =>  NUMBER := NUMBER + 1 ;
             when T_COLON          =>  exit ;
             when others           =>  CHECK(FALSE, "Unexpected input entry in first"
                                                 & " line of the Verilog table");
            end case ;
        end loop ;
        CHECK(L'LENGTH > 0,"Illegal first line of the Verilog table (':' missing)!");
        CHECK(NUMBER >= 1, "Zero input in Verilog Table!") ;
        P.NUMBER_OF_INPUTS := NUMBER ;
        -- PRINT("Found number of inputs equal to " & IMAGE(NUMBER) & ".");
        NUMBER := 0 ;
        NEXT_TABLE_VALUE(F, L, TRUE, TOKEN, LEVEL, EDGE) ;
        CHECK(TOKEN = T_LEVEL,
                  "Illegal state or output specification in first line of the table!") ;
        NEXT_TABLE_VALUE(F, L, TRUE, TOKEN, LEVEL, EDGE) ;
        -- decide if primitive is combinational or sequential
        case TOKEN is
         when T_SEMICOL    => P.PRIMITIVE_IS_SEQUENTIAL := FALSE ;
                              NUMBER := 1 ;
         when T_COLON      => P.PRIMITIVE_IS_SEQUENTIAL := TRUE ;
         when others       => CHECK(FALSE,"Unexpected end of first line in the table!");
        end case ;
        -- count number of lines in the table
        while not ENDFILE(F) loop
            GET_NEXT_CHAR(F, L, C) ;
            case C is
             when ';'       =>  NUMBER := NUMBER + 1 ;
             when 'E' | 'e' =>  KEYWORD(1) := C ;
                                L2 := new STRING'(L.all) ;
                                -- looking for keyword ENDTABLE
                                READ(L2, KEYWORD(2 to 8));
                                DEALLOCATE(L2) ;
                                if STRING_EQ(KEYWORD ,"ENDTABLE") then
                                    exit ;
                                end if ;
             when others =>     null;
            end case ;
        end loop ;
        P.NUMBER_OF_TABLE_LINES := NUMBER ;
        CHECK(not ENDFILE(F),"Keyword ENDTABLE not found during Verilog file analysis!");
        return P ;
end GET_PARAMETERS ;

function TABLE_DEFINITION(FILENAME : STRING) return TABLE_MATRIX is
        constant P : PARAMETERS := GET_PARAMETERS(FILENAME) ;
        constant NBINPUT : POSITIVE := P.NUMBER_OF_INPUTS ;
        constant NBLINE : POSITIVE := P.NUMBER_OF_TABLE_LINES ;
        constant IS_SEQ : BOOLEAN := P.PRIMITIVE_IS_SEQUENTIAL ;
        variable TABLE : TABLE_MATRIX(1 to NBLINE) ;
```

```
        -- content of the Verilog primitive table
        file VERILOG_FILE : TEXT is in FILENAME ;
        variable CURRENT_LINE : LINE ;
        variable TOKEN : TABLE_TOKEN ;
        variable LEVEL : DXBIT ;
        variable EDGE : DXBIT_VECTOR(1 to 2) ;
    begin
        SEARCH_TABLE(VERILOG_FILE,CURRENT_LINE) ;
        PRINT("----------------------------------------") ;
        if IS_SEQ then
            PRINT("The Verilog primitive is SEQUENTIAL.") ;
        else
            PRINT("The Verilog primitive is COMBINATIONAL.") ;
        end if ;
        PRINT("Number of inputs: " & IMAGE(NBINPUT));
        PRINT("Number of lines: " & IMAGE(NBLINE));
        PRINT("----------------------------------------") ;
        -- read the table content and store it into variable TABLE
        for I in 1 to NBLINE loop
            TABLE(I).INPUTS := new TABLE_INPUT_VECTOR(1 to NBINPUT) ;
            for J in 1 to NBINPUT loop
                NEXT_TABLE_VALUE(VERILOG_FILE,CURRENT_LINE,
                                        IS_SEQ,TOKEN,LEVEL,EDGE) ;
                CHECK(TOKEN = T_LEVEL or TOKEN = T_EDGE, "Unexpected " &
                        IMAGE(J) & "th input entry at line " & IMAGE(I) & " of the table!") ;
                if TOKEN = T_LEVEL then
                    TABLE(I).INPUTS(J).LEVEL := LEVEL ;
                elsif TABLE(I).HAS_EDGE then  -- TOKEN = second T_EDGE
                    CHECK(FALSE,"Error in table: two edges at line " & IMAGE(I) & "!") ;
                else  -- TOKEN = first T_EDGE
                    TABLE(I).INPUTS(J).IS_EDGE := TRUE ;
                    TABLE(I).INPUTS(J).EDGE := EDGE ;
                    TABLE(I).HAS_EDGE := TRUE ;
                end if ;
            end loop ;
            NEXT_TABLE_VALUE(VERILOG_FILE,CURRENT_LINE,
                                    IS_SEQ,TOKEN,LEVEL,EDGE) ;
            CHECK(TOKEN = T_COLON,"Number of inputs not correct at line " &
                                    IMAGE(I) & " of the table (expecting ':')!") ;
            if IS_SEQ then
                NEXT_TABLE_VALUE(VERILOG_FILE,CURRENT_LINE,
                                        IS_SEQ,TOKEN,LEVEL,EDGE) ;
                CHECK(TOKEN = T_LEVEL,"Illegal state value at line " & IMAGE(I) &
                                        " of the table (expecting level value)!") ;
                TABLE(I).STATE := LEVEL ;
                NEXT_TABLE_VALUE(VERILOG_FILE,CURRENT_LINE,
                                        IS_SEQ,TOKEN,LEVEL,EDGE) ;
                CHECK(TOKEN = T_COLON,"State field not correct at line " & IMAGE(I) &
                                        " of the table (expecting ':')!") ;
            end if ;
            NEXT_TABLE_VALUE(VERILOG_FILE,CURRENT_LINE,
```

```
                                    IS_SEQ,TOKEN,LEVEL,EDGE) ;
        if IS_SEQ then
            CHECK(TOKEN=T_MINUS or
                    (TOKEN=T_LEVEL and LEVEL>='X' and LEVEL<='1'),
                    "Illegal output value at line " & IMAGE(I) & " of the table!") ;
            if TOKEN = T_MINUS then
                TABLE(I).OUTPUT.NO_CHANGE := TRUE ;
            else
                TABLE(I).OUTPUT.LEVEL := LEVEL ;
            end if ;
        else
            CHECK(TOKEN=T_LEVEL and LEVEL>='X' and LEVEL<='1',
                    "Illegal output value at line " & IMAGE(I) & " of the table!") ;
            TABLE(I).OUTPUT.LEVEL := LEVEL ;
        end if ;
        NEXT_TABLE_VALUE(VERILOG_FILE,CURRENT_LINE,
                                    IS_SEQ,TOKEN,LEVEL,EDGE) ;
        CHECK(TOKEN = T_SEMICOL,"Unexpected end of line " & IMAGE(I) &
                                        " of the table (expecting ';')!") ;
    end loop ;
    return TABLE ;
end TABLE_DEFINITION ;

function IMAGE(V : PRIMITIVE_BIT) return CHARACTER is
-- converts PRIMITIVE_BIT into CHARACTER
begin
    case V is
     when '?'    =>  return '?' ;
     when 'B'    =>  return 'B' ;
     when 'X'    =>  return 'X' ;
     when '0'    =>  return '0' ;
     when '1'    =>  return '1' ;
     when 'Z'    =>  return 'Z' ;
    end case ;
end IMAGE ;

procedure DISPLAY_TABLE(     FILENAME          :    STRING ;
                             variable TABLE    :    in TABLE_MATRIX ;
                             IS_SEQUENTIAL     :    BOOLEAN      ) is
-- debug procedure used to write the table content into a file
    file F : TEXT is out FILENAME ;
    variable L : LINE ;
    procedure SWRITE(L : inout LINE ; S : STRING) is
    begin  -- to resolve ambiguities of the TEXTIO WRITE procedures
        WRITE(L,S) ;
    end;
begin
    SWRITE(L,"--- Verilog Table  ---") ; WRITELINE(F,L) ;
    SWRITE(L," ") ; WRITELINE(F,L) ;
    for I in TABLE'RANGE loop
        for J in TABLE(I).INPUTS'RANGE loop
```

```
                if TABLE(I).INPUTS(J).IS_EDGE then
                    SWRITE(L," (" & IMAGE(TABLE(I).INPUTS(J).EDGE(1))
                            & IMAGE(TABLE(I).INPUTS(J).EDGE(2)) & ") ") ;
                else
                    SWRITE(L,"   " & IMAGE(TABLE(I).INPUTS(J).LEVEL) & "  ") ;
                end if ;
            end loop ;
            SWRITE(L, " : ") ;
            if IS_SEQUENTIAL then
                SWRITE(L, IMAGE(TABLE(I).STATE) & " : ") ;
            end if ;
            if TABLE(I).OUTPUT.NO_CHANGE then
                SWRITE(L, "- ;") ;
            else
                SWRITE(L, IMAGE(TABLE(I).OUTPUT.LEVEL) & " ;") ;
            end if ;
            WRITELINE(F,L) ;
        end loop ;
        SWRITE(L," ") ; WRITELINE(F,L) ;
        SWRITE(L, "--- End of Table ---") ; WRITELINE(F,L) ;
end DISPLAY_TABLE ;

end VERILOG_PRIMITIVE ;
```

Much simpler, the generic entity PRIMITIVE is given below:

```
library VLIB ; use VLIB.VERILOG_PRIMITIVE.all ;
entity PRIMITIVE is
        generic ( FILENAME    :    STRING ;
                  -- name of the ASCII file containing the primitive table
                  DELAY       :    TIME := 1 ns
                  -- output delay
        ) ;
        port (    OUTPUT      :    out XBIT ;
                  -- primitive output
                  INPUT       :    in  XBIT_VECTOR
                  -- primitive inputs
        ) ;
end PRIMITIVE ;
```

The first generic parameter is the name of the Verilog file, and the second is a propagation delay applied to any output change. As the original Verilog primitive, this entity has only one output port. The input port is an unconstrained array: the constraint applied will be the number of input ports of the Verilog primitive. The VHDL designer may thus use this generic entity for any Verilog primitive.

The architecture of this entity is the following:

```
architecture VERILOG of PRIMITIVE is
        constant P : PARAMETERS := GET_PARAMETERS(FILENAME) ;
        constant IS_SEQ : BOOLEAN := P.PRIMITIVE_IS_SEQUENTIAL ;
        -- true if primitive is sequential, false if combinational
        constant NBINPUT : POSITIVE := P.NUMBER_OF_INPUTS ;
        -- number of primitive inputs
        constant NBLINE : POSITIVE := P.NUMBER_OF_TABLE_LINES ;
        -- number of lines specified in the table
begin
        process
            variable TABLE : TABLE_MATRIX(1 to NBLINE) :=
                                        TABLE_DEFINITION(FILENAME) ;
            -- primitive table
            variable STATE : XBIT ;
            -- current state (for sequential primitive)
            variable LAST_INPUT : XBIT_VECTOR(1 to NBINPUT) ;
            -- input values at the last process activation
            variable OUTPUT_VALUE : XBIT ;
            -- new output value
            variable MATCH : BOOLEAN ;
            -- TRUE if INPUT matches a table entry
            variable WAIT_LEVEL_ENTRY : BOOLEAN ;
            -- TRUE if a matching edge-sensitive entry of the table has been found
        begin
          LAST_INPUT := INPUT ;
         assert NBINPUT = INPUT'LENGTH
            report "Number of inputs not consistent with the Verilog table!"
            severity FAILURE ;
          DISPLAY_TABLE("check.out",TABLE,IS_SEQ) ;  -- for debug
         loop
            MATCH := FALSE ;
            WAIT_LEVEL_ENTRY := FALSE ;
            -- scan all the lines of the primitive table
            for I in 1 to NBLINE loop
                -- check if line I is matching the real input and state values
                if EQ(INPUT,LAST_INPUT,TABLE(I).INPUTS.all,TABLE(I).HAS_EDGE)
                and (not IS_SEQ or EQ(STATE,TABLE(I).STATE)) then
                    if not IS_SEQ or not TABLE(I).HAS_EDGE then
                        OUTPUT_VALUE := NEXT_VALUE(TABLE(I).OUTPUT,STATE);
                        MATCH := TRUE ;
                        exit ;
                    elsif not WAIT_LEVEL_ENTRY then
                        OUTPUT_VALUE := NEXT_VALUE(TABLE(I).OUTPUT,STATE);
                        MATCH := TRUE ;
                        WAIT_LEVEL_ENTRY := TRUE ;
                    end if ;
                end if ;
            end loop ;
            if not MATCH then
                OUTPUT_VALUE := 'X' ;
            end if ;
```

```
            if IS_SEQ then
                STATE := OUTPUT_VALUE ;
            end if ;
            OUTPUT <= OUTPUT_VALUE after DELAY ;
            LAST_INPUT := INPUT ;
            wait on INPUT ;
        end loop ;
    end process ;
end VERILOG ;
```

As explained previously, this architecture consists of a single process that first analyzes the Verilog file at the initialization and then, at any change of an input port, compares the input values to each line of the table. In case of conflicting table entries, according to the Verilog reference manual, a level-sensitive entry always dominates an edge-sensitive entry: this explains the use of the boolean WAIT_LEVEL_ENTRY. Finally, the default value 'X' is applied to the output if no match is encountered, and the state variable is updated if the primitive is sequential.

A first example of the use of these design units has been given above with the combinational primitive ANDOR. Below are listed two other examples that have already been discussed in the previous paragraph: the primitives LATCH and FLIPFLOP. For each of them, the Verilog file, the corresponding VHDL testing unit, and simulation results are listed.

- **The sequential level-sensitive LATCH**

Verilog File	VHDL Testing Unit

// File latch.ver
primitive LATCH(Q,CLOCK,D);
    output Q ;
    reg Q ;
    input CLOCK, D ;
    table
    // clock d state q
        1 1 :?: 1 ;
        1 0 :?: 0 ;
        0 ? :?: - ;
    endtable
endprimitive
	library VLIB ; use VLIB.VERILOG_PRIMITIVE.all ;
	entity ENT is end;
	architecture ARC of ENT is
	    component LATCH
	        port (Q : out XBIT; CLOCK, D : in XBIT);
	    end component;
	    for all : LATCH use entity VLIB.PRIMITIVE(VERILOG)
	        generic map ("latch.ver",1 ns)
	        port map (OUTPUT => Q,
	                INPUT(1) => CLOCK, INPUT(2) => D);
	    signal Q : XBIT ;
	    signal D, CK : XBIT ;
	begin
	    I : LATCH port map (Q,CK,D) ;
	    D <= '0' after 5 ns, '1' after 15 ns, 'X' after 25 ns;
	    CK <= '0' after 1 ns, '1' after 10 ns,
	            '0' after 20 ns, 'X' after 30 ns;
	end ;

```
# ** Note: ----------------------------------------
# ** Note: The Verilog primitive is SEQUENTIAL.
# ** Note: Number of inputs : 2
# ** Note: Number of lines  : 3
# ** Note: ----------------------------------------
list ck d q
run
  ns  ck d q
   0  X X X
   1  0 X X
   5  0 0 X
  10  1 0 X
  11  1 0 0
  15  1 1 0
  16  1 1 1
  20  0 1 1
  25  0 X 1
  30  X X 1
  31  X X X
```

• **The sequential edge-sensitive FLIPFLOP**

Verilog File	VHDL Testing Unit

// File flipflop.ver
primitive FLIPFLOP(Q,CLOCK,D);
    output Q ;
    reg Q ;
    input CLOCK, D;
    table
    // clock d state q
      (01) 1 : ? : 1 ;
      R 0 : ? : 0 ;
      (0X) 1 : 1 : 1 ;
      (0x) 0 : 0 : 0 ;
      (?0) ? : ? : - ;
      ? (??) : ? : - ;
    endtable
endprimitive
	library VLIB ; **use** VLIB.VERILOG_PRIMITIVE.**all**;
	entity ENT **is end**;
	architecture ARC **of** ENT **is**
	    component FLIPFLOP
	        port (Q : **out** XBIT; CLOCK, D : **in** XBIT);
	    end component;
	    for all : FLIPFLOP **use entity**
	        VLIB.PRIMITIVE(VERILOG)
	        generic map ("flipflop.ver",1 ns)
	        port map (OUTPUT => Q, INPUT(1) => CLOCK,
	                INPUT(2) => D);
	    signal Q : XBIT ;
	    signal D, CK : XBIT ;
	begin
	    I : FLIPFLOP **port map** (Q,CK,D) ;
	    D <= '0' **after** 5 ns, '1' **after** 15 ns, 'X' **after** 25 ns,
	        '0' **after** 30 ns, '1' **after** 45 ns, '0' **after** 65 ns,
	        '1' **after** 80 ns , '0' **after** 88 ns ;
	    CK <= '0' **after** 2 ns, '1' **after** 10 ns, '0' **after** 20 ns,
	        'X' **after** 35 ns, '0' **after** 40 ns, '1' **after** 50 ns,
	        'X' **after** 55 ns , '0' **after** 58 ns, '1' **after** 60 ns,
	        'X' **after** 70 ns, '0' **after** 73 ns, '1' **after** 75 ns,
	        '0' **after** 80 ns, '1' **after** 85 ns, '0' **after** 90 ns,
	        'X' **after** 95 ns ;
	end;

```
# ** Note: -----------------------------------
# ** Note: The Verilog primitive is SEQUENTIAL.
# ** Note: Number of inputs : 2
# ** Note: Number of lines  : 6
# ** Note: -----------------------------------
list ck d q
run
 ns ck d q
  0  X X X
  2  0 X X
  5  0 0 X
 10  1 0 X
 11  1 0 0
 15  1 1 0
 20  0 1 0
 25  0 X 0
 30  0 0 0
 35  X 0 0
 40  0 0 0
 45  0 1 0
 50  1 1 0
 51  1 1 1
 55  X 1 1
 56  X 1 X
 58  0 1 X
 60  1 1 X
 61  1 1 1
 65  1 0 1
 70  X 0 1
 71  X 0 X
 73  0 0 X
 75  1 0 X
 76  1 0 0
 80  0 1 0
 81  0 1 X
 85  1 1 X
 86  1 1 1
 88  1 0 1
 90  0 0 1
 95  X 0 1
 96  X 0 X
```

A means to use the Verilog table syntax in VHDL has been presented in this paragraph. Besides its practical aspect, the detailed methodology also highlights some powerful programming features of VHDL: the predefined TEXTIO package allows the VHDL designer to read all kinds of ASCII files, even written in other languages such as Verilog, and to extract modeling information from them like truth-table or ROM specifications.

Of course, the goal was not to write a Verilog analyzer in VHDL: procedures and functions of the package VERILOG_PRIMITIVE are much simpler than subprograms written for a real compiler. But VHDL provides enough features to make it possible.

Finally, this paragraph also illustrates the benefits to the designer of working with predefined libraries (such as VLIB). These libraries may include sophisticated design units that needed some modeling investment: their reusability is thus of great interest.

9.8.2. Gate- and Switch-Level Modeling, Path Delay Specifications

Verilog has a certain number of predefined features that make it more adapted to gate- and switch-level modeling than VHDL:

- twenty-six predefined unidirectional and bidirectional logic gates and switch devices (see paragraph 9.5.6) ;
- strength definitions using a scale of eight values:

Description	Keyword	Level
high impedance (for gate output, continuous assign)	highz0 - highz1	0
small capacitor (for **trireg** net)	small0 - small1	1
medium capacitor (for **trireg** net)	medium0 - medium1	2
weak drive (for gate output, continuous assign)	weak0 - weak1	3
large capacitor (for **trireg** net)	large0 - large1	4
pull drive (for gate output, continuous assign)	pull0 - pull1	5
strong drive (for gate output, continuous assign)	strong0 - strong1	6
supply drive (for gate output, continuous assign)	supply0 - supply1	7

- a resolution mechanism based on strength comparisons and on the use of range of possible strengths and states (*ambiguous strengths*).

Modeling these constructs and mechanisms in VHDL is possible but will not lead to the same simulation efficiency (speed): this is due to the fact that Verilog predefined gates and strength definitions are part of the language itself, so they can be directly handled and optimized by a Verilog simulation kernel. In order to obtain comparable performances, it could be a good strategy to develop VHDL packages adapted for such modeling, especially if these packages become standardized (e.g., the STD_LOGIC_1164 package: see chapter 4, "Structuring the Environment"). In that case, VHDL simulators could directly "recognize" these specific features and optimize gate output computations and strength resolutions.

Verilog also provides a very concise way to assign delays to paths across a module. The Verilog designer can use a *specify block* in order to describe paths in a module and to assign delays to these paths. Two kinds of path description are available:

• parallel module paths (=>), establishing parallel bit-to-bit connections;
• full module paths (*>), establishing full bit-to-vector connections.

For example, if D and Q are two vectors of four bits (index 0 to 3), then (D => Q) specifies the four paths D[0] to Q[0], D[1] to Q[1], D[2] to Q[2], and D[3] to Q[3], whereas (D *> Q) specifies the 16 possible paths D[0] to Q[0], D[0] to Q[1],..., D[3] to Q[2], D[3] to Q[3].

Considering a module LATCH having D[0:3] and CLK as input ports and Q[0:3] as output port, the following specify block may be written in Verilog:

```
specify
        (CLK *> Q) = (TRISE_CLK_Q, TFALL_CLK_Q) ;
        // four paths : CLK to Q[0], CLK to Q[1], CLK to Q[2] and CLK to Q[3].
        (D => Q) = (TRISE_D_Q, TFALL_D_Q) ;
        // four paths : D[0] to Q[0], D[1] to Q[1], D[2] to Q[2] and D[3] to Q[3].
endspecify
```

If this module also contains internal delays (*distributed delays*) — due to gate propagations, for example — then the highest value between the sum of distributed internal delays and the corresponding port-to-port path delay (if any) is used for simulation. Delays used in a specify block have the same syntax as the one explained in 9.6.2.2.

This mechanism is very convenient and powerful to specify delays across complex modules. Using similar delay specifications in VHDL is not straightforward: VHDL timing information can only be included in signal assignments or in "**wait for**" statements, and they cannot be completely separated from output computations, as is the case in Verilog with the specify block, which can also override internal delays. Finally, as stated previously for gate- and switch-level modeling, introducing a complex mechanism in VHDL in order to implement these Verilog features will inevitably lead to some simulation performance drawbacks.

9.9. CONCLUSION

Verilog and VHDL share some common characteristics: for example, clear separation between concurrent and sequential domains, concurrent semantic (delta-delay mechanism), and software constructs (subprograms). Many descriptions using either always statements and continuous assignments (in Verilog) or process statements and concurrent signal assignments (in VHDL) will seem very similar at first glance.

Nevertheless, there are some fundamental differences between the two languages. First of all, Verilog has many more predefined constructs than VHDL: predefined operators, predefined basic gates, predefined resolution functions, and so forth. The use of Verilog will perhaps look simpler for the IC designer who can find in this language all the basic constructs he needs. In VHDL, even the basic gates like the nand or not gates have to be explicitly defined. As is often stated in this book, every modeling methodology based on VHDL must first include the writing of commonly used packages and entities: the designer must start circuit modeling only when he has this set of precompiled design units at his disposal. Modeling from scratch is possible in Verilog, but really not efficient in VHDL.

Secondly, Verilog can be used to model devices at a lower level than VHDL. Indeed, this language includes switch-level modeling, while the description of bidirectional gates (although possible) is really not efficient and usable in VHDL. Verilog also provides a very convenient truth-table syntax for primitive descriptions: this syntax is very concise and straightforward for the IC designer, but also allows some speed optimizations in the simulator. With such features, the simulation of logic descriptions will be quicker in Verilog than in VHDL.

All these Verilog capabilities also make this language easier to use as a synthesis description language. The *don't-care* notion is provided by Verilog only (symbol '?' in case statements and primitive descriptions). VHDL allows many ways to describe the same devices; therefore it needs conventions (and subset definitions) to be used by synthesis tools (see section 3.10 of chapter 3).

Finally, and this is the counterpart of the previous differences, VHDL provides more constructs for high-level modeling: abstract datatypes for variables and signals, textio procedures, packages, etc. VHDL is thus more appropriate than Verilog for functional or system-level modeling. The VHDL designer may define signals of any abstract type, use powerful sequential constructs, and so on. Of course, at a lower level, all these user-defined constructs may be less efficient in terms of simulation speed than the predefined Verilog constructs, but some types and operators could be *recognized* by an optimized VHDL simulator (this is already the case for some VHDL simulators: they provide VHDL packages defining logic types that match the *native* simulator logic types).

There is always a trade-off between the power of a language (high-level description capabilities, overloading, type verifications, etc.) and the simulation speed of descriptions written in this language. Restricting descriptions to the use of (many) predefined constructs or allowing (many) user-defined constructs leads to different modeling possibilities as well as to different simulation performances. Thus, giving a definitive verdict regarding a comparison between the two languages really makes no sense.

Verilog has already been used by many designers, but the effective use of VHDL as well as its acceptance as a standard are increasing rapidly. How both languages are and will be used, and for what kind of application (ICs, PCBS, System, etc.), may be very different from one company (or design department) to another.

10. UDL/I AND VHDL

10.1. INTRODUCTION

In the previous chapters, VHDL has been compared to M and Verilog, two hardware description languages sharing three main characteristics:

- they have been designed as languages for proprietary simulators;
- they are currently widely used by IC and/or PCB designers;
- they are native to the U.S.

From these points of view, the language UDL/I — Unified Design Language for Integrated circuits — is totally different from the other two. Indeed, this hardware description language is

- intended to be a standard, independent from any specific product;
- not currently used, its design being not completed;
- native to Japan.

It is certainly not necessary to recall that VHDL is an international standard, that its use by designers is increasing rapidly but started only a few years ago, and that the standardization process is mainly supported by a join effort of the U.S. and Europe.

The comparison of UDL/I with VHDL thus seems very interesting: both languages have the same original target, i.e., they were designed to become an exchange standard and do not depend on a specific CAD vendor, which was not true for M, and for Verilog in its first stage.

This common philosophy in their design could initially make the reader believe that these languages are very similar: as this chapter will show, they are, on the contrary, very different.

UDL/I has been developed since 1989 by the Japan Electronic Industry Development Association, depending on some major Japanese companies, such as NTT. A UDL/I committee is in charged of writing a language reference manual: the draft version number 1.0h4 of May 14, 1991 (in translation) is the basis of this chapter [UDL91].

One of the main characteristics of the *Unified Design Language for Integrated circuits* is that it is indeed very specific to IC modeling. Some features provided by VHDL have been customized in UDL/I to be particularly convenient for integrated circuit modeling. Some other VHDL features are largely or even totally missing in UDL/I.

The floor plan of this comparison is thus a mixture of VHDL and UDL/I notions: the first sections compare the design units, the sequential and concurrent domains, and the objects and types handled by both languages; the following sections are divided according to the different UDL/I description bodies (structural and behavior); and the last sections attempt to give equivalent VHDL templates for typical UDL/I modeling structures, such as the primitive description section and the automaton.

Note: the only reference for this comparative study has been the draft of the UDL/I language reference manual. No simulators are available for the time being. The reader must thus be aware that the UDL/I models written in this chapter have not been simulated and may become obsolete with a new release of the reference manual. Nevertheless, the reader may presume that the spirit and the possibilities of the UDL/I language will remain roughly unchanged.

10.2. DESIGN UNIT

The only design unit available in UDL/I is the *design description*, comprising several (sub)entities, which are the *module descriptions*. The design description contains the model of one integrated circuit, which consists of several subunits, or modules, each of them described by a module description.

A typical example of UDL/I design description is given below:

```
IDENT :      MICROPROCESSOR_32_BIT ;
DATE :       08/19/91 ;
VERSION :    AB1234.01 ;
AUTHOR :     Serge MAGINOT ;
COMMENT :    This is a comment which is stored in the database after compilation. ;

"This is a volatile comment: not stored in the database"
< module description >
...
< module description >
CEND ;
```

The first statement is compulsory and indicates the name of the chip. The four next statements, called common management statements in the reference manual, are optional: they provide a means to store the date, the version number, the author name, and any comment in the design database. Then follows any number of module descriptions.

In VHDL, there is no design unit specific to the whole integrated circuit. The *entity* is used to model the chip interface, an *architecture* to model its structure or behavior, and other entities and architectures (instantiated via components) to model the submodules of this chip.

The UDL/I common management statements (date, version, author, comment statements) are not implemented in VHDL. It is true that the VHDL designer may (and is strongly advised to) write such comments in his VHDL files, but these comments are not stored in the database after compilation. On the other hand, UDL/I management statements are real statements, and their information (including the comment) is kept in the database. As illustrated by the previous example, UDL/I provides also volatile comments (neither processed nor stored in the design database): such comments are surrounded by double quotes (").

Nevertheless, the UDL/I common management statements can be very properly implemented in VHDL via user-defined attributes. Indeed, if the following package is defined:

```
package MANAGEMENT_DEFINITIONS is
    type TIMESTAMP is record
        M : INTEGER range 1 to 12 ;
        D : INTEGER range 1 to 31 ;
        Y : INTEGER range 1980 to INTEGER'HIGH;
    end record ;
    subtype DECIMAL_DIGIT is CHARACTER range '0' to '9' ;
    type REVISION is array(1 to 2) of DECIMAL_DIGIT ;
    type VERSION_NUMBER is record    -- to be consistent with the UDL/I definition,
        V : STRING(1 to 6) ;         -- but the version number could be defined simpler.
        R : REVISION ;
    end record ;
    attribute DATE : TIMESTAMP ;
    attribute VERSION : VERSION_NUMBER ;
    attribute AUTHOR : STRING ;
    attribute COMMENT : STRING ;
end MANAGEMENT_DEFINITIONS ;
```

then the following VHDL design unit may be declared:

```
use WORK.MANAGEMENT_DEFINITIONS.all ;
entity MICROPROCESSOR_32_BIT is
    attribute DATE of MICROPROCESSOR_32_BIT : entity is (8, 19, 1991) ;
    attribute VERSION of MICROPROCESSOR_32_BIT : entity is ("AB1234","01") ;
    attribute AUTHOR of MICROPROCESSOR_32_BIT : entity is "Serge MAGINOT" ;
    attribute COMMENT of MICROPROCESSOR_32_BIT : entity is
        "This is a comment which is stored in the database after compilation." ;
end MICROPROCESSOR_32_BIT ;
```

These VHDL statements are more or less equivalent to the previous UDL/I common management statements because the corresponding attribute values are stored in the design database. Furthermore, they can be processed from within the design unit.

Writing comments is recommended in any methodology for any programming and description language, but these comments are often not processed. Comments such as version number and author name are compulsory in a good design methodology. UDL/I common management statements, as well as relevant VHDL user-defined attributes, are processed and stored in the database: it is then possible to think about miscellaneous verifications and management processings.

A typical UDL/I module description that can be found in the previous design description titled MICROPROCESSOR_32_BIT could be

```
NAME :        RAM_32_BIT ;
PURPOSE :     LOGSIM, LOGIC_SYNTHESIS, FORMAL_PROOF ;
PROCESS :     CMOS, TTL ;
LEVEL :       MACRO_CELL ;
DATE :        08/10/91 ;
VERSION :     RAM101.01 ;
AUTHOR :      Serge MAGINOT ;
COMMENT :     This is a comment which is stored in the database after compilation. ;

OUTPUTS :               READY /0/ ;
INPUTS ADDRESS :        ABUS <0:7> ;
BUS DATA :              DBUS <31:0> /(f, z)/ ;
CLOCK :                 CLK ;
RESET :                 RST ;
POWERS :                VDD, VSS ;

TYPES :  BUFFER, DECODER, RAM_CELL ;
        BUFFER :        B<0:7> ;
        DECODER :       DEC ;
        RAM_CELL :      R<0:255, 31:0> ;
END_TYPES ;

<body of description> ::=
        <behavior description> | <structural description> | <primitive description section>
[ <assertion section> ]

END ;
```

The first statement indicates the name of the module. The second one gives the purpose(s) of the module description, i.e., if the description is for logic simulation, synthesis, formal proof, circuit analysis, test, or other verifications. Of course, one module (identified by its name) may have several descriptions corresponding to different purposes.

This notion of purpose is very important. VHDL is defined in the language reference manual with a simulation semantics: there is neither synthesis nor formal proof semantics. The only unambiguous purpose of a VHDL model is the logic simulation: using VHDL as an input language for synthesis tools often leads to different interpretations. Most of today's synthesis vendors have defined a VHDL subset with their own synthesis semantics.

In contrast, the UDL/I reference manual handles different purposes, such as simulation, synthesis, or verifications. The first chapter of the manual specifies that *"[UDL/I is] good for logic synthesis as well as simulation. [There is a] clear separation between the hardware to be synthesized and additional information for simulation or verification"*.

The purpose statement shows this *clear separation,* which is indeed well implemented in the language. Furthermore, the reader will notice in the next sections that UDL/I is well suited to synthesis, most of the language features having a synthesis semantics (leading sometimes to strong limitations in behavioral description capabilities, as compared to VHDL).

In the example of module description, the purpose statement lists three different purposes. All the purpose names are user defined, except *LOGSIM,* which is predefined for the logic simulation purpose.

The third statement specifies the fabrication process or technology to be used for the module (this statement is optional); process names are user defined. The fourth statement is the level statement, which provides the nesting level of the module in the design hierarchy; level names are user defined, except the keyword *END,* which indicates the bottom level of the design.

Next, optional common management statements may be found in the module description. When some of them are missing, corresponding management statements of the design description are taken; otherwise, module management statements override the corresponding statements found in the design description.

After the management statements come the external pin definitions, which correspond to the VHDL port declarations. Actually, these pin definitions are more complex than the VHDL ports declarations and will be detailed in paragraph 10.4.1.

Next are the optional types statements. The name is confusing, because it has nothing in common with the VHDL types. Actually, the UDL/I types statements are more or less equivalent to VHDL component declarations. These statements declare the submodule names, optionally with their pin description, and the names of the instances of each submodule. This statement is described in paragraph 10.5.1.

After these types statements, the body of the description is written. It may be of three kinds: a structural description (section 10.5), a behavioral description (section 10.6), or a primitive description section (paragraph 10.9.1). Only one type of description is allowed in a particular body of a module: *a mixed style as in VHDL is not allowed in UDL/I!*

Finally, the module description may include an optional assertion section, described in section 10.7: this section includes statements that are very similar to the VHDL assertion statements.

Figure 10.1 shows the UDL/I design description compared to the five VHDL design units. Three main parts may be distinguished in each UDL/I module:

- a first part including the purpose, process, level, and common management statements, which can be modeled with user-defined attributes in VHDL (UDL/I comments stored in the design database);
- a second part including the name statement and the external pin definitions, which corresponds to the VHDL entity declaration (without *generic parameters, which are not implemented in UDL/I);*
- a last part including the types statements and the body of the description, which corresponds to the VHDL architecture body.

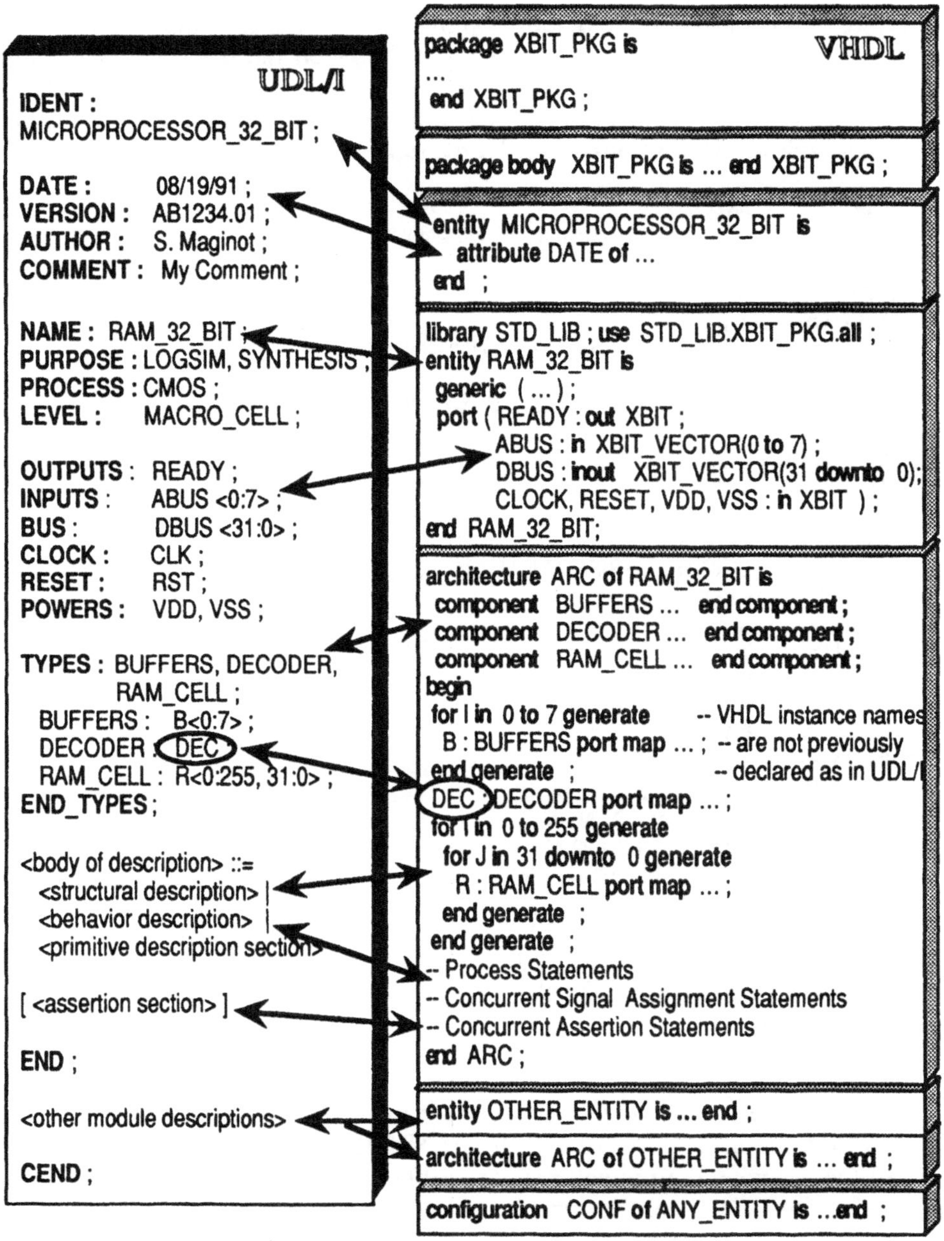

Fig 10.1 Design Units

Thus, any UDL/I description module (encapsulated into a design description) is more or less equivalent to a VHDL couple entity and architecture. Although there is no separation between the interface description (entity) and the internal description (architecture), it is possible to specify different descriptions for the same module if each of these descriptions is associated with a different purpose. For example:

```
NAME :          MUX ;
PURPOSE :       LOGSIM ; "description for logic simulation"
LEVEL :         END ;
INPUTS :        A, B, SEL ;
OUTPUTS :       Y ;
BEHAVIOR_SECTION ; "Behavioral description"
      BEGIN
          Y := IF SEL THEN A ELSE B END_IF ;
      END ;
END_SECTION ;
END ;
"--------------------------------------------------------------"
NAME :          MUX ;
PURPOSE :       LAYOUT_VERIFICATION; "description for layout vs schematic verifications"
LEVEL :         CELL ;
INPUTS :        A, B, SEL ;
OUTPUTS :       Y ;
TYPES :         NAND2, INV ;
                NAND2 : N1, N2, N3 ;
                INV : I ;
END_TYPES ;
NET_SECTION ; "Structural description"
      NETA = FROM (.A) TO (N1.A) ;
      NETB = FROM (.B) TO (N2.A) ;
      NETS = FROM (.SEL) TO (I.A, N1.B) ;
      NETSB = FROM (I.Y) TO (N2.B) ;
      NETNA = FROM (N1.Y) TO (N3.A) ;
      NETNB = FROM (N2.Y) TO (N3.B) ;
      NETY = FROM (N3.Y) TO (.Y) ;
END_SECTION ;
END ;
```

These different UDL/I descriptions are not exactly of the same nature as different architectures associated with one VHDL entity. Indeed, in the VHDL case, all the associated architectures are usable for the logic simulation, whereas only one of the UDL/I descriptions may have the *LOGSIM* purpose.

As will be detailed in section 10.5, UDL/I structural descriptions do not implement the configuration mechanism of VHDL (instantiation of components further configured with entities). There is thus no UDL/I equivalent to the VHDL configuration declaration design unit.

Likewise, as illustrated by the previous figure, *software* design units comparable to the VHDL package declaration and package body are not provided by UDL/I. As the next sections will underline, UDL/I does not implement many programming language constructs and is really more focused on hardware modeling.

10.3. SEQUENTIAL AND CONCURRENT DOMAINS

From a naive point of view, we might say (as in the previous chapters) that the concurrent world is basically modeling the interconnection of black boxes, whereas the sequential world is intended to implement sequential algorithms.

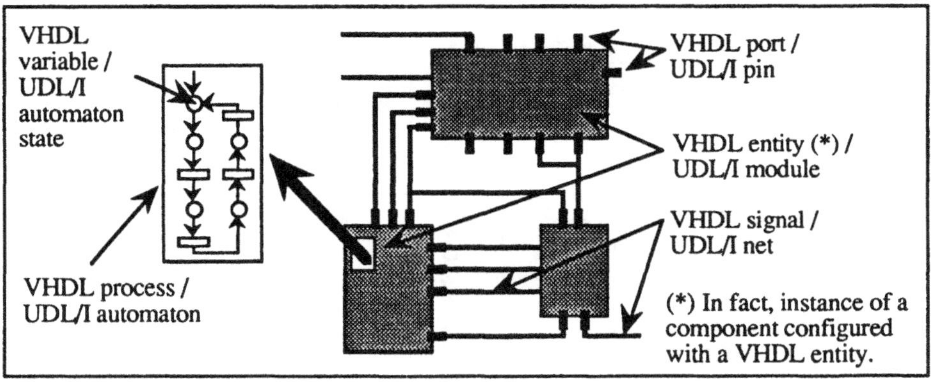

Fig 10.2 Sequential and Concurrent Domains

As illustrated above, black boxes are modeled by VHDL entities and UDL/I modules and interconnected by VHDL signals and UDL/I nets. This netlist description is only one kind of concurrent domain in UDL/I, the other one being the behavior description, which cannot be mixed with the previous one, as explained above.

The sequential domain, implemented by processes in VHDL, is restricted to the automaton model in UDL/I: this device is said to be sequential in the sense that it executes *sequences* of commands.

The following figure identifies sequential and concurrent domains in both languages:

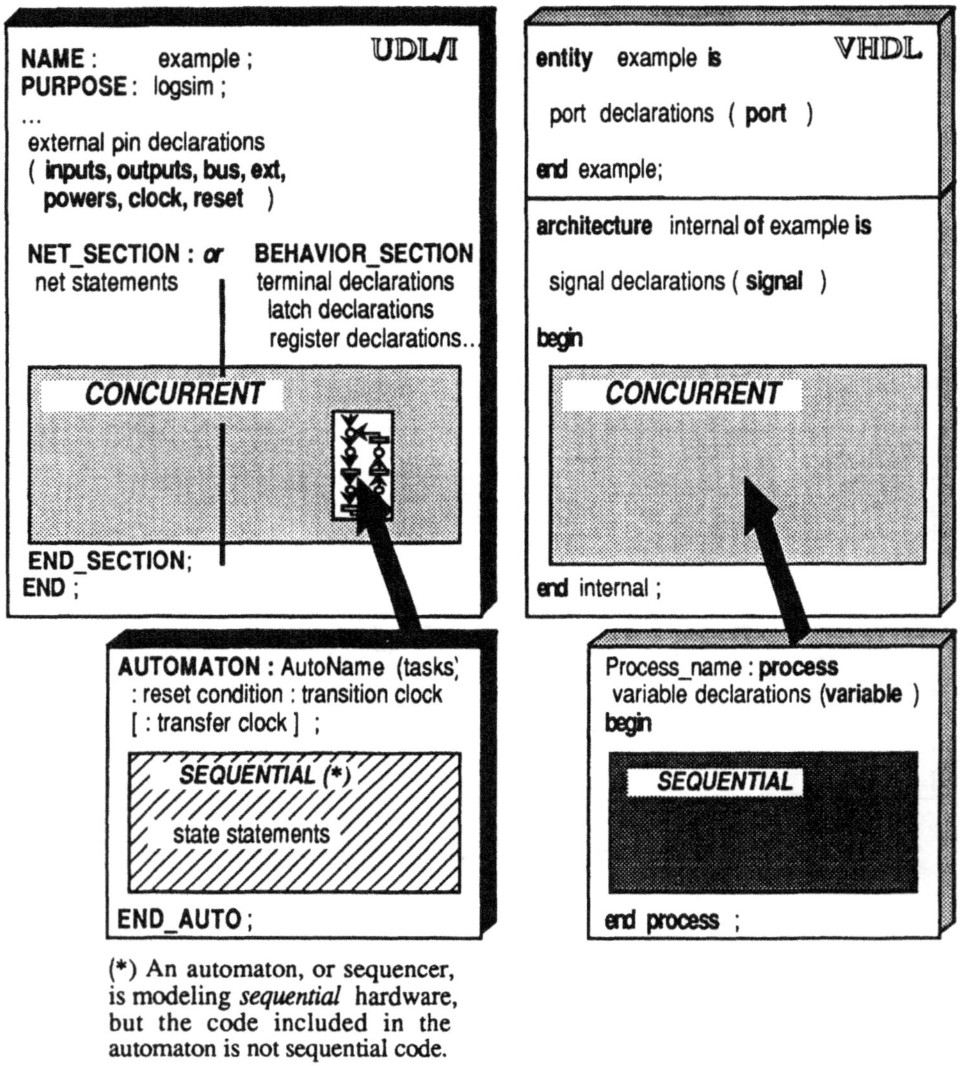

Fig 10.3 Sequential and Concurrent Domains

As the above schema shows, the domain of a UDL/I module description is the concurrent domain. Indeed, the body of a module description can be a structural, behavior, or primitive description (remember that only one of these three styles is allowed in a particular module: no mixed style, as in VHDL, is permitted).

The case of the primitive description is a special case that will be covered in paragraph 10.9.1. Let us consider the two others types of description:

- A structural body only describes the interconnection of other modules (no other statements are allowed). According to the reference manual, pins are connected together by *nets* (which are not declared and are only used in net statements: see sections 10.4 and 10.5). Of course, such netlist description is part of the concurrent domain.
- A behavior body has declarative and executable parts. Some objects, or *facilities*, that can be declared in this body (and only in this one) are *terminals, registers, latches, RAMs,* and *ROMs,* which are more or less equivalent to VHDL signals (but with specific behaviors: see section 10.4). The executable part comprises concurrent behavioral statements (see section 10.6) and automaton definitions, being thus in the concurrent domain like any VHDL architecture body with concurrent signal assignments and process statements.

These UDL/I automatons may be seen as the only feature modeling sequential hardware. Indeed, an automaton is defined by a given number of states, transitions between states, and commands associated with each state or transition. The UDL/I designer may thus describe any sequence of statements, even if the written code is not sequentially executed.

Actually, there is *no sequential code (domain) in UDL/I* like those in the VHDL processes or in any software programming language. In fact, we may say in some sense that *UDL/I only provides concurrent statements* (no sequential statements). These statements will be detailed, with their semantics, in paragraph 10.6.3.

More precisely, the UDL/I reference manual defines the semantics of the language by using an equivalent subset of the whole UDL/I behavior syntax. This subset is called the *core subset*. Translation rules are given to define all statements (like automaton definitions) in this subset. The UDL/I designer may thus read in the manual that all the statements associated with a given place or state are equivalent to structured statements found in the behavior section of the module, i.e., are concurrent statements.

Finally, the UDL/I reference manual does not define any subprogram constructs: only standard and system functions may be called in expressions and statements (see paragraph 10.6.1). This is an important particularity (and simplification) of UDL/I: *no sequential process, no sequential subprogram.* This lack of sequential domain may be considered as the price paid to get a hardware description language that has a straightforward interpretation for synthesis, but it will definitively prohibit any attempt to make system-level descriptions with the UDL/I language.

On the other hand, the concurrent world (i.e., the hardware world!) is very well defined in UDL/I, which is actually the first priority of a hardware description language. The semantics of this concurrent domain is more classic: it is based on an event-driven simulation algorithm (assignment statements are executed each time any signal value appearing in the right-hand side changes).

But UDL/I simulation behaviors also present some interesting originalities in the case of conflicts or simultaneous events (see paragraph 10.6.3 on non-deterministic behavior).

10.4. OBJECTS AND TYPES

As mentioned in the previous sections, there is a clear separation in UDL/I between netlist and behavior descriptions. Thus, almost all the named objects handled in UDL/I are specific to the behavior section. There is no equivalent to the VHDL signals, which are used both in netlist descriptions and in behavioral statements. The following figure lists these objects (and their types) in both languages.

	UDL/I
PINS	**inputs, outputs, bus, ext, powers, clock, reset**
NETS	not declared
FACILITIES	**terminal, register, latch, rom, ram rompat, const**
AUTOMATON OBJECTS	**state, task**

	VHDL
signal (**port**)	Any VHDL type except **access** and **file**
variable	Any VHDL type
constant	Any VHDL type except **access** and **file**
file	file containing objects of any VHDL type, except **access** and **file**

Fig 10.4 Objects and Types

With regard to the objects of the tables above, the following comparisons may be stated: UDL/I pins are equivalent to VHDL ports, UDL/I nets and facilities are comparable to VHDL signals, and UDL/I automaton states and tasks, since they can be accessed from outside the automaton (see paragraph 10.9.2), may be implemented in VHDL with signals as well. Thus, all the UDL/I objects are more or less equivalent to VHDL ports and signals, i.e., objects declared in the concurrent world.

This remark is consistent with the fact that UDL/I does not implement the real sequential world (i.e., encapsulation of sequential code) and that all its statements are concurrent statements. Thus, as they are defined in the sequential domain only (in processes and subprograms), VHDL variables have no equivalent in UDL/I.

CONST and ROMPAT facilities are, respectively, equivalent to VHDL constants of type BIT_VECTOR and of array type of BIT_VECTOR elements.

The UDL/I reference manual also does not define files. In VHDL, files are very convenient for specifying complex parameter sets (such as ROM patterns) or for system modeling.

Once again, we may notice that the number of UDL/I software constructs is very restricted (no files, no variables, no subprograms). The hardware constructs are very specific to integrated circuit modeling: it is true that this specificity is the purpose of this Unified Design Language for... Integrated circuits!

This specialization appears, for example, in the types of the UDL/I objects. Whereas VHDL lets the user define signals of any type (predefined type bit, but also any enumeration, array or record types), *all the UDL/I signals have a bit logic type*. Actually, they may have only one of the following values:

- 0 for logic 0
- 1 for logic 1
- Z for high impedance
- X for unknown value

Of course, such a logic type can be modeled in VHDL by the following enumeration type:

```
type UDLI_BIT is ('X', '0', '1', 'Z') ;
```

Different resolution functions and associated subtypes of UDLI_BIT may then be declared in VHDL that model the UDL/I bit logic. Actually, UDL/I signals are defined either by external pin definitions or facilities declarations, which correspond to VHDL port declarations and VHDL signal declarations, respectively.

As mentioned previously, *nets* used in the UDL/I netlist descriptions *are not declared objects* (as VHDL signals used in structural descriptions are), but they may be considered as terminals (one kind of facility: see below) that inherit properties from the attached external pins of instantiated modules.

The following correspondences may thus be stated:

- UDL/I external pins are equivalent to VHDL ports;
- UDL/I (undeclared) nets are equivalent to *structural* VHDL signals;
- UDL/I facilities are equivalent to *behavioral* VHDL signals.

This means that *structural* signals are used in port maps of component instantiations, whereas *behavioral* signals are targets of signal assignments. Of course, the reader must be aware that one particular VHDL signal can be used for both descriptions: there is no separation as in UDL/I, where structural and behavioral descriptions cannot be mixed.

Finally, as the next declarations will show, UDL/I wires may be grouped together to make *buses:* DATA_BUS <15:0> is equivalent to the VHDL signal DATA_BUS(15 downto 0).

10.4.1. UDL/I External Pin Definitions

Following is a table listing all the possible UDL/I external pin definitions:

UDLI Pins	UDL/I Syntax	VHDL mode
INPUTS	**INPUTS** [<pin type>, {<pintype>}] : <pin definition name> [/ <input default> /] { , <pin definition name> [/ <input default> /] } ;	**in**
OUTPUTS	**OUTPUTS** [<pin type>, {<pintype>}] : <pin definition name> [/ <output default> /] {, <pin definition name> [/ <output default> /] } ;	**out**
BUS	**BUS** [<pin type>, {<pintype>}] : <pin definition name> [/ <default> /] {, <pin definition name> [/ <default> /] } ;	**inout**
EXT	**EXT** [<pin type>, {<pintype>}] : <pin definition name> {, <pin definition name>} ;	**linkage**
POWERS	**POWERS** : <pin definition name> { , <pin definition name> } ;	**linkage** or **in**
CLOCK	**CLOCK** : <pin definition name> [/ <input default> /] { , <pin definition name> [/ <input default> /] } ;	**in**
RESET	**RESET** : <pin definition name> [/ <input default> /] { , <pin definition name> [/ <input default> /] } ;	**in**

Seven pin classes are available in UDL/I. The first four are general-purpose pins corresponding to VHDL ports of different modes: UDL/I *outputs, inputs, bus,* and *ext* pins are, respectively, equivalent to VHDL ports of mode *out, in, inout,* and *linkage* Actually, UDL/I type *ext* defines external pins having an unknown direction, used, for example, to model transistor terminals. Therefore, it is more or less equivalent to the mode *linkage*, even if this mode has no real semantics in VHDL.

To each one of these four pins may be attached a *pin type*: this pin type is a user-defined simple name that may be used to classify pins with another criterion (there are no predefined type names in UDL/I). For example, the designer may define the following pins:

```
INPUTS   ADDRESS   :   ABUS <0:15> ;
BUS      OPERAND   :   DBUS <0:31> ;
```

In this example, ADDRESS and OPERAND have no predefined semantics, but any meaning may be given to these names by the designer.

Of course, VHDL ports are defined with types, but these types specify the structure of these ports: scalar, array, record, and so on. Subtypes may also be defined that associate constraints with the parent type or that simply rename the type without additional meaning. For example, the following subtypes may be declared:

```
type UDLI_BIT_VECTOR is array ( NATURAL range <> ) of UDLI_BIT ;
subtype ADDRESS is UDLI_BIT_VECTOR ( 0 to 15 ) ;
-- index constraint
subtype DATA is UDLI_BIT_VECTOR ( 0 to 31 ) ;
-- index constraint
subtype OPERAND is DATA ;
-- no constraint, OPERAND and DATA are equivalent (OPERAND only renames DATA)
-- OPERAND has no more meaning than in the above UDL/I declaration
```

Thus, previous UDL/I pin declarations are more or less equivalent to the following VHDL port declarations:

```
port ( ABUS : in ADDRESS ;  DBUS : inout DATA ) ;
```

But in this case, ADDRESS and DATA are VHDL subtypes that define, for example, the range of the ports (they are not just names as in UDL/I).

Finally, the four pin declarations seen up to now may have an input default field (INPUTS and BUS) or an output default field (OUTPUTS and BUS).

The input default field defines the assumed value of the pin when it is left unconnected. Possible input default values are the four logic values 0, 1, X, Z plus the value F, which means *FORCE*; i.e., the pin should *not* be left unconnected. When no explicit default value is given, value F is assumed. In VHDL, when no explicit default value is given for an input port, then this port must be connected. Thus, the following UDL/I and VHDL declarations are equivalent:

```
INPUTS : D /z/ ;        <=>        port ( D : in UDLI_BIT := 'Z' ) ;
INPUTS : D /f/ ;        <=>        port ( D : in UDLI_BIT ) ;
INPUTS : D ;            <=>        port ( D : in UDLI_BIT ) ;
```

In the two last cases, the input port must be connected in the corresponding module or component instantiation.

The output pins may also have default values, but their meaning is different. Indeed, the output default field defines the resolution function, named *dot function* in UDL/I, which is used to resolve the conflict of two or more OUTPUT pins connected together.

Possible output default values are A, O, 1, 0, X, H, Z, and F. The seven first values correspond, respectively, to the standard functions WAND, WOR, TRIREC1, TRIREC0, TRIRECX, TRIRECH, TRIRECX used as dot functions. The value F means that no dot function is used. When no output default is specified, the output default value X (dot function TRIRECX) is assumed.

Actually, the UDL/I syntax defines the mechanism of conflict resolution (wired-X, wired-and, tri-state, etc.) in the external pin definitions (and behavioral facility assignments: see paragraph 10.6.2), but not for the structural nets. In VHDL, the resolved subtype indication of the declared signal specifies its conflict resolution both in behavioral and structural descriptions.

In VHDL, a resolution function may be associated with ports *and* signals. Usually, the VHDL designer defines resolved subtypes to further declare ports and signals with these subtypes. As will be detailed in 10.6.1.5, all the UDL/I standard functions (and among them those used as dot functions) can be programmed in VHDL. For clarity, these functions are defined with the same UDL/I name, thereby allowing the following resolved subtype declarations:

```
subtype WAND_BIT     is  WAND UDLI_BIT ;     -- all these VHDL functions have
subtype WOR_BIT      is  WOR UDLI_BIT ;      -- a single input parameter which is an
subtype TRIREC1_BIT  is  TRIREC1 UDLI_BIT ;  -- unconstrained array of type
subtype TRIREC0_BIT  is  TRIREC0 UDLI_BIT ;  -- UDLI_BIT_VECTOR (see 10.6.1.5)
subtype TRIRECX_BIT  is  TRIRECX UDLI_BIT ;  -- their use as resolution function for
subtype TRIRECH_BIT  is  TRIRECH UDLI_BIT ;  -- type UDLI_BIT is thus legal...
subtype TRIRECZ_BIT  is  TRIRECZ UDLI_BIT ;
```

Thus, the following UDL/I and VHDL declarations can be written:

```
OUTPUT : Q /h/ ;    <=>    port ( Q : out TRIRECH_BIT ) ;
OUTPUT : Q /o/ ;    <=>    port ( Q : out WOR_BIT ) ;
OUTPUT : Q /f/ ;    <=>    port ( Q : out UDLI_BIT ) ; -- unresolved port
```

But the UDL/I and VHDL declarations listed in each of these three lines are not equivalent. The hierarchical resolution mechanism is indeed completely different in both languages:

- in UDL/I, the dot (resolution) function associated with the output port of a given module is used to resolve conflicts on the wire, located *outside* the given module, to which the port is connected;
- in VHDL, the resolution function associated with the output port of a given entity is used to resolved conflicts *inside* the architecture of the given entity (if the output port is assigned by more than one driver inside the architecture).

Actually, the VHDL hierarchical resolution uses port and signal resolution subtypes. In figure 10.5 below, the resolution function TRIREC0_BIT associated with the output *port* P is used to resolve the conflict between the outputs of the two gates G1 and G2. Then, the resolution function

TRIRECX_BIT associated with the *signal* S is used to resolve the conflict between the two output ports P and Q, and so on.

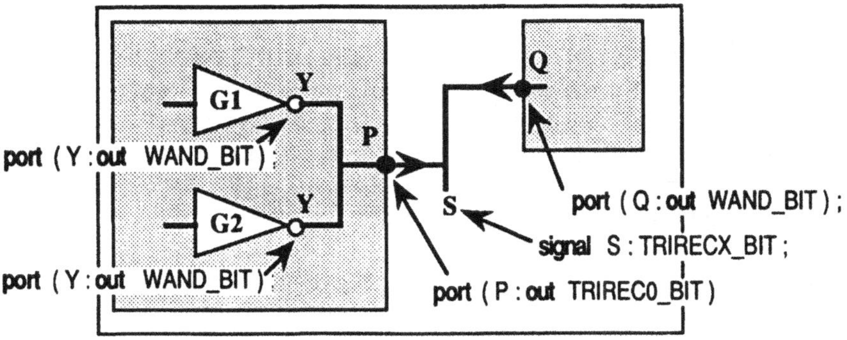

Fig 10.5 VHDL Hierarchical Conflict Resolution

In UDL/I, resolution specifications of *structural* descriptions are only defined for output (or bus) pin definitions, and not for nets. As stated before, the dot function of a given port resolves the conflict on the wire, which is connected to this port but outside the module: thus, all ports connected to the same wire *must have the same dot functions* (this should be checked by the simulator).

In figure 10.6 below, the dot function WAND associated with both output ports Y of the two gates G1 and G2 is used to resolve the conflict between them. Then, the dot function TRIREC0 associated with both ports P and Q is used to resolve the conflict on the net S, and so on.

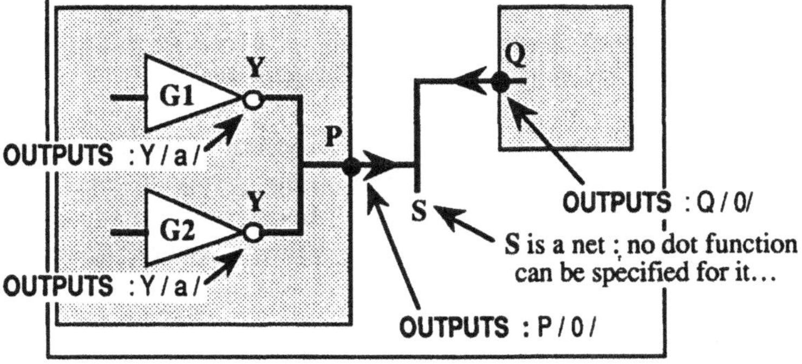

Fig 10.6 UDL/I Hierarchical Conflict Resolution

Thus, connecting two ports associated with different dot functions is not possible in UDL/I. Some could argue that such connections are not realizable with today's technologies in integrated circuits. Such connections may have some sense at a system-level design, but UDL/I is not intended to be used at this modeling level.

Nevertheless, the VHDL possibility of associating resolution functions with ports and signals seems easier and safer to use, mostly because the resolution function is specified in the same design unit in which it is used.

Finally, UDL/I conflict resolution between facilities (such as terminals and registers) uses a completely different mechanism — the OFFSTATE clause — which will be detailed in paragraph 10.6.2.

Coming back to the table showing all the UDL/I external pin declarations, the reader notices that the last three pins are specific pins having no equivalent in VHDL. They model the *power, clock, and reset pins*. Power pins are meaningless for simulation, but they may be used by place-and-route tools; no semantics is defined for clock and reset pins by the UDL/I reference manual, but their definition may be linked to the clock expressions and reset statements.

This UDL/I terminology is consistent with the philosophy of the language: it is very IC oriented and includes a complete synthesis semantics.

10.4.2. UDL/I Facility Declarations

According to the UDL/I reference manual, seven kinds of facilities can be declared: *terminal, latch, register, RAM, ROM, ROMPAT,* and *constant.* Up to now, most of these facilities have been compared to VHDL signals with some specific behaviors. Actually, *UDL/I facilities are intended to model the hardware devices denoted by their name* (latch, register, etc.), thus leading to straightforward meanings for synthesis tools.

Following are listed all these UDL/I facilities with a *first intuitive* modeling in VHDL (more accurate equivalences will be given in paragraph 10.6.2 and 10.6.3, the latter focusing on the statements themselves).

• The *terminal* is used to model wires inside combinational circuits or blocks. It more or less corresponds to the output pin of a gate: it is referenced in expressions or assigned by statements without clock expression (without *AT* statement or expression that introduce triggering conditions: see 10.6.3.6). Thus:

```
TERMINAL : SUM, MUXOUT ;
...
SUM := A @ B @ CIN ;
MUXOUT := IF CONTROL THEN A ELSE B END_IF ;
```

A terminal is equivalent to a VHDL signal that is only used in assignments modeling combinational devices, such as

```
signal SUM, MUXOUT : UDLI_BIT ;
...
SUM <= A xor B xor CIN ;  -- "xor" being the overloaded operator on type UDLI_BIT
MUXOUT <=    A when CONTROL = '1' else
             B when CONTROL = '0' else
             'X' ;
```

• The *latch* models the level-sensitive memory device of the same name: it should be assigned by *AT* statements or expressions (see 10.6.3.6) with clock expressions using level-sensitive conditions.

A level-sensitive clock expression is an explicit *HIGH* or *LOW* clock statement (using keywords **HIGH** and **LOW**) or a simple expression in which case HIGH is assumed. For example:

```
LATCH : Q ;
...
AT HIGH (CLK) DO
      Q := D ;
 END_DO ;
```

The UDL/I latch is equivalent to a VHDL signal assigned by statements including level-sensitive expressions, i.e.,

```
signal Q : BIT ;
...
Q <= D when CLK = '1'
      else Q ;
```

```
signal Q : BIT ;
...
process
begin
      if CLK = '1' then
            Q <= D ;
      end if ;
      wait on D, CLK ;
end process ;
```

```
signal Q : BIT ;
...
B : block (CLK = '1')
begin
      Q <= guarded D ;
end block ;
```

These three models underline various modeling capabilities of VHDL. This is an advantage from the simulation point of view, but it may be considered as a weakness from the synthesis point of view: too many descriptions are possible for the same device (to be synthesized), thus leading to ambiguities. Unlike VHDL, UDL/I always provides unambiguous synthetizable descriptions.

• The *register* models a memory device that is either edge-sensitive or level-sensitive or both. It is the only facility that can use edge clocking. Its level-sensitive behavior is similar to the latch or RAM behaviors.

The register may be assigned by one of four statements: assignment by AT statement with *RISE* or *FALL* clock statement (edge-sensitive), assignment by

AT statement with *HIGH* or *LOW* clock statement (level-sensitive, like the latch), *PRESET* statement (asynchronous set to 1), and *RESET* statement (asynchronous set to 0). **RISE, FALL, HIGH, LOW, PRESET**, and **RESET** are UDL/I keywords (see paragraph 10.6.3).

```
REGISTER : REG<0:3> ;
...
IF ENABLE THEN
      AT RISE(CLK) DO REG := A ; END_DO ;
END_IF ;
IF CLEAR THEN
      RESET(REG) ;
END_IF ;
```

The UDL/I register is equivalent to a VHDL signal assigned in VHDL statements corresponding to the four specified UDL/I statements, i.e., including edge- and level-sensitive latches, reset, and preset.

```
signal REG : UDLI_BIT_VECTOR(0 to 3) ;
...
REG <=      "0000" when CLEAR = '1' else                  -- Level sensitive
            A when not CLK'STABLE and CLK = '1' else      -- Edge sensitive
            REG ;
```

Of course, although modeling the same device (a register), the above VHDL code is not really equivalent to the UDL/I one: the VHDL signal REG is the target of one assignment, whereas the UDL/I register REG is the target of two assignments.

Writing two assignments in VHDL would have necessitated defining a resolved signal (being assigned by two drivers) with resolution rules implementing those used by UDL/I, which are: priority of the asynchronous transfer over the synchronous one, and non-deterministic resolution in the case of two transfers having the same priority. This important aspect of UDL/I will be explained in paragraph 10.6.3.

• The *RAM* models the classical level-sensitive memory element. Multiple ports can be defined. The simultaneous read and write access to the same port is allowed only if both operations are performed at the same RAM address. Similarly, a simultaneous write operation on different ports is allowed only if data and address are the same for all ports. This behavior is a classical RAM behavior (although some could argue that it is only *one* possible RAM behavior among many).

The RAM has the same asynchronous behavior as the latch, and thus is accessed using the same statements:

```
INPUTS : WR, RE ;
INPUTS : ADDRESS<0:10> ;
BUS : DATA<0:31> ;
RAM : MEM<0:2047,0:31> ;
...
IF WR THEN
        AT HIGH(CLK) DO MEM<ADDRESS> := DATA ; END_DO ;
END_IF;
IF RE THEN
        DATA := MEM<ADDRESS> ;
END_IF;
```

The UDL/I RAM may be modeled in VHDL by using a signal of type array (it is also possible to use a variable within a process):

```
library UDLI_LIBRARY ; use UDLI_LIBRARY.UDLI_TYPES.all ; -- cf 10.6.1.5
entity RAM_ENTITY is
        port (   WR, RE      :   in UDLI_BIT ;
                 ADDRESS     :   in UDLI_BIT_VECTOR(0 to 10) ;
                 DATA        :   inout UDLI_BIT_VECTOR(0 to 31)   ) ;
end RAM_ENTITY ;

use WORK.SOME_PKG.NATURAL_VALUE ;
architecture A of RAM_ENTITY is
        type MY_RAM is array (0 to 2047) of UDLI_BIT_VECTOR(0 to 31) ;
        signal MEM : MY_RAM ;
begin
        process
        begin
            wait until WR = '1' ;
            MEM(NATURAL_VALUE(ADDRESS)) <= DATA ;
        end process ;
        process
        begin
            wait until RE = '1' ;
            DATA <= MEM(NATURAL_VALUE(ADDRESS)) ;
        end process ;
end ;
```

In this model, NATURAL_VALUE is a user-defined function converting UDLI_BIT_VECTOR into NATURAL (which is not predefined, as the conversion between UDL/I buses and integers is).

• The *ROM* is, of course, modeling the Read-Only Memory. The *ROMPAT* is the UDL/I facility used to specify the content of this ROM.

```
ROM : MEM<0:1023, 0:7> ;
ROMPAT MEM : HFF, H02, H45, B00110101, B11111111 ;
```

The ROMPAT statement uses patterns that are simply constants having the data size of the declared ROM. This ROM can be modeled in VHDL by a constant of type array.

```
type MY_ROM is array (0 to 1023) of UDLI_BIT_VECTOR(0 to 7) ;
-- type declaration
constant MEM : MY_ROM := (     UDLI(X"FF"), UDLI(X"02"), UDLI(X"45"),
                               UDLI(B"00110101"), UDLI(B"11111111"),
                               others => "XXXXXXXX") ;
```

Here, UDLI is a function used to convert bit string literals into arrays of type UDLI_BIT_VECTOR (for which such a format is not defined):

```
function UDLI (B : BIT_VECTOR) return UDLI_BIT_VECTOR is
        variable U : UDLI_BIT_VECTOR(B'range) ;
begin
        for I in B'range loop
            U(I) := UDLI_BIT'VAL( BIT'POS( B(I) ) + 1 ) ;
        end loop;
        return U ;
end;
```

Actually, a better way to specify a ROM content in VHDL is to read a file containing these data. This can be easily achieved with the READ procedures exported by the predefined TEXTIO package. As UDL/I does not implement any file or I/O procedures, such a powerful and convenient ROM initialization is impossible to model in this language.

• The UDL/I constant declaration is very simple and can be modeled in VHDL by the declaration of a constant of type UDLI_BIT_VECTOR:

```
CONST : C1 = 4B0101, C2 = 8H0F, C3 = 8B01011111 ;
```

The first digit of each UDL/I value is the bit length of the constant. It is not necessary in VHDL to specify this bit length, because it is automatically extracted from the array itself:

```
constant C1 : UDLI_BIT_VECTOR := UDLI (B"0101") ;
constant C2 : UDLI_BIT_VECTOR := UDLI (X"0F") ;
constant C3 : UDLI_BIT_VECTOR := UDLI (B"01011111") ;
-- previous conversion function (UDLI) has been used.
```

Among all the above-listed facilities, the terminal, register, latch, RAM and ROM may be declared with delay specifications:

```
REGISTER : REG<0:3> DELAY 1.3NS ;
TERMINAL : TERM DELAY 0.5NS ;
```

Such a delay is associated with the facility itself and is added to delays that can appear in assignments:

```
TERM := A & B DELAY 1.2NS ;
```

In that case, the total delay between an edge on A and a resulting edge on TERM is equal to 0.5 + 1.2 = 1.7 ns.

In VHDL, delays may only be specified in assignment statements. It is possible to attach an intrinsic delay to the signal itself with a user-defined attribute (which is static, such as the UDL/I facility delay), but the total delay will be explicit in each statement:

```
architecture EXAMPLE of ANY_ENTITY is
        attribute DELAY : TIME ;
        signal TERM : BIT ;
        attribute DELAY of TERM : signal is 0.5 ns ;
        -- signal TERM declared in this architecture will have an intrinsic delay of 0.5 ns
begin
        TERM <= A and B after TERM'DELAY + 1.2 ns ;  -- A and B are supposed to be inputs
end ;
```

All these facilities may also be defined with aliases. The UDL/I syntax is the following ("!!" is the symbol of concatenation):

```
REGISTER : R<31:0>          = OP<0:1> !! ADDR<0:29>
                            = LEFTWORD<15:0> !! RIGHTWORD<15:0> ;
```

Such aliases may also be declared in VHDL:

```
signal R : UDLI_BIT_VECTOR(31 downto 0) ;
alias OP : UDLI_BIT_VECTOR(0 to 1) is R(31 downto 30) ;
alias ADDR : UDLI_BIT_VECTOR(0 to 29) is R(29 downto 0) ;
alias LEFTWORD : UDLI_BIT_VECTOR(15 downto 0) is R(31 downto 16) ;
alias RIGHTWORD : UDLI_BIT_VECTOR(15 downto 0) is R(15 downto 0) ;
```

The UDL/I syntax is more compact. Furthermore, it allows more possibilities than the VHDL syntax, such as

```
TERMINAL : M<1:3,1:4>  = L<1:12> ;
```

A 2D array has been renamed into a 1D array, with the following convention: M<1:1> = L<1>, M<1:2> = L<2>, ... M<3:4> = L<12>. Such alias definition is impossible in VHDL: a conversion function has to be used between two objects of different types.

As a conclusion for this section, it may be observed that all the UDL/I objects (external pins, facilities) may be modeled into VHDL objects (ports, signals). But the difference implied by the UDL/I semantics between all the facilities no longer exists between the corresponding VHDL signals: those are fundamentally the same VHDL objects (signals), their only difference consisting in the way they are assigned. Any specialization is thus left to the designer, and no such semantics checking is performed by the simulator. This makes no difference for the simulation, but increases the difficulty of a synthesis interpretation of a VHDL model.

On the other hand, the UDL/I reference manual defines different semantics for terminals and registers: terminals are evaluated when any value of the right side of an assignment changes, registers only when asynchronous or synchronous clock conditions are met. These clock conditions are not explicit in VHDL: clocks are just user-defined signals that control the activation of processes, like any other input signal of combinational devices (see previous models).

UDL/I and VHDL have in common their event-driven mechanism, which is sufficient to define the way facilities or signals are computed, i.e., sufficient to define their *simulation semantics*. From the simulator point of view, terminals and registers are indeed not so different as they seem to be (excepting real differences in conflict resolutions: see paragraph 10.6.3). Thus, as VHDL only includes simulation semantics, such a distinction is not fundamental.

But terminals and registers also correspond to very different hardware devices: wires in combinational circuits in the former case, memory elements in the latter. In other words, their *synthesis semantics* is quite different. As UDL/I also defines a clear synthesis meaning for each object, this separation between all kinds of facilities seems necessary in this language. If a synthesis semantics has to be defined for VHDL, then adding features that clearly identify equivalent terminals and registers could be one good strategy.

10.5. UDL/I STRUCTURAL DESCRIPTION

10.5.1. Types or Components Declarations

In UDL/I, modules can be directly instantiated into other modules. The modules are referenced (and their instances declared!) in the module into which they are instantiated. Following is the example of the multiplexer of section 10.2:

```
NAME :        MUX ;                   "Name of the module"
...
INPUTS :      A, B, SEL ;             "Input pins of module MUX"
OUTPUTS :     Y ;                     "Output pin of module MUX"
TYPES :   NAND2, INV ;                "Two Module References: NAND2, INV."
          NAND2 : N1, N2, N3 ;        "Declaration of three instances of module NAND2"
          INV : I ;                   "Declaration of a single instance of module INV"
END_TYPES ;
...
END ;
```

Modules can be referenced with the list of their pins (and the attributes of them): the order of these pins may be different from that of the original module description. It is also possible to declare arrays of instances.

```
TYPES :   DECODER( I<0:2>#INPUTS, O<0:7>#OUTPUTS ) ;
          DECODER : INST, D<0:3> ;
END_TYPES ;
```

In the above example, the module DECODER has been referenced; it has one single input pin I of size three-bit and one output pin O of size eight-bit. Five instances of this module have been declared, namely, INST, D<0>, D<1>, D<2>, and D<3>.

In VHDL, components (and not entities) are instantiated inside architectures and linked to couples entity-architecture via the configuration mechanism. These components must be declared before their instantiation, but there is no explicit declaration of the instances: the instance names are the labels written in the instantiation statements. The labels are implicitly declared at the beginning of the corresponding declarative part (and can be used in configuration specifications).

```
architecture STRUCTURE of ANY is
    component NAND2 -- the component (and not the entity)
        port (A, B : in BIT ; Y : out BIT ) ;
    end component ;
    component INV -- the component (and not the entity)
        port (A : in BIT ; Y : out BIT ) ;
    end component ;
    component DECODER_COMP -- the component (and not the entity)
        port ( I : in BIT_VECTOR(0 to 2) ; O : out BIT_VECTOR(0 to 7) ) ;
    end component ;
    for all : NAND2 use entity WORK.NAND2(STRUCT) ; -- configuration of N1, N2, N3
    for I : INV use entity WORK.INV(DATAFLOW) ; -- configuration of I
    for INST : DECODER_COMP use entity WORK.DECODER(BEHAVIOR) ;
    -- configuration of component DECODER_COMP: the component is
    -- linked to the couple entity(architecture): DECODER(BEHAVIOR).
    -- INST is the name of an instance which is implicitly declared after keyword "is"
    ...
```

```
begin
        INST : DECODER_COMP port map (...) ;
        -- instantiation of DECODER_COMP : INST is the label (name) of the instance
        N1 : NAND2 port map (...) ; -- instance N1
        N2 : NAND2 port map (...) ; -- instance N2
        N3 : NAND2 port map (...) ; -- instance N3
        I : INV port map (...) ; -- instance I
end;
```

UDL/I does not implement the configuration mechanism, and therefore there is no component. Components must be declared in VHDL (there are new objects, different from the entities), but modules are just referenced in UDL/I (they already exist). As explained in the next paragraph, it is also possible in VHDL to have arrays of instances or conditional instance(s) with the generate statement.

10.5.2. Netlist Description

In UDL/I, the instances declared in the types statement are connected together in the *net section*. Such a net section is one possible body of the module description and excludes any other behavioral section in the same module (no mixed style of description).

The net section may include two kinds of statements: the *net statement* and the *delay statement*. The net statement may have three possible syntaxes:

```
Net_Name = FROM ( Pin_Name ) TO ( Pin_Name ) ;  -- direction specified
Net_Name = ( Pin_Name ) ( Pin_Name ) ;           -- direction specified (implicit FROM...TO)
Net_Name = Pin_Name  Pin_Name ;                  -- direction not specified
```

The first two forms of the statement specify the *direction* of the net; the third does not. Here again is the example of section 10.2:

```
NET_SECTION ;                                  "Net statements specifying the direction"
        NETA = FROM (.A) TO (N1.A) ;           ".A is the input pin A of global module MUX."
        NETB = FROM (.B) TO (N2.A) ;           "N2.A is the input pin A of the instance N2"
        NETS = FROM (.SEL) TO (I.A, N1.B) ;    "of submodule NAND2."
        NETSB = FROM (I.Y) TO (N2.B) ;         "NETSB is the name of the net"
        NETNA = FROM (N1.Y) TO (N3.A) ;        "connecting I.Y to N2.B"
        NETNB = FROM (N2.Y) TO (N3.B) ;
        NETY = FROM (N3.Y) TO (.Y) ;
END_SECTION ;
```

The name at the left of each statement is the *net name*. This net name is *not* declared (as terminals and pins are). It can be an array name:

```
NAME : SEQ_DEVICE ;
PURPOSE : LOGSIM ;
INPUTS : DBUS<1:8> ;
INPUTS : SM ;
CLOCK : CLK ;
OUTPUTS : DATA<1:2> ;
TYPES :   FLIPFLOP( CK#CLOCK, D#INPUTS, Q#OUTPUTS ),
              MUX4TO1( I<0:3>, SEL, O) ;
                  FLIPFLOP : F<1:8> ;  "8 instances of FLIPFLOP"
                  MUX4TO1 : M<1:2> ;  "2 instances of MUX4TO1"
END_TYPES ;
NET_SECTION ;
      NETIN<1:8> = (.DBUS<1:8>) (F<1:8>.D) ;         "8 nets"
      NETCK = (.CLK) (F<1:8>.CK) ;                   "1 net"
      NETSM = (.SM) (M<1:2>.SEL) ;        "1 net (M<1>.SEL and M<2>.SEL tied together)"
      NET1<1:4> = (F<1:4>.Q) (M<1>.I<0:3>) ;         "4 nets"
      NET2<1:4> = (F<5:8>.Q) (M<2>.I<0:3>) ;         "4 nets"
      NETOUT<1:2> = (M<1:2>.O) (.DATA<1:2>) ;        "2 nets"
END_SECTION ;
END ;
```

In VHDL, direction cannot be specified in the netlist as in UDL/I. At any rate, in most of the cases, both UDL/I and VHDL simulators are able to deduce the direction of a particular net from the direction of the ports or pins connected to it. This is not true when only VHDL ports of mode **linkage** or UDL/I pins of type EXT are connected to a given net: in that case, the direction of the net could be useful.

Arrays of instances are modeled in VHDL with the generate statement. A VHDL translation of the previous UDL/I model is as follows:

```
entity SEQ_DEVICE is
      port (   DBUS          : in BIT_VECTOR(1 to 8) ;
               SM, CLK       : in BIT ;
               DATA          : out BIT_VECTOR(1 to 2) ) ;
end SEQ_DEVICE ;
architecture STRUCTURE of SEQ_DEVICE is
      component FLIPFLOP
          port ( CK, D : in BIT ; Q : out BIT ) ;
      end component ;
      component MUX4TO1
          port ( I : in BIT_VECTOR(0 to 3) ; SEL : in BIT ; O : out BIT ) ;
      end component ;
      signal SIG : BIT_VECTOR(1 to 8) ; -- a signal of type array of bits (8 bits)
begin
      F : for I in 1 to 8 generate  -- generation of an array of 8 instances
          FF : FLIPFLOP port map (CLK, DBUS(I), SIG(I)) ;
      end generate ;
      M1 : MUX4TO1 port map (SIG(1 to 4), SM, DATA(1)) ; -- first port is connected to
      M2 : MUX4TO1 port map (SIG(5 to 8), SM, DATA(2)) ; -- an array of 4 bits
end STRUCTURE ;
```

In UDL/I, the netlist is described net by net, whereas in VHDL it is described instance by instance. Each VHDL instance has a port map specifying the connection between ports and signals, the last mentioned items being declared in the architecture (UDL/I nets are *not* declared objects).

The VHDL generate statement allows the description of arrays of components (as in UDL/I with the array notation INSTANCE<1:8>), but also of conditional instance, which is not possible in UDL/I:

```
-- assuming that N is a generic parameter of the corresponding entity:
F : if N mod 2 = 1 generate  -- conditional generation of a single instance
        FF : FLIPFLOP port map (CLK, DBUS(N), SIG(N)) ;
end generate ;
```

Actually, VHDL conditional generate statements usually (but not necessarily!) depend on generic parameters or deferred constants, which are very useful for the description of parameterized blocks (such as those generated by synthesis tools). But since UDL/I implements neither generic parameters nor deferred constants, the necessity of this conditional statement becomes very relative.

10.5.3. Delay Specifications

As stated earlier, the UDL/I netlist section may include delay statements. These statements may also be written in a primitive section (see paragraph 10.9.1).

The full syntax of this delay statement is very complex: a specific delay can be specified *for each path* (from pin to pin), *for each edge* (rising, falling, from or to high-impedance state), and also with a *loading factor*: the loading factor is multiplied by the total loading capacitance of the pin to give the loading delay, which is added to the basic delay to produce the final delay. Finally, all these parameters may be specified with *minimum, typical* and *maximum* values. For example:

```
NAME : COMBINATIONAL_DEVICE ;
<Purpose and Level Statements>
INPUTS : IN1, IN2, IN3 ;
OUTPUTS : OUT1, OUT2 ;
<Modules References, Instances Declarations>
NET_SECTION ;    or PRIMITIVE_SECTION ;
<Net Statements>   or <Truth Table Description>

 DELAY .OUT1 :            "Beginning of delay statement related to output pin OUT1:"
       LOAD = 0.2PF ,     "Load capacitance due to this module on pin OUT1"
       MAXLOAD = 3.0PF ,  "Maximum total capacitance admitted on pin OUT1: "
                          "otherwise warning message issued by simulator"
```

```
        RISE =   (               "Delay specification for edge 0 to 1 on output pin OUT1:"
            FROM (.IN1)  = (2.8NS, 3.2NS, 3.6NS) ,          "Delay from IN1 to OUT1"
            FROM (.IN2)  = (1.9NS, 2.1NS, 2.4NS) ,          "Delay from IN2 to OUT1"
            DEFAULT      = (2.0NS, 2.2NS, 2.4NS) ,          "Delay from others to OUT1"
            DRIVE   = (0.15NS/PF, 0.17NS/PF, 0.19NS/PF)     "Loading factor on pin OUT1"
                 ) ,
        EDGE ( 1 Z, 0 Z ) = (  "Delay specification for edges 1 to Z and 0 to Z on pin OUT1"
            FROM (.IN1)  = (1.9NS, 2.0NS, 2.1NS) ,          "Delay from IN1 to OUT1"
            DEFAULT      = (1.7NS, 1.8NS, 1.9NS) ,          "Delay from others to OUT1"
            DRIVE   = (0.15NS/PF, 0.18NS/PF, 0.19NS/PF)     "Loading factor on pin OUT1"
                  ) ,
        ... ;                    "End of delay statement related to output pin OUT1"

    DELAY .IN1 :                 "Delay statement related to input pin IN1 :"
        LOAD = 0.5PF ;           "Load capacitance due to this module on pin IN1"
        ...
  END_SECTION ;
  END ;
```

Such delay specifications are not easy to implement in VHDL, especially loading delays: this important backannotation feature takes into account the total capacitance introduced on a net by all pins connected to it (and specified in different modules by the LOAD value).

All the other features — choice between minimum, typical, and maximum delays, addition of the loading delay to the basic delay corresponding to the current edge — may be modeled in VHDL using packages of constants and functions (principles of such modeling are given in the chapter concerning Verilog).

It is not clear in the UDL/I reference manual whether delays specified in a netlist section override delays included in the instantiated modules or not (to prevent designer headaches, it should certainly be the case!). Since a module containing a primitive section is always at the bottom of the hierarchy (level *END*), delays specified in a primitive section are absolute delays.

At any rate, UDL/I delay specifications are very well defined and should satisfy the IC designer. Even if the philosophy is quite different, there is certainly a lack in VHDL in this area. This lack can be partially fixed with some packages, except for the backannotation mechanism (loading capacitor effects).

10.6. UDL/I BEHAVIORAL DESCRIPTION

10.6.1. Predefined Operators and Functions

UDL/I provides many predefined functions and operators, which are listed below and compared to equivalent VHDL subprograms (when these exist). Of course, the VHDL designer always has the possibility to define any procedure or function he needs, whereas the UDL/I designer can only use the predefined ones: UDL/I does not implement user-defined subprograms.

10.6.1.1. Bit Logic and Boolean Operators

Bit Logic Operators	UDL/I	VHDL
unary bitwise invert	^	**not**
unary two's complement	-	*not predefined*
binary bitwise and	&	**and**
binary bitwise nand	^&	**nand**
binary bitwise or	!	**or**
binary bitwise nor	^!	**nor**
binary bitwise xor	@	**xor**
binary bitwise xnor	^@	*not predefined*
binary bitwise equality	=	*not predefined*
binary bitwise inequality	^=	*not predefined*
unary reduce and	&	*not predefined*
unary reduce nand	*not predefined*	*not predefined*
unary reduce or	!	*not predefined*
unary reduce nor	*not predefined*	*not predefined*
unary reduce xor	@	*not predefined*
unary reduce xnor	*not predefined*	*not predefined*

These UDL/I operators are defined for the bit values X, 0, 1, Z, whereas VHDL logical operators are only predefined on type BIT ('0','1') and must be redefined on type UDLI_BIT. UDL/I provides more unary operators (reduce and, or, xor) and also a logical equality that may produce an unknown result (X). This equality (as well as the inequality) operator also accepts the don't-care value (?) as an input. Following is the truth table of the UDL/I equality:

=	X	0	1	Z	?
X	X	X	X	X	1
0	X	1	0	X	1
1	X	0	1	X	1
Z	X	X	X	X	1
?	1	1	1	1	1

Such an operator can be easily implemented in VHDL on an enumerated type including the five elements 'X', '0', '1', 'Z', and '?'. The result of this new operator will be of type UDLI_BIT, and therefore it *cannot* be used instead of the boolean equality as a condition in *IF* statements (which must be of type BOOLEAN).

On the other hand, UDL/I does not include a boolean equality producing only TRUE or FALSE value, as in VHDL: the equality and inequality operators used in UDL/I are only those defined here (= and ^=). The evaluation of expressions using these operators may give as a result 1, 0, or X. When used as the condition of *IF* statements, the value 1 will lead to the execution of the *THEN* clause, 0 of the *ELSE* clause, and X of the *OFFSTATE* clause (see 10.6.3.1).

10.6.1.2. Relational Functions/Operators

Relational Operators	UDL/I	VHDL
less than	LT	<
less than or equal to	LE	<=
greater than	GT	>
greater than or equal to	GE	>=

Relational functions and operators do not have the same semantics in both languages: UDL/I functions are predefined on bit arrays, which are actually considered as unsigned integers, whereas VHDL operators are predefined on all scalar types (integers, reals, physicals, enumerations) and discrete array types (such as BIT_VECTOR).

Thus, UDL/I bit array comparisons are unsigned integer comparisons, whereas VHDL bit vector comparisons are based on the order of the enumeration literals listed in the subtype declaration. For example:

```
type BIT is ('0', '1') ;
-- predefined type BIT, '0' < '1'
type UDLI_BIT is ('X' , '0', '1', 'Z') ;
-- user-defined type, 'X' < '0' < '1' < 'Z'
type CRAZY_BIT is ('1', '0') ;
-- user-defined type, '1' < '0'
```

If bit arrays include X or Z values, the result of UDL/I comparison is implementation dependent: this is dangerous for portability reasons (and a standard is supposed to be portable). In VHDL, the comparison is always deterministic (for the previous UDLI_BIT type, 'X' < '0' is always true).

10.6.1.3. Arithmetic Functions/Operators

Arithmetic Operators	**UDL/I for bit arrays (treated as unsigned integers)**	**VHDL for integers and reals (but *not* bit vectors)**
addition	ADD, ADDC	+
subtraction	SUB, SUBC	-
incrementation	INC	*not predefined*
decrementation	DEC	*not predefined*
multiplication	MULT	*
division	DIV	/
modulus	*not predefined*	**mod**
remainder	MOD	**rem**
exponentiation	*not predefined*	**
absolute value	*not predefined*	**abs**

ADD performs the addition of two bit arrays of the same length L and of one bit carry-in, and returns a bit array of identical length L, without generating an output carry. ADDC performs the addition on the same inputs but returns a bit array of length L+1, generating the output carry at the MSB. SUB and SUBC have the same differences.

There is a fundamental difference between arithmetic subprograms in UDL/I and VHDL: UDL/I functions are predefined on bit arrays, which are again considered as unsigned integers, whereas VHDL operators are only predefined on integers and reals, but *not* on bit vectors for which there is no implicit conversion into integers.

Thus, explicit conversion functions have to be written in VHDL to convert BIT_VECTOR or UDLI_BIT_VECTOR into INTEGER and vice versa. The VHDL modeling of such functions, as well as those corresponding to the UDL/I functions listed above, is straightforward. Using the operator overloading will enhance readability:

```
library UDLI_LIBRARY ;
use UDLI_LIBRARY.UDLI_TYPES.all ; -- package detailed in 10.6.1.5
package UDLI_ARITHMETIC is
  function TO_INT( V : UDLI_BIT_VECTOR) return INTEGER ;
  -- function converting UDLI_BIT_VECTOR into INTEGER
```

```
function TO_UDLI( I : INTEGER) return UDLI_BIT_VECTOR ;
-- function converting INTEGER into UDLI_BIT_VECTOR
function ADD (A, B : UDLI_BIT_VECTOR ; CIN : UDLI_BIT) return UDLI_BIT_VECTOR ;
-- addition corresponding to the UDLI function ADD
function ADDC (A,B : UDLI_BIT_VECTOR ; CIN : UDLI_BIT) return UDLI_BIT_VECTOR;
-- addition corresponding to the UDLI function ADDC
function "+"(A, B : UDLI_BIT_VECTOR) return UDLI_BIT_VECTOR ;
-- addition corresponding to the UDLI function ADD with CIN = '0' (for example)
-- overload of the predefined operator "+" having integer or real arguments
 ...
end UDLI_ARITHMETIC ;
```

Complementing these declarations, as well as writing the corresponding bodies, will constitute a good exercise for the reader.

10.6.1.4. Bit Manipulation and Miscellaneous Functions

Some very useful bit manipulation functions are predefined in the UDL/I reference manual. These functions are as follows:

Miscellaneous Operators	M	VHDL
concatenation	!!	&
shift right logical	SRL	*not predefined*
shift left logical	SLL	*not predefined*
shift right arithmetic	SRA	*not predefined*
shift left arithmetic	SLA	*not predefined*
circular shift right	CIR	*not predefined*
circular shift left	CIL	*not predefined*
MSB through LSB reverse	REFLECT	*not predefined*
decoder	DECODE	*not predefined*
copy (repeat)	COPY	*not predefined*

Except for the concatenation operator, none of the above UDL/I functions is predefined in VHDL. Seven of them perform classical bit manipulations: logical, arithmetic, and circular shifts and reflection. The DECODE function returns the decoded value of the input expression (DECODE(3b011) = 8b00001000) ; the COPY function repeats the input expression a given number of times (COPY(3b011,3) = 9b011011011).

As with the previous arithmetic functions, these functions are not predefined but are easy to implement in VHDL. It is interesting to notice the different approaches taken by both languages (both of which are or intend to become a standard): VHDL has a very poor number of predefined subprograms, whereas UDL/I seems to provide most of the required functions for IC design.

Would these functions have been sufficient for system modeling? Certainly not, but that was not the purpose of UDL/I, which is focused on integrated circuits. Once again, because VHDL is more general, it is important to emphasize the absolute need in VHDL to define the designer goals related to his specific application in order to *build* the VHDL environment (packages) that the designer needs to perform efficient and comfortable modeling. *Building an environment* means that some designers or developers have to *write* specific VHDL packages, but it also means that a lot of designers are going to *reuse* these packages.

10.6.1.5. UDL/I Standard Functions

Standard functions may be called in primitive description sections (see paragraph 10.9.1). They can thus model basic modules (for which LEVEL is END) and allow efficient simulation: they are indeed often implemented in the kernel of the simulators based on a four-value logic (X, 0, 1, Z).

These functions are the logical functions AND, NAND, OR, NOR, DELAY (buffer), NOT, POWER (output = 1), GRND (output = 0), XOR, XNOR, BUFIF1, BUFIF0, NOTIF1, NOTIF0, the dot functions (see paragraph 10.4.1) WAND, WOR, TRIRECX, TRIREC1, TRIREC0, TRIRECZ, TRIRECH, and two memory functions ROM and RAM.

The implementation of these functions in VHDL is given below. They are all declared in a package called UDLI_TYPES, which also includes the UDLI_BIT types and subtypes declarations seen in paragraph 10.4.1. The corresponding package body follows.

Warning: in UDL/I, the standard functions are actually procedures (the output pin is specified as an argument). It is more convenient in VHDL to define them as functions, so that some of them can be used as resolution functions.

```
package UDLI_TYPES is
        -- unresolved enumeration type corresponding to the UDL/I logic type
        type UDLI_BIT is ('X', '0', '1', 'Z') ;
        -- unresolved array type
        type UDLI_BIT_VECTOR is array (NATURAL range <>) of UDLI_BIT ;
        -- resolution functions
        function WAND ( S : UDLI_BIT_VECTOR ) return UDLI_BIT ;
        function WOR ( S : UDLI_BIT_VECTOR ) return UDLI_BIT ;
        function TRIREC1 ( S : UDLI_BIT_VECTOR ) return UDLI_BIT ;
        function TRIREC0 ( S : UDLI_BIT_VECTOR ) return UDLI_BIT ;
        function TRIRECX ( S : UDLI_BIT_VECTOR ) return UDLI_BIT ;
        function TRIRECZ ( S : UDLI_BIT_VECTOR ) return UDLI_BIT ;
        function TRIRECH ( S : UDLI_BIT_VECTOR ) return UDLI_BIT ;
        -- subtype declarations
        subtype WAND_BIT        is    WAND UDLI_BIT ;
        subtype WOR_BIT         is    WOR UDLI_BIT ;
```

```
        subtype TRIREC1_BIT      is   TRIREC1 UDLI_BIT ;
        subtype TRIREC0_BIT      is   TRIREC0 UDLI_BIT ;
        subtype TRIRECX_BIT      is   TRIRECX UDLI_BIT ;
        subtype TRIRECH_BIT      is   TRIRECH UDLI_BIT ;
        subtype TRIRECZ_BIT      is   TRIRECZ UDLI_BIT ;
        -- logical functions
        function F_AND ( S : UDLI_BIT_VECTOR ) return UDLI_BIT ;
        function F_NAND ( S : UDLI_BIT_VECTOR ) return UDLI_BIT ;
        function F_OR ( S : UDLI_BIT_VECTOR ) return UDLI_BIT ;
        function F_NOR ( S : UDLI_BIT_VECTOR ) return UDLI_BIT ;
        function F_XOR ( S : UDLI_BIT_VECTOR ) return UDLI_BIT ;
        function F_XNOR ( S : UDLI_BIT_VECTOR ) return UDLI_BIT ;
        function DELAY ( S : UDLI_BIT ) return UDLI_BIT ;
        function F_NOT ( S : UDLI_BIT ) return UDLI_BIT ;
        function POWER return UDLI_BIT ;
        function GRND return UDLI_BIT ;
        function BUFIF1 ( CTRL, INPUT : UDLI_BIT ) return UDLI_BIT ;
        function BUFIF0 ( CTRL, INPUT : UDLI_BIT ) return UDLI_BIT ;
        function NOTIF1 ( CTRL, INPUT : UDLI_BIT ) return UDLI_BIT ;
        function NOTIF0 ( CTRL, INPUT : UDLI_BIT ) return UDLI_BIT ;
end UDLI_TYPES ;
```

The body of these functions is given below:

```
package body UDLI_TYPES is
------------------------------------------------------------------------
        function WAND ( S : UDLI_BIT_VECTOR ) return UDLI_BIT is
            variable FOUND_X : BOOLEAN ;
        begin
            for I in S'range loop
                case S(I) is
                  when 'X' => FOUND_X := TRUE ;
                  when '0' => return '0' ;
                  when others => null ;
                end case ;
            end loop ;
            if FOUND_X then
                return 'X' ;
            else
                return '1' ;  -- value also returned if S'LENGTH = 0
            end if ;
        end WAND ;
------------------------------------------------------------------------
        function WOR ( S : UDLI_BIT_VECTOR ) return UDLI_BIT is
            variable FOUND_X : BOOLEAN ;
        begin
            for I in S'range loop
                case S(I) is
                  when 'X' => FOUND_X := TRUE ;
                  when '1' => return '1' ;
                  when others => null ;
```

```
            end case ;
        end loop ;
        if FOUND_X then
            return 'X' ;
        else
            return '0' ;  -- value also returned if S'LENGTH = 0
        end if ;
    end WOR ;
-------------------------------------------------------------------------------
    function TRIREC ( S : UDLI_BIT_VECTOR ; DEF : UDLI_BIT) return UDLI_BIT is
    -- function used only internally: thus not declared in package UDLI_TYPES.
    -- function used for the definition of TRIREC0, TRIREC1, TRIRECX,
    -- TRIRECX and TRIRECH resolution functions.
        variable FOUND_1, FOUND_0 : BOOLEAN ;
    begin
        for I in S'range loop
            case S(I) is
             when 'X' => return 'X' ;
             when '0' => FOUND_0 := TRUE ;
             when '1' => FOUND_1 := TRUE ;
             when others => null ;
            end case ;
        end loop ;
        if FOUND_0 and FOUND_1 then
            return 'X' ;
        elsif FOUND_0 then  -- only '0's
            return '0' ;
        elsif FOUND_1 then  -- only '1's
            return '1' ;
        else  -- only 'Z's or S'LENGTH = 0
            return DEF ;
        end if ;
    end TRIREC ;
-------------------------------------------------------------------------------
    function TRIREC1 ( S : UDLI_BIT_VECTOR ) return UDLI_BIT is
    begin
        return TRIREC( S, '1' ) ;
    end TRIREC1 ;
-------------------------------------------------------------------------------
    function TRIREC0 ( S : UDLI_BIT_VECTOR ) return UDLI_BIT is
    begin
        return TRIREC( S, '0' ) ;
    end TRIREC0 ;
-------------------------------------------------------------------------------
    function TRIRECX ( S : UDLI_BIT_VECTOR ) return UDLI_BIT is
    begin
        return TRIREC( S, 'X' ) ;
    end TRIRECX ;
-------------------------------------------------------------------------------
    function TRIRECZ ( S : UDLI_BIT_VECTOR ) return UDLI_BIT is
    begin
```

```
        return TRIREC( S, 'Z' ) ;
    end TRIRECZ ;
--------------------------------------------------------------------------------
    function TRIRECH ( S : UDLI_BIT_VECTOR ) return UDLI_BIT is
    begin
        return TRIREC( S, 'Z' ) ;       -- does not correspond to the UDL/I function
    end TRIRECH ;                       --  (see below)
--------------------------------------------------------------------------------
    function F_AND ( S : UDLI_BIT_VECTOR ) return UDLI_BIT is
        variable FOUND_X_OR_Z : BOOLEAN ;
    begin
        for I in S'range loop
            case S(I) is
             when 'X' | 'Z' => FOUND_X_OR_Z := TRUE ;
             when '0' => return '0' ;
             when others => null ;
            end case ;
        end loop ;
        if FOUND_X_OR_Z then
            return 'X' ;
        else
            return '1' ;
        end if ;
    end F_AND ;
--------------------------------------------------------------------------------
    function F_NAND ( S : UDLI_BIT_VECTOR ) return UDLI_BIT is
    begin
        return F_NOT(F_AND(S)) ;
    end F_NAND ;
--------------------------------------------------------------------------------
    function F_OR ( S : UDLI_BIT_VECTOR ) return UDLI_BIT is
        variable FOUND_X_OR_Z : BOOLEAN ;
    begin
        for I in S'range loop
            case S(I) is
             when 'X' | 'Z' => FOUND_X_OR_Z := TRUE ;
             when '1' => return '1' ;
             when others => null ;
            end case ;
        end loop ;
        if FOUND_X_OR_Z then
            return 'X' ;
        else
            return '0' ;
        end if ;
    end F_OR ;
--------------------------------------------------------------------------------
    function F_NOR ( S : UDLI_BIT_VECTOR ) return UDLI_BIT is
    begin
        return F_NOT(F_OR(S)) ;
    end F_NOR ;
```

```
--------------------------------------------------------------------------------
    function F_XOR ( S : UDLI_BIT_VECTOR ) return UDLI_BIT is
        variable NB_1 : NATURAL ;
    begin
        for I in S'range loop
            case S(I) is
             when '0' => null ;
             when '1' => NB_1 := NB_1 + 1 ;
             when others => return 'X' ;
            end case ;
        end loop ;
        if NB_1 mod 2 = 1 then
            return '1' ;
        else
            return '0' ;
        end if ;
    end F_XOR ;
--------------------------------------------------------------------------------
    function F_XNOR ( S : UDLI_BIT_VECTOR ) return UDLI_BIT is
    begin
        return F_NOT(F_XOR(S)) ;
    end F_XNOR ;
--------------------------------------------------------------------------------
    function DELAY ( S : UDLI_BIT ) return UDLI_BIT is
    begin
        return S ; -- in UDL/I, the DELAY function does not change 'Z' into 'X'
    end DELAY ;
--------------------------------------------------------------------------------
    function F_NOT ( S : UDLI_BIT ) return UDLI_BIT is
    begin
        case S is
         when '0' => return '1' ;
         when '1' => return '0' ;
         when others => return 'X' ;
        end case ;
    end F_NOT ;
--------------------------------------------------------------------------------
    function POWER return UDLI_BIT is
    begin
        return '1' ;
    end POWER ;
--------------------------------------------------------------------------------
    function GRND return UDLI_BIT is
    begin
        return '0' ;
    end GRND ;
--------------------------------------------------------------------------------
    function BUFIF1 ( CTRL, INPUT : UDLI_BIT ) return UDLI_BIT is
    begin
        case CTRL is
         when '0' =>
```

```
            return 'Z' ;
        when '1' =>
            if INPUT = 'Z' then
                return 'X' ;
            else
                return INPUT ;
            end if ;
        when others =>
            return 'X' ;
    end case ;
end BUFIF1 ;
--------------------------------------------------------------------------------
function BUFIF0 ( CTRL, INPUT : UDLI_BIT ) return UDLI_BIT is
begin
    return BUFIF1 (F_NOT(CTRL), INPUT) ;
end BUFIF0 ;
--------------------------------------------------------------------------------
function NOTIF1 ( CTRL, INPUT : UDLI_BIT ) return UDLI_BIT is
begin
    return F_NOT(BUFIF1(CTRL,INPUT)) ;
end NOTIF1 ;
--------------------------------------------------------------------------------
function NOTIF0 ( CTRL, INPUT : UDLI_BIT ) return UDLI_BIT is
begin
    return F_NOT(BUFIF0(CTRL,INPUT)) ;
end NOTIF0 ;
--------------------------------------------------------------------------------
end UDLI_TYPES ;
```

The UDL/I dot (resolution) functions are very close to the Verilog resolution functions (see chapter 9, "Verilog and VHDL"). Equivalent VHDL functions have already been given for Verilog, and the interested reader may notice that two variants of equivalent VHDL descriptions have been given in the UDL/I and Verilog chapters. But with reference to Verilog, an important warning must be given: the VHDL TRIREC1, TRIREC0,...,TRIRECH functions are *not associative*, and this could lead to strange conflict resolutions.

For example, using the TRIREC1 resolution, a first conflict between 'Z' and 'Z' will give '1': if this '1' value is propagated outside the entity and is then in conflict with a '0', the second conflict resolution would give a final value 'X': it should have been '0', because the first '1' was a *charged* (weak) '1' resulting from the high-impedance state, whereas '0' is a *driven* value.

To model associative resolution functions (giving more accurate results), the four-value logic type is thus not sufficient: it is necessary to differentiate a charged '1' from a driven '1' value, and thus to introduce *strength levels*. Strength modeling is beyond the scope of this paragraph. But similar questions have already been discussed in the Verilog chapter: the reader may refer to the corresponding paragraphs for further details.

Another limitation is in the VHDL TRIRECH function, which does not correspond to the UDL/I one, for which specification the last value of the signal remains when all driver values are equal to 'Z'. Alas, there is no way with the VHDL resolution mechanism to assign the last value of the signal (no change), because only the new values proposed by the drivers are passed as arguments to the resolution function. The VHDL modeling of such tristate signals will use other constructs: for example, they can be declared as guarded signals of the register kind, and having all the inputs equal to 'Z' will be modeled with the disconnection of the corresponding drivers. Thus, the same behavior can be modeled, but not with resolution functions.

Finally, the ROM and RAM standard functions can be translated into VHDL procedures having access to global signals (declared in a package) that represent the memory data. To prevent simultaneous access to data stored at the same address, a correct resolution function should be defined. Of course, multiple ROM and RAM devices (i.e., signals) can be declared. A package including such types and procedures could be

```
library UDLI_LIBRARY ;
use UDLI_LIBRARY.UDLI_TYPES.all ;
package MEMORY is
        type  RES_BIT_VECTOR is array(NATURAL range <>) of RESOLVED UDLI_BIT ;
        -- RESOLVED is any user-defined resolution function
        type UDLI_ACCESS is access RES_BIT_VECTOR ;
        type MEM_DEV is array (NATURAL range <>) of UDLI_ACCESS ;
        -- using access type is a trick to declare "arrays of unconstrained arrays"...
        procedure INIT ( VALUE : MEM_DEVICE ; signal THE_MEM : inout MEM_DEVICE ) ;
        -- procedure initializing the content of the ROM or the RAM
        procedure READ_ROM (     signal CS : in UDLI_BIT ;
                                 signal ADDRESS: in UDLI_BIT_VECTOR ;
                                 signal DATA : out UDLI_BIT_VECTOR ;
                                 signal THE_ROM : inout MEM_DEVICE     ) ;
        -- procedure reading any ROM signal
        procedure READ_WRITE_RAM (     signal CS, RW : in UDLI_BIT ;
                                       signal ADDRESS: in UDLI_BIT_VECTOR ;
                                       signal DATA : inout UDLI_BIT_VECTOR ;
                                       signal THE_RAM : inout MEM_DEVICE      ) ;
        -- procedure reading or writing any RAM signal
        signal ROM256x4 : MEM_DEV(0 to 255) := (others=>new UDLI_BIT_VECTOR(0 to 3));
        -- signal modeling a ROM of 256 data of 4 bits
        signal RAM64x8 : MEM_DEV(0 to 63) := (others=>new UDLI_BIT_VECTOR(0 to 7));
        -- signal modeling a ROM of 256 data of 4 bits
end MEMORY ;
```

10.6.2. UDL/I Conflict Resolution: the OFFSTATE Clause

Before looking at the behavioral constructs of UDL/I, it is necessary to understand the resolution mechanism introduced by the *OFFSTATE* clause. As explained in paragraph 10.4.1, structural descriptions associate resolution specifications only with output or bus pin definitions, and not with net declarations (structural nets are not declared). On the other hand, resolution specifications can be attached to some facilities in a behavioral section: these facilities are the *terminals*, the *external assignable pins*, and the *assignable submodule pins*.

The *OFFSTATE* clause specifies the value of suc.ı a facility when no driver is assigning a value to the facility or when two or more drivers are assigning different values to the same facility (conflict). This clause is thus adapted to terminals: it cannot be used for registers and latches because their behavior is specified differently (with clock conditions: see next paragraph). The UDL/I reference manual gives the following table:

Driving Values	*OFFSTATE* Value			
	0	1	Z	D
(-, -)	**0**	**1**	**Z**	**X**
(-, 0)	0	0	0	0
(-, 1)	1	1	1	1
(0,0)	0	0	0	0
(0,1)	**1**	**0**	**X**	**X**
(1,1)	1	1	1	1

("-" means that the driver is not assigning a value to the terminal (disconnected))

The value of a facility in the two cases stated above is specified by the *OFFSTATE* value, which can be 0, 1, Z, or D. To each of these values corresponds a different behavior of the terminal. As the reader may notice, the behavior is not fully specified by the previous table: what is the resolved value if one of the driving values is equal to X or Z? The UDL/I language reference manual needs to be revised (as does that of VHDL, of course).

An *OFFSTATE* clause can be specified in three statements: *IF* statement, *CASE* statement, and *CASEOF* statement (or, equivalently, in *IF*, *CASE*, and *CASEOF* expressions). All these statements are conditional statements: the *OFFSTATE* value is used to determine the value of the terminals when all associated conditions are false. Further details will be given in the next paragraphs.

In VHDL, conflict resolutions are specified with the resolution function: this function takes as parameters the array of driving values and returns the resolved value of the signal. The resolution function can also be used to specify the value of a signal when all its drivers are disconnected, provided that this signal has been declared as a *guarded signal of kind BUS* (in this case, the resolution function is called without parameter after the last driver disconnection and returns the *off-state* value).

Thus, four resolved subtypes that have been declared in the package UDLI_TYPES (see 10.6.5.1) may be *renamed* in order to model the UDL/I terminals:

```
library UDLI_LIBRARY ;
use UDLI_LIBRARY.UDLI_TYPES.all ;
package UDLI_TERMINAL is
                                                     -   when (-,-)   when (0,1)
    subtype TERMINAL_0 is WOR_BIT ;                  -      '0'          '1'
    subtype TERMINAL_1 is WAND_BIT ;                 -      '1'          '0'
    subtype TERMINAL_Z is TRIRECZ_BIT ;              -      'Z'          'X'
    subtype TERMINAL_D is TRIRECX_BIT ;              -      'X'          'X'
    -- the resolved subtypes WOR_BIT,... have been declared in the package
    -- UDLI_TYPES to model the UDL/I dot functions used in output pin definitions
end ;
```

It is easy to verify that the associated resolution functions *WOR, WAND, TRIRECZ,* and *TRICRECX* satisfy the UDL/I *OFFSTATE* specifications given by the previous table (of course, this translation is valid only if the guessed specifications for X and Z values are correct).

Then, the declaration of any VHDL signal modeling a UDL/I terminal specified by the OFFSTATE value 1 (for example) will be

```
signal ONE_TERM : TERMINAL_1 bus ;
-- guarded signal of kind bus (resolution function called at the last driver disconnection)
```

Such a VHDL modeling of the UDL/I terminals will be extended in the next paragraph (in the parts concerning the UDL/I conditional statements).

10.6.3. Statements

The behavior description section of a module consists of the statements presented here. Most of these statements model the behavior of the facilities seen in paragraph 10.4.2: terminal, latch, register, and so on. UDL/I defines other statements that control the activation and synchronization of the state machines and will be detailed in paragraph 10.9.2 on the UDL/I automaton.

The statements assigning facilities may be of six types: assign statement, begin_end statement, *IF* statement, *CASE* statement, *CASEOF* statement, and *AT* statement. *All these UDL/I statements are concurrent statements* .

The semantics of these statements depends on the kinds of facilities that are assigned. The UDL/I semantics is defined in the reference manual with a core subset. In this core subset, all the facilities are translated into one of two facilities: terminals (*combinational* elements: terminals) and registers (*memory* elements: latches, registers, RAMs). Thus, two semantics rules are defined:

- A statement assigning a *terminal* is executed whenever a signal value of the right side of the assignment changes. Such a statement models *combinational* devices. When this statement is a conditional statement and if all the associated conditions are false, then the value of the terminal is given by the *OFFSTATE* clause (see previous paragraph).

- A statement assigning a *register* may have a synchronous or asynchronous behavior (the right side of this statement must thus include a clock expression). In the case of an edge-sensitive clock expression, the assignment is executed only when the edge condition occurs. In the case of a level-sensitive clock expression, the assignment is executed whenever the level condition becomes true or the input value changes while this level condition is true. Both kinds of statement model *memory* devices. The *OFFSTATE* clause is ignored by the register.

The VHDL semantics defines all concurrent statements with the same terms as the first UDL/I rule. Actually, the VHDL translation of the second kind of UDL/I statements will use processes that executes sequential code (not implemented in UDL/I: see section 10.3), thus having a behavior similar to the UDL/I one.

10.6.3.1. Assign Statement

The assign statement is of the form

```
<concatenation> := <expression> ;
```

where the expression may be of six kinds: simple expression with optional delay, begin_end expression, *IF* expression, *CASE* expression, *CASEOF* expression, and *AT* expression. The statements using the five last expressions (begin_end, *IF, CASE, CASEOF,* and *AT*) are equivalent to the five begin_end, *IF, CASE, CASEOF,* and *AT* statements: they will be explained in the next paragraphs.

The first form of the assign statement is

```
<concatenation> := <simple expression> [ <delay> ] ;
```

It contains no clock expression and can thus only assign terminals. For example, using the standard function ADDC

```
COUT !! ACC<0:7> := ADDC( A<0:7>, B<0:7>, CIN ) DELAY 8.25NS ;
```

which is equivalent to the VHDL concurrent assignment:

```
COUT & ACC <= UBIT9'( ADDC(A, B, CIN) ) after 8.25 ns ;
-- where ACC, A and B are of subtype UDLI_BIT_VECTOR(0 to 7)
-- and UBIT9 is the subtype UDLI_BIT_VECTOR(0 to 8)
-- this subtype qualification is compulsory because VHDL cannot deduce the type
-- of an aggregate from the type of the elements of this aggregate...
```

Like the UDL/I one, the corresponding VHDL statement is executed each time a signal value of the right side changes.

10.6.3.2. Begin_End Statement

The begin_end statement is just a construct that enables a group of statements to become syntactically equivalent to one statement. This statement has the following syntax:

```
BEGIN
        <statement>
        { <statement> }
END ;
```

All the statements included between BEGIN and END keywords are concurrent statements. In VHDL, such statements could be written in a block statement (without a guard condition and declaration). Actually, this UDL/I construct is not really a block, because it is not a declarative region.

The begin_end expression is similarly used to represent a blocked expression:

```
BEGIN
        <expression>
        { <expression> }
END ;
```

The following two syntaxes are equivalent:

TERM := **BEGIN** <expression> { <expression> } **END** ;	**BEGIN** TERM := <expression> ; { TERM := <expression> ; } **END** ;

10.6.3.3. IF Statement

The syntax of this statement is the following:

```
IF <simple expression>   THEN  <statement> {<statement>}
                         [ ELSE  <statement> {<statement>} ]
                         [ OFFSTATE  <off state value> ]
END_IF ;
```

where <simple expression> is *one-bit* wide.

The meaning of the *IF* statement, which is a *concurrent* statement, is the following: if <simple expression> is 1, the *THEN* statements are *valid*; if

<simple expression> is 0, the *ELSE* statements are *valid*; if <simple expression> is X or Z, or if <simple expression> is 0 and there is no *ELSE* clause, the *OFFSTATE* clause specifies the value assigned to all the terminals appearing in the previous statements.

Valid statements means that the statements are executed with the semantics of concurrent statements, i.e., according to one of the two rules mentioned at the beginning of paragraph 10.6.3. If all signals appearing on the right side of these statements remain quiet, then no statement may be executed. Thus, the selected statements were said to be *valid* instead of *executing* in order to distinguish their behavior from the execution of sequential statements.

Furthermore, the UDL/I reference manual specifies that the *IF* statement is converted in the core subset into these two statements:

```
IF <simple expression>    THEN  <statement of the THEN clause>
                                { <statement of the THEN clause> }
                          [ OFFSTATE  <same off state value>  ]
END_IF ;
IF ^<simple expression> THEN  <statement of the ELSE clause>
                                { <statement of the ELSE clause> }
                          [ OFFSTATE  <same off state value>  ]
END_IF ;
```

Thus, the previous UDL/I *IF* statement is more or less equivalent to two VHDL *guarded blocks*:

```
LABEL : block ( TRUE(<VHDL translation of simple expression>) )
begin
      <VHDL guarded assignment corresponding to statement of the THEN clause>
      { <VHDL guarded assignment corresponding to statement of the THEN clause> }
end block;
LABEL_ : block ( FALSE(<VHDL translation of simple expression>) )
begin
      <VHDL guarded assignment corresponding to statement of the ELSE clause>
      { <VHDL guarded assignment corresponding to statement of the ELSE clause> }
end block;
```

TRUE and FALSE are functions overloading the elements of the enumeration type BOOLEAN and defined on type UDLI_BIT:

```
function TRUE (S : UDLI_BIT)
      return BOOLEAN is
begin
      if S = '1' then
          return TRUE ;
      else
          return FALSE ;
      end if ;
end ;
```

```
function FALSE (S : UDLI_BIT)
      return BOOLEAN is
begin
      if S = '0' then
          return TRUE ;
      else
          return FALSE ;
      end if ;
end ;
```

Such a translation is consistent with the modeling of the UDL/I terminals by VHDL guarded signals of kind *BUS* (see paragraph 10.6.2). The two guarded blocks above have guarded conditions that use the functions TRUE and FALSE: therefore, these guarded conditions cannot be true simultaneously (and thus, the resolution function is never called to resolve a meaningless conflict between the THEN and the ELSE assignments). On the other hand, the conditions are both false when the result of the simple expression is either 'X' or 'Z': at the time when this simple expression becomes unknown, both drivers are disconnected, and the resolution function is called without parameters (signals of kind *BUS*), returning the value of the signal corresponding to the UDL/I *OFFSTATE* clause. An example:

```
TERMINAL : A, B, C, SEL ;
TERMINAL : TERM1, TERM2 ;
...
IF SEL THEN
      TERM1 := A ;
      TERM2 := C ;
ELSE
      TERM1 := B ;
OFFSTATE
      Z
END_IF ;
```

```
signal A, B, C, SEL : UDLI_BIT ;
signal TERM1, TERM2 : TERMINAL_Z bus ;
...
T : block ( TRUE(SEL) )
begin
      TERM1 <= guarded A ;
      TERM2 <= guarded C ;
end block;
F : block ( FALSE(SEL) )
begin
      TERM1 <= guarded B ;
end block;
```

In the VHDL example, when SELECT becomes equal to 'X' (or 'Z'), both guarded conditions become false, all drivers are disconnected, and the resolution function TRIRECZ is called without argument, returning the value 'Z', which is assigned to the signal TERM1. This corresponds exactly to the UDL/I behavior.

As stated in the beginning of paragraph 10.6.3, the UDL/I registers ignore the *OFFSTATE* clause: if the simple expression associated with the *IF* statement is X or Z, or if this simple expression is 0 and there is no *ELSE* clause (i.e., cases where the *OFFSTATE* clause is used for terminals), the UDL/I registers simply keep their last value (neither synchronous nor asynchronous transfers can occur). At first glance, the UDL/I registers may thus be modeled in VHDL by *guarded signals of the register kind*, which keep their last value after the last driver disconnection. The modeling of these facilities will be detailed in 10.6.3.6 (*AT* statements).

Finally, UDL/I provides an *IF* expression that can be assigned:

```
TERM := IF <simple expression>   THEN <expression>
                                 [ ELSE <expression> ]
                                 [ OFFSTATE  <off state value> ]
          END_IF ;
```

This is a shortcut notation totally equivalent to the following *IF* statement:

```
IF <simple expression>    THEN TERM := <expression> ;
                          [ ELSE TERM := <expression> ; ]
                          [ OFFSTATE  <off state value>  ]
END_IF ;
```

which can be then translated into the corresponding VHDL guarded blocks. Of course, if the previous terminal TERM is assigned only by one *IF* expression (one driver), it can also be modeled in VHDL by a concurrent conditional signal assignment:

```
TERM <=  <THEN expression> when TRUE(<simple expression>) else
         <ELSE expression> when FALSE(<simple expression>) else
         <'0', '1', 'Z' or 'X' corresponding to the offstate value 0, 1, Z or D> ;
```

10.6.3.4. CASE Statement

The UDL/I syntax of the *CASE* statement is

```
CASE <simple expression>     OF
      # <expanded range> {, <range>}       <statement> {<statement>}
      { # <expanded range> {, <range>}     <statement> {<statement>} }
      [ ELSE                               <statement> {<statement>}  ]
      [ OFFSTATE  <off state value>  ]
END_CASE ;
```

where <simple expression> and the <expanded range>s should have the same bit size. The meaning of this statement can be extrapolated from the meaning of the *IF* statement. Actually, the UDL/I reference manual provides the following example of conversion:

```
CASE S OF
      # R1, R2       <statements1>
      # R3 : R4      <statements2>
      ELSE           <statements3>
      OFFSTATE Z
END_CASE ;
```

which is equivalent to the following UDL/I statement:

```
BEGIN
      IF S=R1 ! S=R2
          THEN <statements1>
          OFFSTATE Z
      END_IF ;
      IF GE(S,R3) & LE(S,R4)
          THEN <statements2>
          OFFSTATE Z
      END_IF ;
      IF ^(S=R1 ! S=R2) & ^(GE(S,R3) & LE(S,R4))
          THEN <statements3>
          OFFSTATE Z
      END_IF ;
END ;
```

which can be then translated into three VHDL guarded blocks (according to 10.6.3.3). Similarly, a *CASE* expression is available in UDL/I:

```
TERM := CASE S OF
            # R1, R2      <expression1>
            # R3 : R4     <expression2>
            ELSE          <expression3>
            OFFSTATE Z
        END_CASE ;
```

This expression can be converted into the corresponding *CASE* statement or, if the terminal TERM is only assigned by this statement (one driver), it can *sometimes* be translated into a VHDL concurrent selected signal assignment:

```
signal TERM : TERMINAL_Z bus ;
...
with S select
      TERM <=     <expression1> when R1 | R2 ,
                  <expression2> when R3 to R4 ,
                  <expression3> when others ;
```

Such a translation is not always so simple, because the different choices in the *CASE* statement can overlap each other: in such a case, all the associated clauses corresponding to the selected choices are executed. This is forbidden in VHDL, where the different choices of a selected assignment must be disjoint: these different VHDL choices would therefore need to be redefined.

10.6.3.5. CASEOF Statement

The *CASEOF* statement has the following syntax:

```
CASEOF
      # <signal name> {, <signal name>}   <statement> {<statement>}
      { # <signal name> {, <signal name>}  <statement> {<statement>} }
      [ OFFSTATE  <off state value> ]
END_CASEOF ;
```

where all the named signals must be one-bit wide. This statement can be converted into the following UDL/I begin_end statement:

```
BEGIN
      IF <signal name> { ! <signal name> }        THEN   <statement> {<statement>}
                                               [ OFFSTATE  <same off state value> ]
      END_IF ;
      { IF <signal name> { ! <signal name> }      THEN   <statement> {<statement>}
                                               [ OFFSTATE  <same off state value> ]
      END_IF ; }
END ;
```

and then translated into the corresponding VHDL guarded blocks.

Of course, UDL/I also provides the *CASEOF* expression (with straightforward meaning):

```
TERM := CASEOF
            # S1, S2      <expression1>
            # S3, S4      <expression2>
            OFFSTATE Z
        END_CASEOF ;
```

The *CASEOF* statement (or expression) has no direct equivalent VHDL construct (the UDL/I converted form with *IF* statements should be used for translation).

10.6.3.6. AT Statement

This statement is the fundamental statement used to model synchronous and/or asynchronous registers. The syntax of this statement is

```
AT <clock expression> DO <statement> { <statement> } END_DO ;
```

Of course, UDL/I also defines the *AT* expression, which can be assigned to registers:

```
REG := AT <clock expression> DO <expression> END_DO ;
```

This is equivalent to the following *AT* statement:

```
AT <clock expression> DO REG := <expression> ; END_DO ;
```

The UDL/I reference manual lists the possible clock expressions:

clock expressions	meaning
<simple expression>	same as RISE_HIGH(<simple expression>)
RISE(<simple expression>)	transfers data at the rising edge (0 to 1) of <simple expression>
FALL(<simple expression>)	transfers data at the falling edge (0 to 1) of <simple expression>
HIGH(<simple expression>)	transfers data while <simple expression> = 1 and latches when <simple expression> goes from 1 to 0
LOW(<simple expression>)	transfers data while <simple expression> = 0 and latches when <simple expression> goes from 0 to 1
RISE_HIGH(<simple expression>)	same as RISE(<simple expression>) for registers and HIGH(<simple expression>) for latches.
RISE_LOW(<simple expression>)	same as RISE(<simple expression>) for registers and LOW(<simple expression>) for latches.
FALL_HIGH(<simple expression>)	same as FALL(<simple expression>) for registers and HIGH(<simple expression>) for latches.
FALL_LOW(<simple expression>)	same as FALL(<simple expression>) for registers and LOW(<simple expression>) for latches.

The expressions using keywords *RISE* and *FALL* model edge-sensitive (synchronous) conditions, whereas the expressions using keywords *HIGH* and *LOW* model level-sensitive (asynchronous) conditions. The other expressions model both edge- and level-sensitive conditions, depending on the type of facility that is assigned (register or latch).

VHDL has neither such keywords nor predefined functions implementing these edge- and level-sensitive clock expressions. A first step to translate these expressions could be to define the four functions RISE, FALL, HIGH, and LOW that work on type UDLI_BIT:

```
library UDLI_LIBRARY ;
use UDLI_LIBRARY.UDLI_TYPES.all ;
package UDLI_CLOCK is
        function RISE (signal S : UDLI_BIT) return BOOLEAN ;
        -- return TRUE is S is rising from '0' to '1'
        function FALL (signal S : UDLI_BIT) return BOOLEAN ;
        -- return TRUE is S is falling from '1' to '0'
        function HIGH (signal S : UDLI_BIT) return BOOLEAN ;
        -- return TRUE is S = '1'
        function FALL (signal S : UDLI_BIT) return BOOLEAN ;
        -- return TRUE is S = '0'
end UDLI_CLOCK ;
```

These functions are easy to implement in VHDL (they will use predefined attributes on signals such as S'EVENT and S'LAST_VALUE). The following portions of VHDL code are supposed to be included in a VHDL architecture that *uses* the previous package UDLI_CLOCK.

- **Edge-sensitive clock expression**

```
REGISTER : REG ;
...
AT RISE (CLK) DO REG := D ; END_DO ;
```

This single assignment can be modeled in VHDL in two slightly different ways:

```
signal REG : UDLI_BIT ;
...
REG <= D when RISE (CLK) else REG ;
-- equivalent process sensitive to D, CLK and REG
```

or even better:

```
signal REG : UDLI_BIT ;
...
process
begin
        wait on CLK ; -- process sensitive to CLK only
        if RISE (CLK) then
              REG <= D ;
        end if ;
end process ;
```

The behavior of this VHDL process is equivalent to that of the UDL/I *AT* statement in the case where the signal (register) REG has a single driver only. The case of multiple drivers (conflicts) will be handled later on.

- **Level-sensitive clock expression**

```
REGISTER : MEM ;  or        LATCH : MEM ;
...
AT HIGH (CLK) DO MEM := D ; END_DO ;
```

This single assignment can be modeled in VHDL in two ways:

```
signal MEM : UDLI_BIT ;
...
MEM <= D when HIGH (CLK) else MEM ;
-- equivalent process sensitive to D, CLK and MEM
```

or even better:

```
signal MEM : UDLI_BIT ;
...
process
begin
        wait on CLK, D ; -- process sensitive to CLK and D only
        if HIGH (CLK) then
            MEM <= D ;
        end if ;
end process ;
```

The previous remark applies here: this VHDL process is equivalent to the UDL/I *AT* statement if the signal (register, latch, RAM) MEM has a single driver only.

In the case of multiple drivers, the conflict resolution of these UDL/I facilities is not at all defined in the same way as the terminals (see paragraph 10.6.2 on the *OFFSTATE* clause). The UDL/I resolution mechanism used for registers depends on the behavior of these multiple assignments: synchronous assignments only, asynchronous assignments only, both synchronous and asynchronous assignments.

• conflict between synchronous assignments:

```
AT RISE (CK1)  DO  REG := D1 ; END_DO ;
AT RISE (CK2)  DO  REG := D2 ; END_DO ;
```

If the two edges RISE(CK1) and RISE(CK2) are not simultaneous, then there is no conflict: the active driver places its value, which remains until the next assignment. If both edges are simultaneous, then both assignments are executed at the same simulation time point, but in an *order determined by a non-deterministic method.*

• conflict between asynchronous assignments:

```
AT HIGH (CK1)  DO  REG := D1 ; END_DO ;
AT HIGH (CK2)  DO  REG := D2 ; END_DO ;
```

If only one level condition HIGH(...) is true, then the corresponding assignment will be executed without any conflict. If both level conditions are true simultaneously, then only one of the two values D1 and D2 will be assigned: the *choice is determined by a non-deterministic method.*

• conflict between synchronous and asynchronous assignments:

```
AT RISE (CK1)  DO  REG := D1 ;  END_DO ;
AT HIGH (CK2)  DO  REG := D2 ;  END_DO ;
```

When both conditions are true, only the asynchronous behavior is executed: the synchronous behavior is only executed when the asynchronous behavior is not active (level condition false). *The asynchronous behavior always overcomes the synchronous behavior.*

Non-determinism is one of the main characteristics of UDL/I: it is not and cannot be implemented in VHDL (see paragraph 10.6.4). Thus, it will not be possible to model exactly the behavior of the UDL/I register assignments in VHDL. Whether implementing this non-determinism in a hardware description language is a good point or not is another question and will be handled in paragraph 10.6.4.

The second difficulty is the priority of the UDL/I asynchronous assignments over the synchronous ones. All the signal assignments in VHDL have the same priority: the notion of priority to solve a conflict can only be used in resolution functions in which parameters are only the driving values. Priority can thus be assigned to some signal values, but not to assignment statements.

One *painful* way to implement this priority in VHDL could be to store the synchronous/asynchronous information in the value assigned to the register, and thus to use signals of record type: one element of this record is the signal value, and the other element stores the kind of assignment (asynchronous, synchronous). The example of such a modeling is given below, although its use seems not very realistic:

```
library UDLI_LIBRARY ; use UDLI_LIBRARY.UDLI_TYPES.all ;
package IT_IS_HARD_TO_BELIEVE is
        type TYPE_OF_TRANSFER is (SYNCHRONOUS, ASYNCHRONOUS) ;
        type UDLI_REGISTER is
         record
             VALUE : UDLI_BIT ;
             ASSIGN : TYPE_OF_TRANSFER ;
         end record ;
        type UDLI_REGISTER_ARRAY is array (NATURAL range <>) of UDLI_REGISTER ;
        function RESOLVE ( S : UDLI_REGISTER_ARRAY ) return UDLI_REGISTER ;
        -- this resolution function first compares the ASSIGN field of the driving values
        -- and forget the values having a SYNCHRONOUS field when at least one value
        -- has an ASYNCHRONOUS field.
end ;
```

An architecture using that package could then include:

```
signal REG : RESOLVE  UDLI_REGISTER ;
...
REG <= (D1, SYNCHRONOUS) when RISE(CK1) else REG ;
REG <= (D2, ASYNCHRONOUS) when HIGH(CK2) else REG ;
-- second statement will always overcome the first statement
```

But such a modeling is very inconvenient and is therefore not recommended. Such a synchronous and asynchronous behavior can be easily modeled with the same VHDL concurrent signal assignment:

```
REG <=   D2 when HIGH(CK2) else
         D1 when RISE(CK1) else
         REG ;
```

This concurrent signal assignment is equivalent to a process sensitive to all signals D1, D2, CK1, CK2, and REG, but the level-sensitive waveform is always evaluated in the first place and thus has priority over the edge-sensitive waveform.

Beside these modeling problems of non-determinism and asynchronous-priority, the basic behavior of all these UDL/I register assignments is that two assignments that are not simultaneous are never in conflict. The reader should remember that a VHDL driver associated with a given assignment is always connected to the assigned signal unless this signal is a guarded signal having a disconnection clause, or the assignment is a guarded assignment of a guarded block whose condition turns off. Thus, the two asynchronous VHDL assignments

```
REG <= D1 when HIGH(CK1) else REG ;
REG <= D2 when HIGH(CK2) else REG ;
```

are always in conflict (the resolution function is always called with the two driving values), even when the two level conditions HIGH(CK1) and HIGH(CK2) are not simultaneously true.

One way to avoid this permanent conflict is to define *guarded signals of kind REGISTER* and to use guarded assignments in guarded blocks:

```
signal REG : TRIRECX_BIT register ;
...
-- first guarded block modeling one asynchronous assignment
LABEL1 : block ( HIGH(CK1) )
begin
        REG <= guarded D1 ;
end block ;
-- second guarded block modeling one asynchronous assignment
LABEL2 : block ( HIGH(CK2) )
begin
        REG <= guarded D2 ;
end block ;
```

Thus, the guarded condition of the first block is evaluated whenever CK1 changes. For the next steps, let us assume that HIGH(CK2) is always false: the driver of the assignment (REG <= guarded D2) is disconnected and does not contribute to the signal value. When HIGH(CK1) becomes true, the signal REG receives the value D1. Then, while HIGH(CK1) remains true, each change of the value of D1 will re-execute the same assignment. Finally, when HIGH(CK1) becomes false (the condition is re-evaluated because CK1 changed), the guarded signal REG is disconnected and, as it is declared to be of kind *REGISTER*, it keeps its last assigned value. Thus, if both guarded conditions are not simultaneously true, there is no conflict between the two assignments.

The same methodology can be adopted to model two synchronous behaviors. Instead of writing

```
REG <= D1 when RISE(CK1) else REG ;
REG <= D2 when RISE(CK2) else REG ;
```

which causes a permanent conflict (resolution function call) between the two assignments (even when both conditions RISE(CK1) and RISE(CK2) are not simultaneously true), it could be better to write

```
signal REG : TRIRECX_BIT register ;
...
-- first guarded block modeling one synchronous assignment
LABEL1 : block ( RISE(CK1) and not CK1'STABLE)
begin
        REG <= guarded D1 ;
end block ;
-- second guarded block modeling one synchronous assignment
LABEL2 : block ( RISE(CK2) and not CK2'STABLE)
begin
        REG <= guarded D2 ;
end block ;
```

The introduction of CK1'STABLE in the guarded condition could look strange to a VHDL beginner, but it is compulsory. Actually, the guarded assignments must be executed only at the rising edge of the corresponding clock and must be disconnected one delta-delay after. Thus, the guarded condition of the first block is evaluated at *each change of CK1 or CK1'STABLE.*

Assuming that CK2 remains '0', when CK1 is rising from '0' to '1', the guarded condition becomes true and value D1 is assigned to REG. One delta-delay after this rising edge of CK1, the signal CK1'STABLE becomes true while CK1 remains at '1': at this time step, the guarded condition becomes false, and thus the driver on REG is disconnected and REG keeps its last assigned value. The reader may check that omitting CK1'STABLE would give to this block the same behavior as the level-sensitive block with condition HIGH(CK1).

Thus, although modeling the UDL/I terminals with VHDL guarded signals of kind *BUS* was satisfactory, modeling the UDL/I registers with VHDL guarded signals of kind *REGISTER* is only half-satisfactory: non-determinism and asynchronous-priority remain unimplemented in VHDL.

10.6.4. UDL/I Non-Deterministic Semantics

As explained in the previous paragraph, one of the main specifications of UDL/I is to include a non-deterministic semantics. In the register assignments, when two asynchronous transfers occur simultaneously, only one of the two values is assigned to the facility: the choice of this value must be non-deterministic according to the UDL/I reference manual.

VHDL does not include this non-deterministic semantics. Furthermore, *it is impossible to model non-deterministic behaviors in VHDL*. This is due to one of the most important constraints of VHDL: this language must be portable. Thus, the simulation of a given VHDL entity must produce exactly the same results on any platform for any number of executions.

But this UDL/I non-deterministic modeling brings other questions. Of course, *hardware is fundamentally non-deterministic*: when a flip-flop latches data that are not stable at the rising edge of its clock, the stored data are uncertain and may change from one load to the other. Thus, a hardware description language that models such a non-determinism is certainly closer to the hardware itself.

But from the designer point of view, is such a non-deterministic behavior so desirable? If the design is wrong because flip-flops latch data while they are unstable, is it better that the simulator produces these non-deterministic (uncertain, changing) results or displays error messages (coming from assertion violations) and/or propagates 'X' values?

The advantage of portability is fundamental: when a given VHDL design produces simulation errors, it is often more convenient if these errors are reproducible on others simulators in order to allow a good communication between designer teams. So what is the solution?

If the hardware description language is so close to the hardware that it reproduces a kind of non-determinism, that could be taken as a proof of perfection of the language. But then the designers would be strongly recommended to write assertions that check setup and hold specifications in order to prevent time-consuming debugging of their design.

10.7. UDL/I ASSERTION SECTION

The UDL/I module has a specific section that includes assertion statements. These assertions are very similar to the VHDL assertions: they provide a means to check some conditions on signals and to generate messages when these conditions are not satisfied.

In VHDL, the assertion statements are handled just as other concurrent (or sequential) statements: they can be written within structural or functional descriptions, although it is often recommended to write them at specific places (such as the entity statement part: see next examples).

The syntax of the UDL/I assertion statement is:

ASSERT <assertion simple expression> [**ELSE** <print statement>] **END_ASSERT** ;

The syntax of the VHDL assertion statement is:

assert <condition> [**report** <string message>] [**severity** <severity level>] ;

VHDL assertions also provides a means to stop simulation if the checked conditions are false. Actually, a predefined enumeration type SEVERITY_LEVEL is defined in the package STD:

type SEVERITY_LEVEL **is** (NOTE, WARNING, ERROR, FAILURE);

Each VHDL assertion has a severity level: the lowest value (NOTE) only prints a message on the standard output of the simulator; the highest value (FAILURE) prints the message and stops the simulation. UDL/I assertions only generate warning messages.

The simple expression in each UDL/I assertion statement must be one-bit wide. In addition to all logic and standard functions, it may use four predefined functions, which are

SETUP(<clock edge>, <input signal>, <time>)	"specifies setup time for a register"
HOLD(<clock edge>, <input signal>, <time>)	"specifies hold time for a register"
SETUP_HOLD(<clock edge>, <input signal>, <setup time>, <hold time>)	"specifies setup and hold times for register"
PULSE_WIDTH(<clock>, <level>, <time>)	"specifies the minimum pulse width of a clock"

The first three functions use an edge specification, which can be

RISE(<clock>)	"clock has changed from 0 to 1"
FALL(<clock>)	"clock has changed from 1 to 0"
TO_ONE(<clock>)	"clock has changed from 0 to 1, X to 1 or Z to 1"
TO_ZERO(<clock>)	"clock has changed from 1 to 0, X to 0 or Z to 0"

Possible UDL/I assertions are therefore

```
ASSERT SETUP(RISE(.CLK)), .D, 2NS)
       ELSE PRINT "Setup Violation..."
END_ASSERT ;
ASSERT PULSE_WIDTH (.CLK, 1, 20NS)
       ELSE PRINT "Pulse Width too small..."
END_ASSERT ;
```

Equivalent timing verifications may, of course, be written in VHDL, even if this language does not predefine functions specifying timing constraints or clock edges, such as the UDL/I functions listed above. Using the signal attributes S'EVENT and S'LAST_VALUE, it is easy to implement equivalent functions in VHDL.

Furthermore, a good methodology could be to define VHDL procedures that execute these assertions. For example, let us define the following package TIMING_CHECK:

```
library UDLI_LIBRARY ;
use UDLI_LIBRARY.UDLI_TYPES.all ;
package TIMING_CHECK is
        procedure CHECK_SETUP (signal CLK, DATA : in UDLI_BIT ; SETUP :  in TIME) ;
        -- procedure checking DATA setup at each rising edge of CLK
        -- other timing check procedure declarations
end TIMING_CHECK ;

package body TIMING_CHECK is
        procedure CHECK_SETUP (signal CLK, DATA : in UDLI_BIT; SETUP :  in TIME) is
            variable DATA_CHANGE : TIME := 0 ns ;
        begin
          loop    -- thanks to this infinite loop, variable DATA_CHANGE is created
                  -- at the elaboration and exists until the end of the simulation
            wait on CLK, DATA ;  -- waiting a change on CLK or DATA
            if DATA'EVENT then
                DATA_CHANGE := NOW ;
            elsif CLK = '1' and CLK'LAST_VALUE = '0' then  -- rising edge
                assert NOW - DATA_CHANGE >= SETUP
                report "Setup violation..."
                severity WARNING ;
            end if ;
          end loop ;
        end CHECK_SETUP ;
        -- other procedure bodies
end TIMING_CHECK ;
```

Such verification procedures are passive, i.e., they do not execute signal assignments. Thus, any entity statement part may include concurrent calls of these procedures. Placed in the visible part of the design unit, these timing verification procedures become part of the specifications of the design unit:

```
library UDLI_LIBRARY ;
use UDLI_LIBRARY.TIMING_CHECK.all ;
entity LATCH is
        port ( D : in UDLI_BIT ; CLK : in UDLI_BIT ; Q : out UDLI_BIT ) ;
begin
        CHECK_SETUP(CLK, D, 3 ns) ;
end LATCH ;
```

Thus, the functionality of the assertions is roughly the same in both languages. Once again, UDL/I has predefined some useful and basic timing functions, whereas VHDL allows the designer to write more powerful timing verification procedures but does not predefine any. Of course, once such appropriate VHDL packages have been written within a designer's team or even a company, they can be and *should* be reused by any designer of that team or company: such a methodology is strongly recommended in order to benefit from the descriptive power of VHDL.

10.8. DESCRIPTION LEVEL

In this section, a typical example of design will be used to illustrate the different levels of description in both languages: structural, dataflow, and behavioral.

The piece of hardware chosen for this purpose is a classical eight-bit accumulator: it includes both combinational and memory devices.

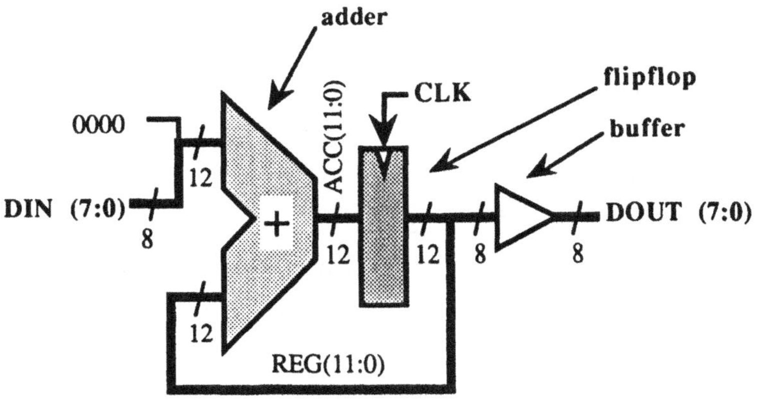

Fig 10.7 Accumulator

10.8.1. Structural Description

The UDL/I structural description of this device is

```
NAME :        ACCUMULATOR;
PURPOSE :     LOGSIM ;
LEVEL :       CELL ;
INPUTS :      DIN<7:0> ;
CLOCK :       CLK ;
POWERS :      GND, VCC ;
OUTPUTS :     DOUT<7:0> ;
...
TYPES :  ADDER (A<11:0>, B<11:0>, S<11:0>), FLIPFLOP(D<11:0>, CK, Q<11:0>),
         BUFFERS (A<7:0>, Y<7:0>) ;        "submodule references"
              ADDER : CADDER ;
              FLIPFLOP : CFLIPFLOP ;
              BUFFERS : CBUFFER ;          "instance declarations"
END_TYPES ;
NET_SECTION ; "Structural description"
       NETIN<7:0> = FROM (.DIN<7:0>) TO (CADDER.A<7:0>) ;
       NETGND = FROM (.GND) TO (CADDER.A<11:8>) ;
       NETREG<11:0> = FROM (FLIPFLOP.Q<11:0>) TO (CADDER.B<11:0>) ;
       NETREG<11:4> = FROM (FLIPFLOP.Q<11:4>) TO (CBUFFER.A<7:0>) ;
       NETACC<11:0> = FROM (CADDER.S<11:0>) TO (FLIPFLOP.D<11:0>) ;
       NETCK = FROM (.CLK) TO (FLIPFLOP.CK) ;
       NETOUT<7:0> = FROM (CBUFFER.Y<7:0>) TO (.DOUT<7:0>) ;
END_SECTION ;
END ;
```

The corresponding VHDL design unit is defined by the following entity declaration:

```
library UDLI_LIBRARY ;
use UDLI_LIBRARY.UDLI_TYPES.all ;
entity ACCUMULATOR is
       port (  DIN    :   in UDLI_BIT_VECTOR(7 downto 0) ;
               CLK    :   in UDLI_BIT ;
               DOUT   :   out UDLI_BIT_VECTOR(7 downto 0)     ) ;
end ACCUMULATOR ;
```

Although it was not compulsory here, the types declared in the package UDLI_TYPES have been used to define ports and signals. The previous entity will be the same for the three descriptions: each description is included in a different architecture. The structural architecture is the following:

```
architecture STRUCTURE of ACCUMULATOR is
        component ADDER
            port (   A, B      :    in UDLI_BIT_VECTOR(11 downto 0) ;
                     S         :    out UDLI_BIT_VECTOR(11 downto 0) ) ;
        end component ;
        component FLIPFLOP
            port (   D         :    in UDLI_BIT_VECTOR(11 downto 0) ;
                     CK        :    in UDLI_BIT ;
                     Q         :    out UDLI_BIT_VECTOR(11 downto 0) ) ;
        end component ;
        component BUFFERS
            port (   A         :    in  UDLI_BIT_VECTOR(7 downto 0) ;
                     Y         :    out UDLI_BIT_VECTOR(7 downto 0) ) ;
        end component ;
        signal ACC, REG : UDLI_BIT_VECTOR(11 downto 0) ;
        signal GND : UDLI_BIT := '0' ;
begin
        CADDER : ADDER port map (    A(11 downto 8) => (others => GND),
                                     A(7 downto 0) => DIN, B => REG, S => ACC) ;
        CFLIPFLOP : FLIPFLOP port map (D => ACC, CK => CLK, Q => REG) ;
        CBUFFER : BUFFERS port map (A => REG(11 downto 4), Y => DOUT) ;
end STRUCTURE ;
```

Differences between the structural descriptions in both languages have been covered deeply in section 10.5.

10.8.2. Dataflow Description

A UDL/I dataflow description of the accumulator is

```
NAME :          ACCUMULATOR;
PURPOSE :       LOGSIM ;
LEVEL :         CELL ;
INPUTS :  DIN<7:0> ;
CLOCK :   CLK ;
OUTPUTS :       DOUT<7:0> ;
BEHAVIOR_SECTION ; "Behavior description"
        TERMINAL :  ACC<11:0> ;
        REGISTER :  REG<11:0> ;
        ACC<11:0>    := ADD(4B0000 !! DIN<7:0>, REG<11:0>, 1B0) ;
        "Terminal assignment using the standard function ADD"
        REG<11:0>    := AT RISE(CLK)  DO  ACC<11:0>  END_DO ;
        "Synchronous assignment (with AT expression) using clock expression RISE(...)"
        DOUT<7:0>    := REG<11:4> ;
END_SECTION ;
END ;
```

The equivalent VHDL architecture will use functions of packages UDLI_ARITHMETIC and UDLI_CLOCK defined, respectively, in 10.6.1.3 and 10.6.3.6. Of course, the entity is the same as before.

```
library UDLI_LIBRARY ;
use UDLI_LIBRARY.UDLI_ARITHMETIC.all , UDLI_LIBRARY.UDLI_CLOCK.all ;
architecture DATAFLOW of ACCUMULATOR is
        signal ACC, REG : UDLI_BIT_VECTOR(11 downto 0) ;
begin
        ACC <= ("0000" & DIN) + REG ;
        -- "+" is the overloaded operator declared in UDLI_ARITHMETIC
        REG <= ACC when RISE(CLK) else REG ;
        -- RISE is the function declared in UDLI_CLOCK
        DOUT <= REG(11 downto 4) ;
end DATAFLOW ;
```

Both descriptions look very similar, but there is one major difference between them: in UDL/I, all the used functions are predefined by the language; in VHDL, the equivalent functions have to be written beforehand in user-defined packages.

10.8.3. Behavioral Description

UDL/I does not provide other behavioral description levels: the last description was both dataflow and behavioral. This is linked to the fact that UDL/I does not provide sequential descriptions: the use of sequential processes to describe algorithms is not possible in UDL/I.

Of course, such a modeling is possible in VHDL. Thus, another architecture of the same device could be

```
library UDLI_LIBRARY ;
use UDLI_LIBRARY.UDLI_ARITHMETIC.all , UDLI_LIBRARY.UDLI_CLOCK.all ;
architecture BEHAVIOR of ACCUMULATOR is
begin
        process
            variable ACC, REG : UDLI_BIT_VECTOR(11 downto 0) ;
        begin
            wait on DIN, CLK ;
            if DIN'EVENT then
                ACC := ("0000" & DIN) + REG ;
                -- "+" is the overloaded operator declared in UDLI_ARITHMETIC
            elsif RISE(CLK) then    -- RISE is the function declared in UDLI_CLOCK
                REG := ACC ;
            end if ;
            DOUT <= REG(11 downto 4) ;
        end process ;
end BEHAVIOR ;
```

Actually, in this case, such a behavioral modeling is not very different from the dataflow one. But for the description of sophisticated sequential algorithms, it is very convenient to encapsulate portions of this sequential code into processes and to simulate its execution on hardware, i.e., in the concurrent world. UDL/I is very limited in that field and is definitely more adapted to IC modeling than to system-level modeling.

10.9. TRANSLATING FROM UDL/I TO VHDL

10.9.1. Primitive Description Section

Apart from the structural and the behavioral sections, the third kind of body allowed in a UDL/I module description is the primitive description section. This kind of section describes simulator primitives by using a truth-table syntax. Following is an example taken from the UDL/I reference manual:

```
PRIMITIVE_SECTION ;
    TABLE :    CLK,   D,  RST   :  Q  :  Q,  QB ;
               ?      ?   0     :  ?  :  0   1  ;
               ?      ?   *     :  0  :  —   —  ;
               R      0   1     :  ?  :  0   1  ;
               R      1   1     :  ?  :  1   0  ;
               N      ?   ?     :  ?  :  —   —  ;
               ?      *   ?     :  ?  :  —   —  ;
               P      0   1     :  0  :  —   —  ;
               P      1   1     :  1  :  —   —  ;
               B      ?   X     :  0  :  —   —  ;
    END_TABLE ;
END_SECTION ;
```

Letters and symbols used in the truth table represent signal levels and edges: '0' means logic value 0, 'B' either 0 or 1, '?' any-value (*don't-care*), 'R' rising edge (from 0 to 1), 'P' positive edge (0 to 1, 0 to X or X to 1), and so on.

In each line, the input values are first listed, then the primitive state (surrounded by colons), and finally the resulting state and output values. The character '—' in an output filed means no-change.

Actually, *the UDL/I primitive section is totally equivalent to the Verilog gate-level primitive* (see paragraph 10.8.1). The characters used in the truth table are the same and have exactly the same meaning. The only difference between the two languages is that UDL/I allows the specification of more than one output values in a primitive section, whereas Verilog primitives may have only one output pin.

Thus, the translation of UDL/I primitive sections into VHDL models can use the same methodology as has been introduced for Verilog primitives. In the corresponding Verilog paragraph 10.8.1, the reader will find many examples of translation, as well as a kind of primitive analyzer written in VHDL that can be easily adapted to UDL/I.

10.9.2. Automaton

As stated in section 10.3, UDL/I does not really implement sequential statements (such as the sequential statements written in a VHDL process or subprogram).

Nevertheless, UDL/I provides a very explicit way to model *finite-state machines*, which may be handled as a kind of sequential device. The description of finite-state machines is achieved in the behavioral section of a module by the automaton definition. The syntax of this automaton definition is the following:

```
AUTOMATON : <automaton name> [ ( <task name> { , <task name> } ) ] :
        <reset condition> : <transition clock> [ : <transfer clock> ] ;
        <state statement> { <state statement> }
END_AUTO ;
```

The automaton is described by a set of *states:* only one of these states is active at a time. Transitions between states are synchronized by a clock. In the above syntax, the reset condition forces the automaton to its initial state (first state declared in the automaton definition), the transition clock synchronizes the state transitions and the (optional) transfer clock specifies data register transfers (when not specified, the transition clock is used instead). The rest of the definition is composed of state statements having the following syntax:

```
<state name> : [ WAIT ( <wait condition> ) : ] [ <statement> { <statement> } ]
```

In each state, a wait clause can be specified: if it is, then the automaton remains in this state until the wait condition becomes true; in the meantime, no statement associated with this state can be executed (including the state transition statement).

Some UDL/I automata can also be controlled by one or several *tasks*. If no task name appears after the automaton name, then the automaton is always active and behaves according to wait conditions and state transition statements (only one state is active at a time). If one or more task name(s) are specified, then the automaton is active if and only if at least one of the tasks is on. If all the tasks are off, then the automaton remains in its active state, but the associated statements are *not* executable. Tasks can be generated (by another automaton), transferred (between automata), and terminated.

VHDL does not provide a specific syntax for finite-state machines. Nevertheless, there are two common ways of describing control parts in VHDL: the first is the process statement with a case statement, and the second uses guarded blocks.

The process statement can be used for modeling state machines having only one state active at a time (like UDL/I automata). A single variable stores the state of the automaton:

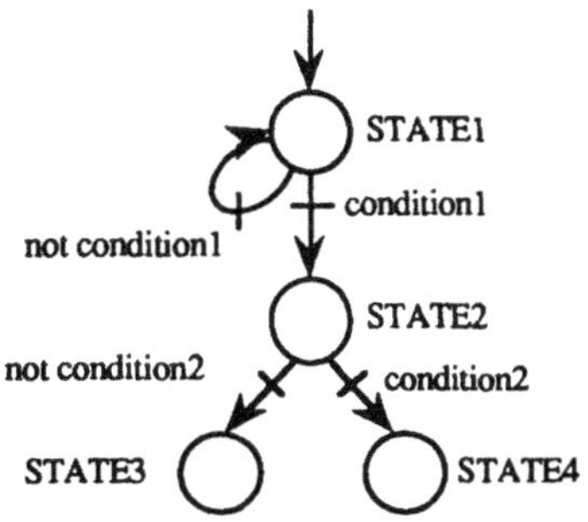

Fig 10.8a Automaton

```
process
    type ENUM_STATE is (..., STATE1, STATE2,...) ;
    variable STATE : ENUM_STATE ;
begin
    case STATE is
     when STATE1 =>
        <statements>
        if CONDITION1 then STATE := STATE2 ; end if ;
     when STATE2 =>
        <statements>
        if CONDITION2 then STATE := STATE4 ;
        else STATE := STATE3 ; end if ;
    ...
    end case ;
    wait until CLOCK = '1' ;
end process ;
```

Guarded blocks can be used for modeling Petri-nets or other automata having more than one active state at a time (these automata can be less easily modeled with UDL/I). Each state can be modeled by a guarded signal of a resolved subtype of type BIT (here, RESOLVE_BIT), with the meaning that the state is active if the corresponding signal is equal to '1'. Each transition is modeled by a guarded block, thus allowing a decentralized control modeling:

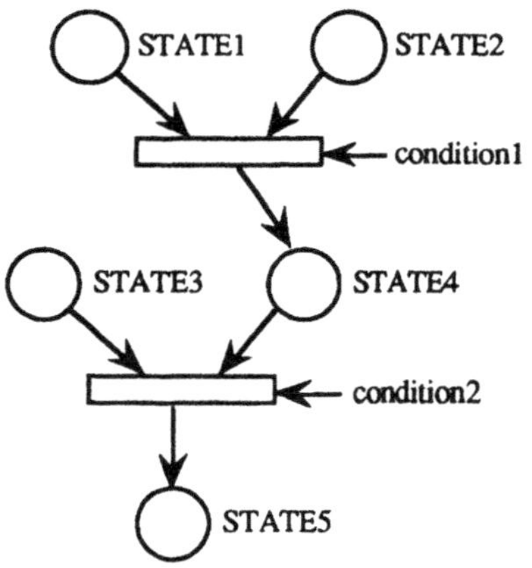

Fig 10.8b Petri-Net

```
signal STATE1, STATE2...: RESOLVE_BIT register;

...

T1 : block (        STATE1 = '1'
                and STATE2 = '1'
                and CONDITION1 )
      STATE4 <= guarded '1' ;  -- becomes active
      STATE1 <= guarded '0' ;  -- becomes inactive
      STATE2 <= guarded '0' ;  -- becomes inactive
end block;

T2 : block (        STATE3 = '1'
                and STATE4 = '1'
                and CONDITION2 )
      STATE5 <= guarded '1' ;  -- becomes active
      STATE3 <= guarded '0' ;  -- becomes inactive
      STATE4 <= guarded '0' ;  -- becomes inactive
end block;
```

Once again, VHDL has a more general approach than UDL/I (this characteristic, of course, has pros and cons). The UDL/I syntax has the advantage of being much more explicit: every designer (and every UDL/I synthesis tool!) is able to recognize an automaton inside a UDL/I module description. This is less obvious in VHDL.

Following is a typical example of a UDL/I automaton (without task):

```
AUTOMATON : FIRST_AUTO : ^RESET : RISE(CLOCK) ;
      STATE1 :   COMMAND1 := 1 ;
                 COMMAND2 := 1 ;
                 -> STATE2 ;          "transition to state STATE2"
      STATE2 :   COMMAND1 := 1 ;
                 COMMAND2 := 0 ;
                 -> STATE3 ;          "transition to state STATE3"
      STATE3 :   WAIT( ENABLE ) :     "automaton stays in STATE3 until ENABLE = 1"
                 COMMAND1 := 0 ;                                 "(or RESET = 0)"
                 COMMAND2 := 1 ;
                 -> STATE1 ;          "transition to state STATE1"
END_AUTO ;
```

This kind of state machine is very common in IC design: it generates a given number of output commands (to some operative part), and these commands are set to 1 or 0 according to the current state or transition. Such an automaton can be modeled in VHDL by the following process statement:

```
FIRST_AUTO : process
      type STATE_TYPE is (STATE1, STATE2, STATE3) ;
      variable STATE : STATE_TYPE ;  -- initialized at STATE1
begin
      wait on RESET, CLOCK ;
      if RESET = '0' then
          -- asynchronous RESET: the automaton is forced in first state
          STATE := STATE1 ;
      elsif CLOCK'EVENT and CLOCK = '1' then
          -- synchronous state transition (rising edge of CLOCK)
          case STATE is
            when STATE1 =>     STATE := STATE2 ;
            when STATE2 =>     STATE := STATE3 ;
            when STATE3 =>     STATE := STATE1 ;
          end case ;
      end if ;
      case STATE is
        when STATE1 =>
          COMMAND1 <= '1' ;
          COMMAND2 <= '1' ;
        when STATE2 =>
          COMMAND1 <= '1' ;
          COMMAND2 <= '0' ;
        when STATE3 =>
          if ENABLE /= '1' then wait until ENABLE = '1' or RESET = '0' ; end if ;
          if ENABLE = '1' then
              COMMAND1 <= '0' ;
              COMMAND2 <= '1' ;
          else -- RESET = '0'
              STATE := STATE1 ;
              COMMAND1 <= '1' ;
              COMMAND2 <= '1' ;
          end if ;
      end case ;
end process ;
```

Although both UDL/I and VHDL descriptions are modeling the same control part, there is a fundamental difference between them: all the statements found inside each UDL/I state description are *concurrent*, whereas all the statements found in the VHDL process statement are *sequential*. As already stated, UDL/I does not have any sequential statement: even in the automaton definition, statements associated with each state have the same semantic as any other concurrent statement (see paragraph 10.6.3). In the previous model, this distinction was not important because only static values (0 or 1) were assigned to the commands: these assignments are thus executed only once at each state, which is the case in both models.

The next example has a different behavior:

```
AUTOMATON : SECOND_AUTO : ^RESET : RISE(CLOCK) ;
      STATE1 :    TERM := A ;
                  -> STATE2 ;
      STATE2 :    TERM := B ;
                  -> STATE1 ;
END_AUTO ;
```

In this example, a terminal TERM is assigned by the value of another terminal A if the automaton is in state STATE1 (and by the value of B if STATE2). Thus, if the automaton remains in state STATE1 (no falling edge on CLOCK), every new value on A will be propagated on TERM. Writing an equivalent VHDL model with a *single* process statement like the one above will not be easy because VHDL process statements only include sequential assignments. Indeed, following is a single VHDL process modeling the same automaton as the previous UDL/I description:

```
SECOND_AUTO : process
      type STATE_TYPE is (STATE1, STATE2) ;
      variable STATE : STATE_TYPE ;  -- initialized at STATE1
begin
      wait on RESET, CLOCK ;
      LOOP1 : loop
          if RESET = '0' then  -- asynchronous RESET: the automaton is forced in first state
              STATE := STATE1 ;
          elsif CLOCK'EVENT and CLOCK = '1' then
              -- synchronous state transition (rising edge of CLOCK)
              case STATE is
                when STATE1 =>    STATE := STATE2 ;
                when STATE2 =>    STATE := STATE1 ;
              end case ;
          end if ;
          case STATE is
            when STATE1 =>
              LOOP2 : loop
                  TERM <= A ;
                  wait on RESET, CLOCK, A ;
                  next LOOP1 when (RESET'EVENT and RESET = '0')
                                or (CLOCK'EVENT and CLOCK = '1') ;
              end loop ;
            when STATE2 =>
              LOOP3 : loop
                  TERM <= B ;
                  wait on RESET, CLOCK, B ;
                  next LOOP1 when (RESET'EVENT and RESET = '0')
                                or (CLOCK'EVENT and CLOCK = '1') ;
              end loop ;
          end case ;
          exit ;
      end loop ;
end process ;
```

In this model, two loops (labeled LOOP2 and LOOP3) are used to implement the concurrent assignments of TERM: if the automaton remains in state STATE1 because there is no falling edge on CLOCK, then signal TERM will be assigned a new value at each event on A. The external loop LOOP1 is necessary to recompute the state value in the case of an edge on RESET or CLOCK.

Such a description is not very convenient and is certainly too verbose: the VHDL process statement is really not adapted (and was not intended!) to implement concurrency. A much better VHDL description would use concurrent statements. Indeed, the previous UDL/I automaton can be modeled in VHDL with guarded blocks:

```
signal TERM : ANY_RESOLUTION_FUNCTION  BIT register ;
...
SECOND_AUTO : block
      type STATE_TYPE is (STATE1, STATE2) ;
      signal STATE : STATE_TYPE ;  -- initialized at STATE1
begin
      S1 : block (STATE = STATE1)
      begin
          TERM <= guarded A ;
      end block S1 ;
      S2 : block (STATE = STATE2)
      begin
          TERM <= guarded B ;
      end block S2 ;
      TRANSITIONS : block (RESET = '0' or (not CLOCK'STABLE and CLOCK = '1'))
      begin
          STATE <= guarded     STATE1 when RESET = '0' or STATE = STATE2 else
                               STATE2 ;
      end block TRANSITIONS ;
end block SECOND_AUTO ;
```

In such a description, the automaton is modeled by a block, and the current state value is stored in an internal signal named STATE. Statements associated with a given state are included inside the same guarded block (S1 or S2): there is one guarded block per state. The guard expression of a given block is the test equality between signal STATE and the state value associated with the given block.

All the signals assigned by the automaton have to be guarded signals of the register kind. As only one block is active at a time (the block corresponding to the current state of the automaton), only one driver is connected to each signal at a time. The resolution function, although compulsory, will thus never be called by the simulator. Furthermore, it could be interesting to include in this resolution function an assertion statement that checks that only one driver is active at a time: this would check that the automaton transition conditions have been properly specified.

Finally, another block has been added at the end of the description, namely the block implementing state transitions (labeled TRANSITIONS). The guard expression of this block is a logical-or between the reset condition and the transition clock expression. This block includes only one statement that computes the new state value.

Such a VHDL description is much closer to the UDL/I automaton semantics than the previous process statement: all the statements used in this description are indeed *concurrent statements*, as in the original UDL/I model. Implementing in VHDL all the capabilities of the UDL/I automaton therefore becomes easier. For example, a wait condition inside a UDL/I state statement will be modeled in VHDL with another condition in the guard expression of the corresponding VHDL guarded block and in the STATE assignment of the TRANSITIONS block:

```
S1 : WAIT( COND1) : ...  =>  B1 : block (STATE=S1 and COND1) ...
                                  ...
                                  STATE <= ... when (STATE=S1 and COND1)...
                                  -- see complete examples below
```

Another typical feature of UDL/I is the use of tasks to control the activity of automata. The following example includes two automata, one being controlled by a task that is generated by the second:

```
AUTOMATON : THIRD_AUTO (TASK) : ^RST : RISE(CK) ;
     S1 :    WAIT( COND1 ) :
             TERM1 := 1 ;
             -> S2 ;
     S2 :    WAIT( COND2 ) :
             FINISH (TASK) ;        "task termination"
             -> S1 ;
END_AUTO ;

AUTOMATON : FOURTH_AUTO : ^RST : RISE (CK) ;
     S1 :    WAIT( COND3 ) :
             TERM2 := A ;
             ->> SECOND_AUTO ** TASK ;      "task generation"
             -> S2 ;
     S2 :    WAIT( COND4 ) :
             TERM2 := B ;
             -> S1 ;
END_AUTO ;
```

These two automata can be modeled in VHDL with two blocks using a boolean signal instead of the UDL/I task:

```
signal TASK : ANY_RESOLUTION_FUNCTION BOOLEAN register ;
signal TERM1, TERM2 : ANY_RESOLUTION_FUNCTION BIT register ;
...
THIRD_AUTO : block
        type STATE_TYPE is (S1, S2) ;
        signal STATE : STATE_TYPE ;  -- initialized at S1
begin
        B1 : block (STATE = S1 and COND1 and TASK)
        begin
            TERM1 <= guarded '1' ;
        end block B1 ;
        B2 : block (STATE = S2 and COND2 and TASK)
        begin
            TASK <= guarded FALSE ;        -- task termination
        end block B2 ;
        TRANSITIONS : block ( (RST = '0' or (not CK'STABLE and CK = '1')) and TASK)
        begin
            STATE <= guarded      S1 when RST = '0' or (STATE = S2 and COND2) else
                                  S2 when (STATE = S1 and COND1) else
                                  STATE  ;
        end block TRANSITIONS ;
end block THIRD_AUTO ;

FOURTH_AUTO : block
        type STATE_TYPE is (S1, S2) ;
        signal STATE : STATE_TYPE ;  -- initialized at S1
begin
        B1 : block (STATE = S1 and COND3)
        begin
            TERM2 <= guarded A ;
        end block B1 ;
        B1TASK : block (STATE = S1 and COND3 and not TASK)
        begin
            TASK <= guarded TRUE ;         -- task generation
        end block B1TASK ;
        B2 : block (STATE = S2 and COND4)
        begin
            TERM2 <= guarded B ;
        end block B2 ;
        TRANSITIONS : block (RST = '0' or (not CK'STABLE and CK = '1'))
        begin
            STATE <= guarded      S1 when RST = '0' or (STATE = S2 and COND4) else
                                  S2 when (STATE = S1 and COND3) else
                                  STATE  ;
        end block TRANSITIONS ;
end block FOURTH_AUTO ;
```

Signals TERM1 and TERM2 are guarded signals of the register kind for the same reason given for signal TERM in the previous example.

The guarded block corresponding to state S1 in the automaton FOURTH_AUTO has been duplicated in order to isolate the task generation. Indeed, the boolean signal TASK is also a guarded signal of the register kind and is assigned by two drivers: one representing the task generation, and the other the task termination. Before the task is generated, the driver corresponding to the task termination is disconnected (corresponding guard expression is false). Before the termination, the driver corresponding to the task generation is disconnected (guard expression false). Thus, only one driver is active at a time, which gives the expected behavior.

UDL/I also implements task transitions: a first automaton can terminate its own task and in the meantime generate the task of another automaton. The syntax is

```
** <first automaton task> --> <second automaton name> ** <second automaton task> ;
```

This task transition is more or less equivalent to a task termination followed by a task generation. Thus, the same methodology as before can be used to implement this statement in VHDL.

As a conclusion for this section, two remarks may be made: VHDL provides many different and powerful ways to model finite state machines, but these ways are not explicit and are often verbose; UDL/I provides a restricted but useful and explicit way to describe finite-state machines.

Indeed, the VHDL designer has the ability to model state machines with one thread of control as well as highly concurrent Petri-nets, but all these models strongly depend on his style of modeling. There is no keyword like *automaton* or *state_machine*, no semantic rule to specify the explicit control part. Synthesis tools usually take a process statement with one case statement and one wait statement as the description for a state machine (like the FIRST_AUTO example), but this is only a convention, not a rule defined in the reference manual.

On the other hand, the UDL/I designer may only describe one specific type of automaton having one thread of control, but the style is forced by the syntax (keywords *automaton* and *end_auto*). As illustrated by this chapter, this syntax allows the description of finite-state machines in a very concise way (equivalent VHDL models are more verbose). All the statements controlled by the automaton are concurrent, which is more realistic than the sequential statements of the VHDL process statement.

10.10. CONCLUSION

The main characteristics of UDL/I that make this language very different from VHDL are the synthesis semantics, the concurrent domain without any sequential statement, and the way all the constructs are specific to integrated circuits (facilities such as registers and latches, standard functions, automaton definition, and so on).

Of course, these three points will be appreciated by the IC designer, who will find in this language all the implicit objects and operators he was expecting to find: addition function on buses (addition on bit vectors must be explicitly defined in VHDL), predefined resolution functions (these also must be defined in VHDL), RAM and ROM facilities, and so on. Once again, although they are not predefined, most of these objects and subprograms may be implemented in VHDL: the IC designer simply needs a set of prewritten packages to achieve efficient (quick and safe) VHDL modeling. If some designers or developers are going to write such packages, many others will just reuse them and thus will find many "predefined" constructs in VHDL — even more than in UDL/I.

The drawback of these characteristics is that UDL/I cannot be used for high-level modeling, such as system-level modeling. Above a given level of abstraction, the use of sequential constructs seems necessary (constructs such as those found in software programming languages: variables, loop statements, abstract datatypes, subprograms, packages, etc.). Of course, at this level, all the constructs cannot have a synthesis meaning: the purpose of such a language is also to model *not-yet-synthesizable* objects.

Of course, synthesis is as important as simulation in IC design: there is certainly a lack in VHDL in the synthesis domain, and UDL/I has the fundamental advantage over VHDL of being defined by both a simulation and a synthesis semantics. In contrast, VHDL may take advantage of its capability of linking high-level descriptions (functional or even specification levels) to synthetizable descriptions (RTL and logic levels).

11. MEMO

Library Units

Entity Declaration

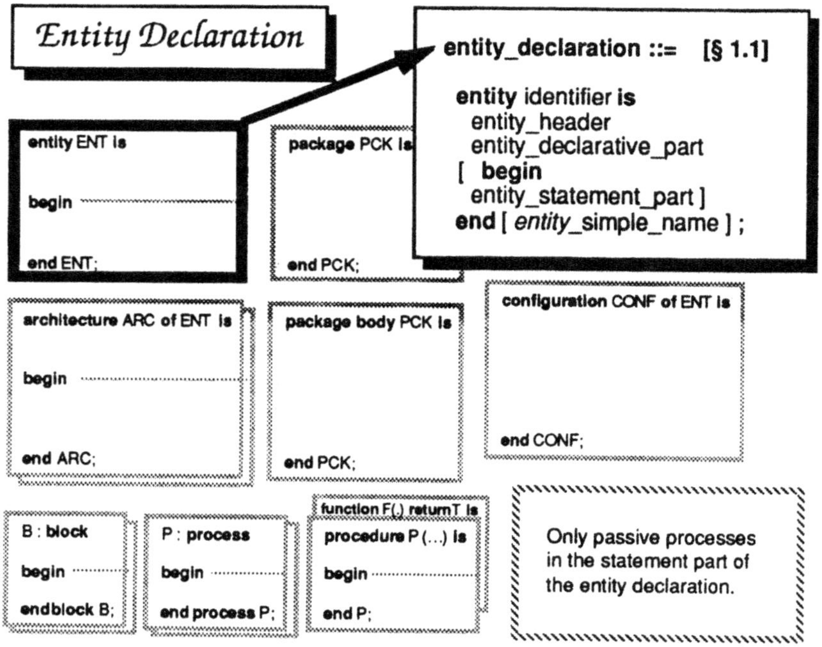


```
entity ENT is
end;
```

```
use WORK.ZBIT_PKG.all; -- for ZBIT
entity ENT1 is
  port (S1 : in BIT; S2 : out ZBIT);
end ENT1;
```

```
entity ENT2 is
  generic (N : in POSITIVE);
  port (S1 : in BIT; S2 : out BIT_VECTOR(1 to N));
end ENT2;
```

```
entity ENT3 is
  generic (N : in POSITIVE);
  port (S1 : in BIT; S2 : out BIT_VECTOR(1 to N));
  use work.TIMING_PKG.all; -- for constant delay
  type bit8 is array(1 to 8) of bit;
  procedure init(signal S: bit8) is
  begin
    S <= ( others => '1') after delay;
  end init;
begin
  assert S1'DELAYED'STABLE(5 ns)
     report "ERROR !"   severity ERROR;
  passive_proc(S1, delay);
end ENT3;
```

Library Unit

Architecture Body

```
architecture_body ::=        [§ 1.2]

  architecture identifier of
  entity_name is
    architecture_declarative_part
  begin
    architecture_statement_part
  end [architecture_simple_name] ;
```

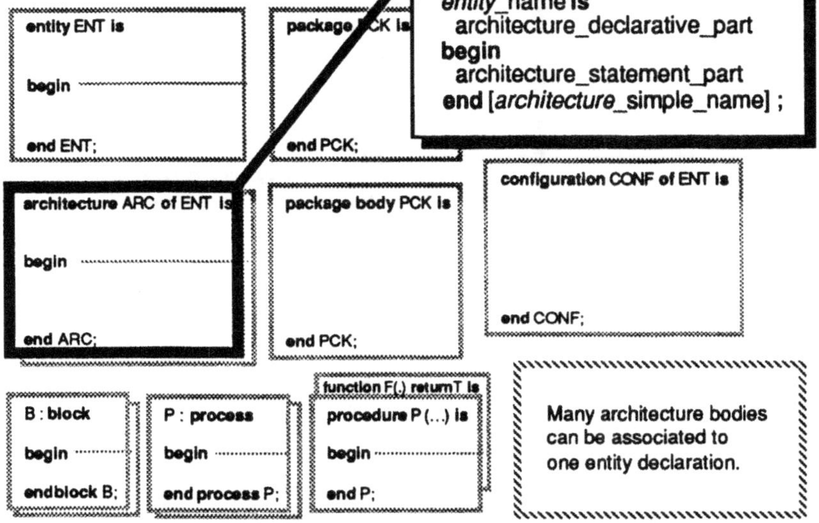


```
architecture ARC of ENT is
begin
end;
```

```
architecture ARC2 of ENT1 is
  constant delay : time :=5 ns;
  signal S : bit;
begin
  S2 <= czbit(S) after 3 ns;
  S <= S1 after delay;
end ARC2;
```

```
architecture ARC1 of ENT1 is
 signal S : bit := '1';
 procedure P ( signal A : in bit;
  signal B : inout bit ) is
 begin
  B <= not B when A='1' else B;
 end P;
begin
  P1 : P( S1, S );
  P2 : S2 <= czbit(S) after 2 ns;
  P3 : process (S)
   variable V : bit;
   begin
   V := S ;
  end process;
end ARC1;
```

```
use work.PARTS. all; -- for CLKGEN
architecture ARC of ENT2 is
 signal S : bit_vector(1 to N);
 component COMP
  port ( A : in bit; B : out bit) ;
 end component;
begin
 S2 <= S after 2 ns;
 C1: for i in 2 to N generate
   C: COMP port map (S1, S(i));
 end generate;
 C2 : CLKGEN port map (S(1));
end ARC;
```

Library Unit

Package Declaration

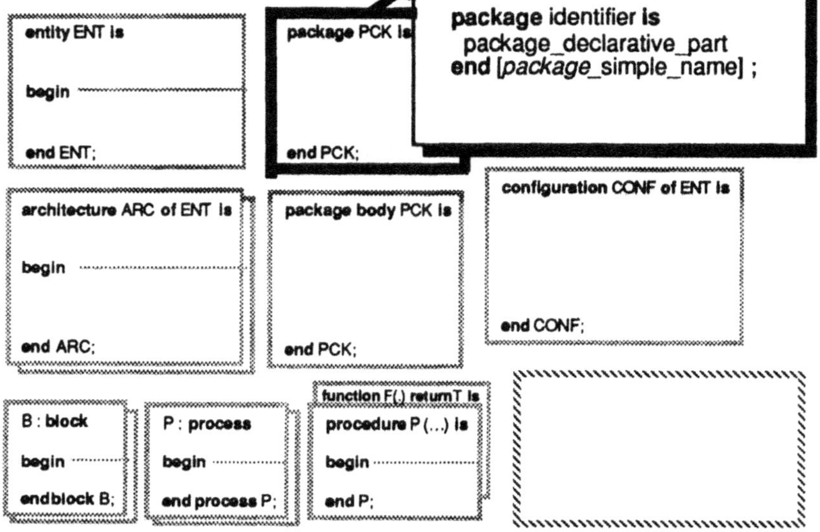

package_declaration ::= **[§ 2.5]**

package identifier **is**
  package_declarative_part
end [*package*_simple_name] ;

```
package PCK is
  constant CINT : integer;
end PCK;
```

```
package PCK1 is
  constant CINT : integer := 10 ;
  subtype T is natural range 0 to CINT;
  type FT is file of T;
  file my_file : FT is in "data.txt";
  function F(S1, S2 : bit) return bit;
end PCK1;
```

```
package PCK2 is
  use work.PARTS. all;
  signal S : resol_bit bus := '1' ;
  disconnect S after 2 ns;
  component LATCH
    generic ( C : natural );
    port ( A : in bit; B : out bit_vector(1 to C) );
  end component;
  component CLOCK
    port ( CK : out bit );
  end component;
end PCK2;
```

```
library LIB;
use LIB.PCK1. all; -- for CINT
package PCK3 is
  constant delay : time ;
  type T is array(1 to CINT) of natural;
  procedure P ( A : in integer; B : out natural);
  function F ( S : bit_vector) return natural;
end PCK3;
```

Library Unit

Package Body

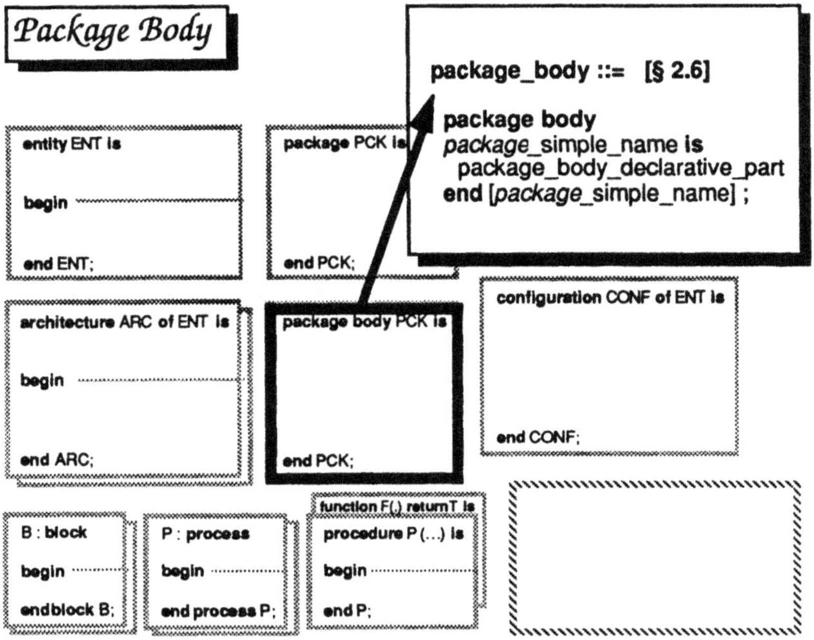


```
use work.PCK0.all;
package body PCK is
  constant CINT: integer := f(6);
end PCK;
```

```
package body PCK1 is
  function G(A,B : bit) return bit is
  begin
    return not (A xor B);
  end G;
  function F(S1,S2 : bit) return bit is
    variable V : bit;
  begin
   V := S1  nand S2;
    return G(V,S2);
  end F;
end PCK1;
```

```
package body PCK3 is
  constant delay : time := 10 fs ;
  procedure P ( A :  in integer; B :  out natural) is
  begin
   B := abs(A);
  end P;
  function F ( S : bit_vector) return natural is
  begin
   for i in S'range loop
     if S(i) = '1' then return 1; end if;
   end loop;
   return 0;
  end F;
end PCK3;
```

Library Unit

Configuration Declaration

configuration_declaration ::= **[§ 1.3]**

configuration identifier **of**
*entity*_name **is**
  configuration_declarative_part
  block_configuration
end[*configuration*_simple_name] ;

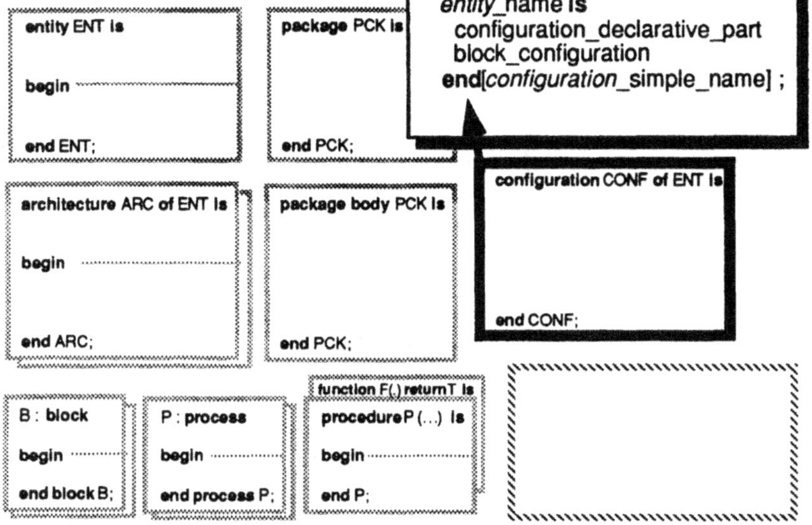


```
configuration CONF of ENT2 is
for ARC
  for all : COMP
    use entity work.LATCH(ARC);
  end for;
end for;
end CONF;
```

```
library LIB;
configuration CONF1 of ENT2 is
use LIB.all;
for ARC
  for C1(2 to N-1)
    for C : COMP
      use entity LATCH(BEHAV);
    end for;                    -- from LIB
  end for;
  for C1(N)
    for C : COMP
      use entity work.REGISTER(STRUCT);   -- from work
       for STRUCT
        for all : BASIC_COMP  use open;
       end for;
    end for;
  end for;
  for C2 : CLKGEN
      use configuration CONFCLOCK  port map (...)  generic map (...) ;
  end for;
end for;
end CONF1;
```

Library Unit

Declarations

Subprogram Declaration

subprogram_declaration ::= [§ 2.1]
subprogram_specification ;

subprogram_specification ::=
  procedure designator
  [(formal_parameter_list)]
| **function** designator
  [(formal_parameter_list)]
  return type_mark

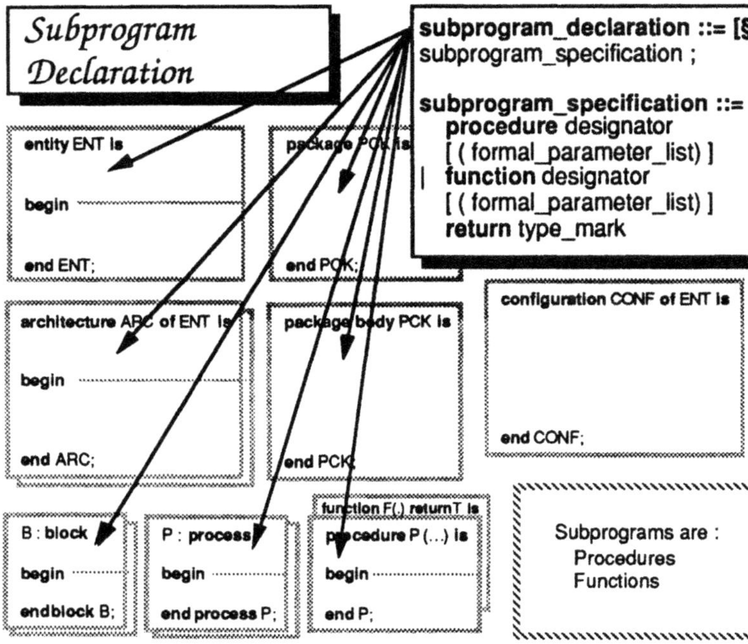

Subprograms are :
- Procedures
- Functions

```
procedure Thank_you ;
```

```
procedure test( A : bit ) ;
```

```
procedure regist( signal D  :  in bit;
                  signal CK :  in bit;
                  signal Q  :  out bit ) ;
```

```
procedure P( variable A : inout color;
             constant B : in choice ) ;
```

```
function Convert( A : bit ) return fuzzy_bit ;
```

```
function equal( signal A, B : zbit )  return boolean ;
```

```
function "+"( A,B : bit_vector )  return bit_vector ;
```

Declaration

Subprogram Body

```
subprogram_body ::=      [§ 2.2]

subprogram_specification is
    subprogram_declarative_part
begin
    subprogram_statement_part
end [ designator ] ;
```

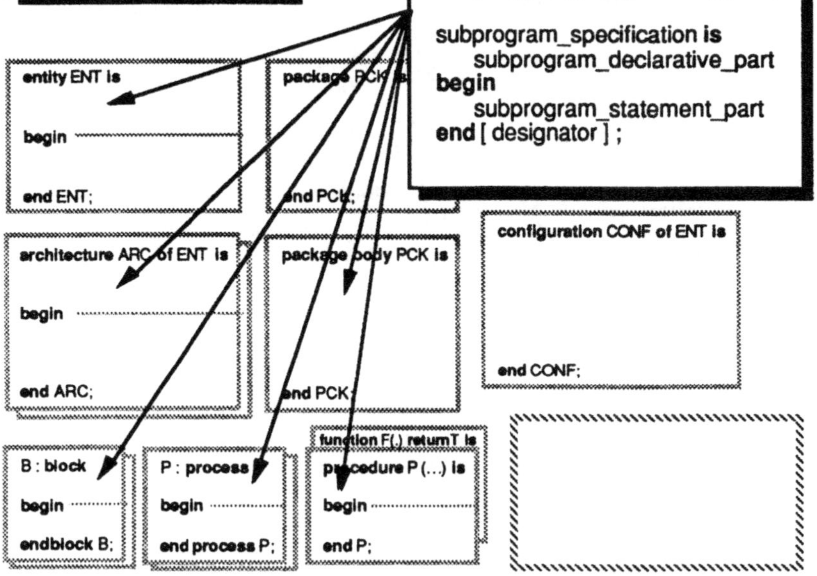


```
procedure Thank_you is
begin
  assert false
  report "Thank You !"
  severity note ;
end Thank_you ;
```

```
procedure regist( signal D  :  in bit;
                  signal CK :  in bit;
                  signal Q  :  out bit ) is
begin
  loop
  wait on CK;
  if CK = '1'  then
    Q <= D  after 1 ns;
  end if;
  end loop;
end regist ;
```

```
function Convert(A : bit)
      return fuzzy_bit is
  variable v : fuzzy_bit;
begin
  if A = '1'  then
    v := High;
  else
   v := Low;
  end if;
  return v;
end Convert;
```

```
procedure P( variable A : inout color;
             constant B :  in choice ) is
 subtype primary_color is color range yellow to red;
 variable X, Y, Z : primary_color;
begin
  ......
end P;
```

```
function  "+"( A,B : bit_vector )   return bit_vector is
begin
  return conv_to_vect( conv_to_int(A) + conv_to_int(B),
                       A'length + B'length );
end "+";
```

Declaration

Type Declaration

type_declaration ::= **[§ 4.1]**
  full_type_declaration
| incomplete_type_declaration

full_type_declaration ::= [§ 4.1]
 type identifier **is** type_definition ;

incomplete_type_declaration ::=
 type identifier ; **[§ 3.3.1]**

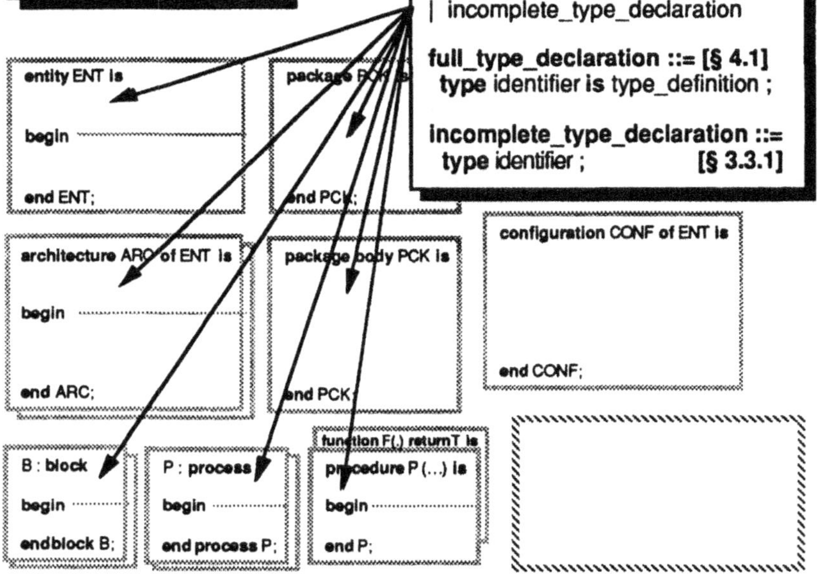


```
type COLOR is (white, yellow, red, blue, green);
type ZBIT is ('X', '0', '1', 'Z');

type BYTE_LENGTH is range 0 to 255;
type WORD_INDEX is range 31 downto 0;
type MY_REAL is range 0.0 to 9.99;
```

```
type CAPACITANCE is
        range 0 to 1E16
  units
    fF;
    pF = 1000 fF;
    nF = 1000 pF;
    uF = 1000 nF;
  end units;
```

```
type COLORS is array ( integer range <> )
                                of COLOR ;
type CTAB is array ( white to red ) of integer;
type MATRIX is array (1 to 5, 0 to 4) of bit;
```

```
type CELL;
type LINK is access CELL;
type CELL is
  record
    value : integer;
    succ, pred : LINK;
  end record;

type COLOR_FILE is
          file of COLOR;
```

In the STANDARD package :

```
type BOOLEAN is (FALSE,TRUE);
type BIT is ('0', '1');
type CHARACTER is (NUL, ...,'}','~', DEL);
type STRING is array (POSITIVE range <>) of CHARACTER;
type BIT_VECTOR is array (NATURAL range <>) of BIT;
```

Declaration

Subtype Declaration

subtype_declaration ::= **[§ 4.2]**
  subtype identifier **is**
    subtype_indication ;

subtype_indication ::= **[§ 4.2]**
  [*resolution_function_*name]
  type_mark [constraint]

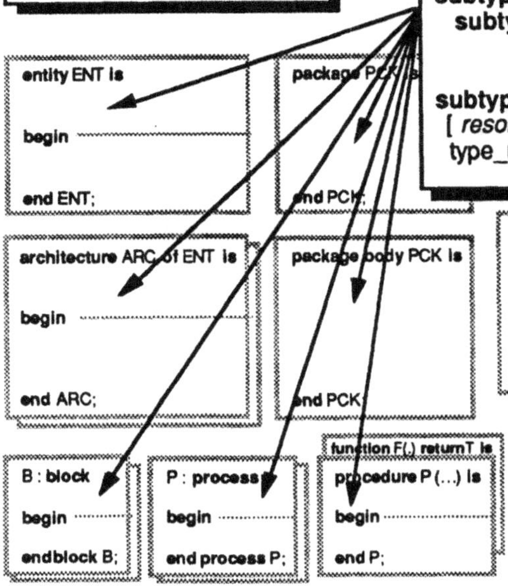

A subtype is
- -A type
- -A constraint

Is compatible with parent type

```
subtype PRIMARY_COLOR is COLOR range yellow to blue ;
subtype SAME_COLOR is PRIMARY_COLOR;

subtype ADDRESS_INTEGER is integer range 0 to 127 ;
subtype HUMAN_SIZE is real range 0.50 to 2.50;

subtype BYTE is bit_vector (7 downto 0);
subtype NAME is string (1 to 31);
subtype COLOR10 is COLORS (1 to 10);
```

```
subtype resolved_bit is resolution_function bit;
subtype SBIT is wiredX ZBIT;
subtype XBIT is wiredX ZBIT range 'X' to '1';
```

In the STANDARD package :

```
subtype NATURAL is integer range 0 to integer'high;
subtype POSITIVE is integer range 1 to integer'high;
```

Declaration

Constant Declaration

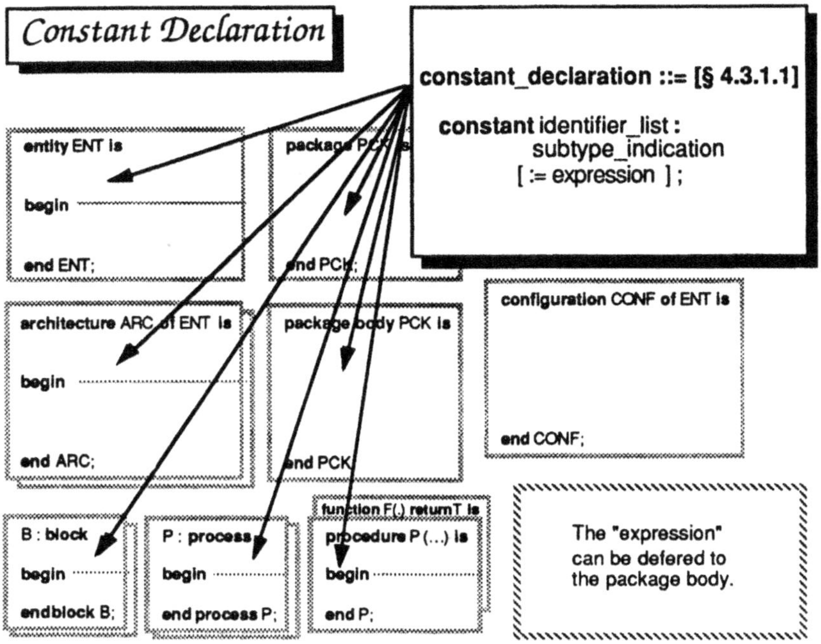


```
constant CTE : integer := 5 ;
constant Vdd, Vcc : bit := '1' ;
constant minimum_setup_time : time := 5 ns;
```

```
constant name : string := "Dupond" ;    -- the constraint comes from the value
constant address : bit_vector := "00111101";   -- the constraint comes from the value
constant mask : bit_vector(1 to 3) := "0101";
constant bittab : bit_vector(1 to 9) := (1 to 3 =>'0', 7 | 9 =>'Z',  others =>'1');
constant tab : table_type(0 to 4) := (2,3,4,-2,0);
```

```
constant deferred : My_Type;
```

```
constant max_address : integer := 2**nb_bit - 1 ;
constant delay : time := tech_param (0.025 pF) ;
```

Declaration

Signal Declaration

signal_declaration ::= **[§ 4.3.1.2]**

```
signal identifier_list :
        subtype_indication
       [ signal_kind ]
       [ := expression ] ;
```

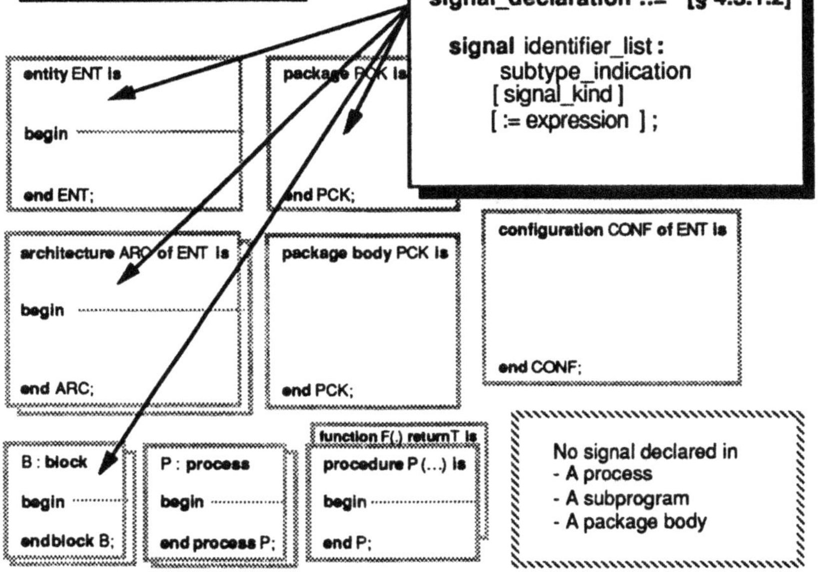


```
signal S1, S2 : bit := '1';
signal status : severity_level ;
signal light_color : Three_Color := Red ;
```

```
signal address : bit8 := "00111101";
signal bittab : bit_vector(1 to 9) := (1 to 3 => '0', 7 | 9 => 'Z',  others => '1');
signal tab : table_type(0 to 4) := (2,3,4,-2,0);
```

```
signal address_bus : resolved_bit_vector bus := "00111101";
```

```
signal resolved_S : resol_function ZBIT register;
```

```
signal integer_address : integer := 2**nb_bit - 1 ;
```

Declaration

Variable Declaration

variable_declaration ::= [§ 4.3.1.3]

```
variable identifier_list :
        subtype_indication
    [ := expression ] ;
```

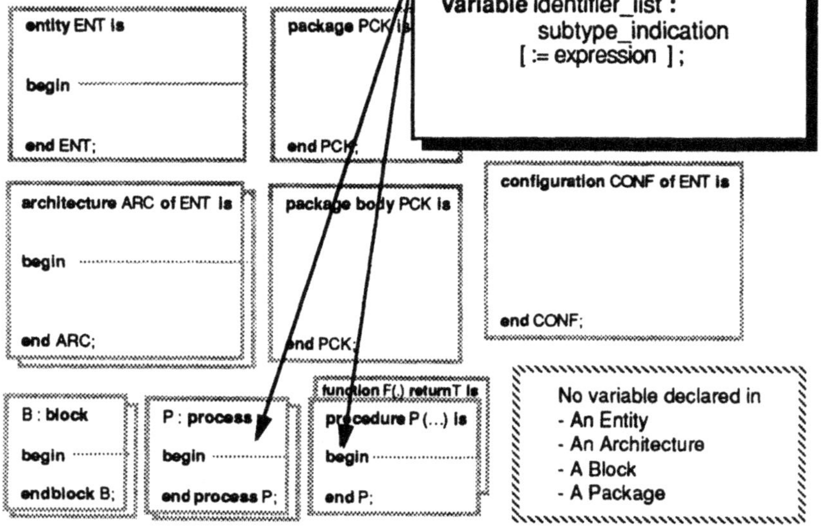


```
variable S1, S2 : bit := '1';
variable status : severity_level ;
variable light_color : Three_Color := Red ;
```

```
variable address_bus : bit8 := "00111101";
variable integer_address : integer := 2**nb_bit - 1 ;
variable bittab : bit_vector(1 to 9) := (1 to 3 => '0', 7 | 9 => 'Z', others => '1');
variable tab : table_type(0 to 4) := (2,3,4,-2,0);
```

```
variable resolved_S : resol_function ZBIT;
```

```
variable Pointer : Access_type := new integer'(10);
```

Declaration

File Declaration

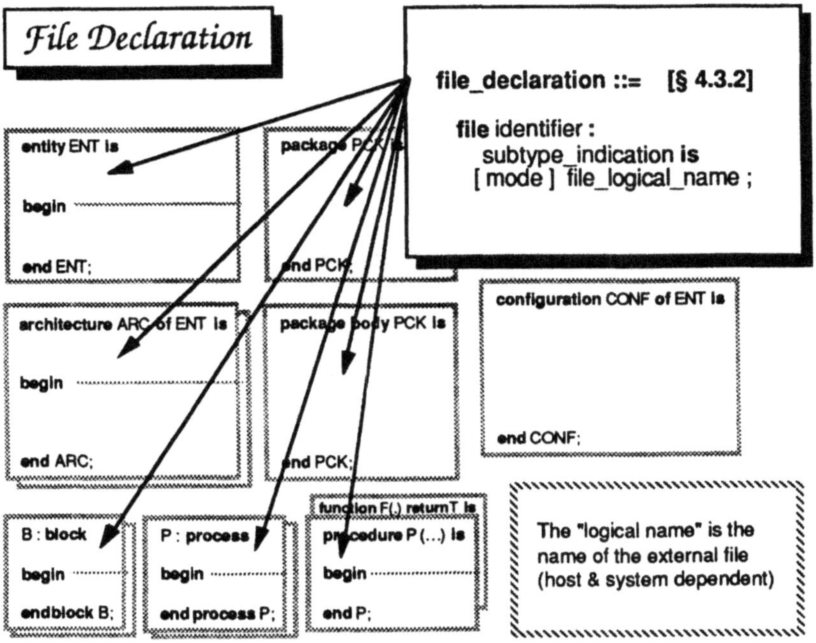


```
file ROM_CONTENT : ROM_File_Type is in "ROM2048.TXT";
```

```
file Sim_Output : My_File_Type is out "/home/user2/sim.res";
```

In the TEXTIO package :

```
file INPUT  : TEXT is in  "STD_INPUT";
file OUTPUT : TEXT is out "STD_OUTPUT";
```

Declaration

Alias Declaration

alias_declaration ::= **[§ 4.3.4]**

alias identifier **:**
    subtype_indication **is** name ;

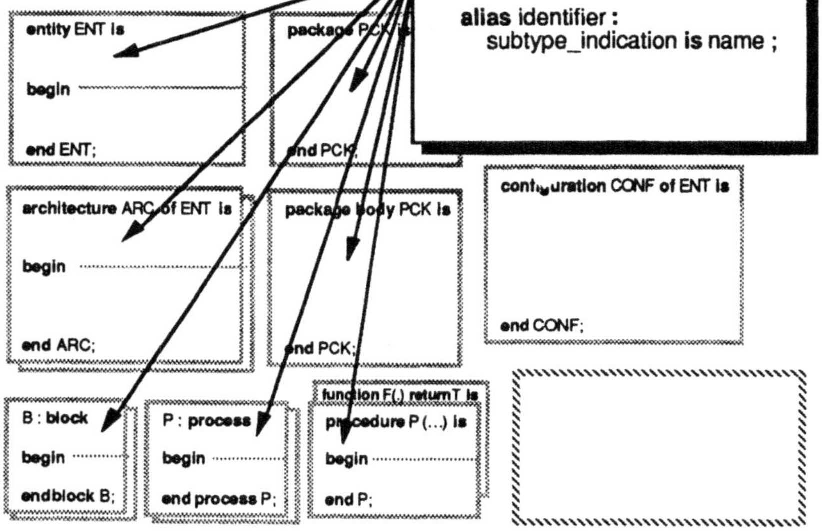


```
constant tcDCK : time := 2.4 ns ;
alias delay : time is tcDCK;
```

```
variable Vect : bit_vector(0 to 7);
alias Rev_Vect : bit_vector(Vect'length downto 1) is Vect ;
```

```
signal REAL_NUMBER : bit_vector(0 to 31);

alias SIGN : bit is REAL_NUMBER(0);
alias MANTISSA : bit_vector(23 downto 0) is REAL_NUMBER(8 to 31);
alias EXPONENT : bit_vector(1 to 7) is REAL_NUMBER(1 to 7);
```

Declaration

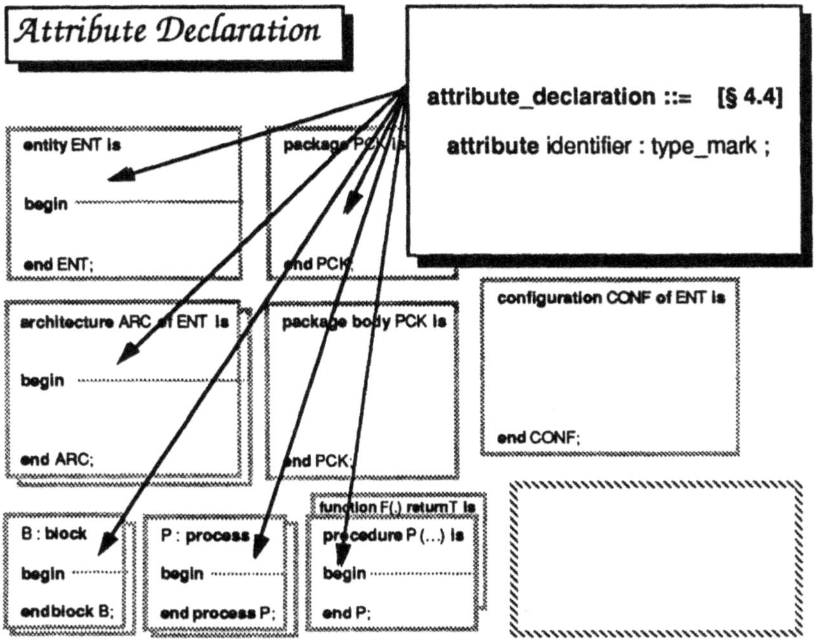

subtype PIN_Natural **is** natural **range** 1 **to** 24;
attribute PIN_Number : PIN_Natural;

attribute capacitance_value : Capacitance ;

attribute instance_location : Coordinate_Type ;

Declaration

Component Declaration

```
component_declaration ::=  [§ 4.5]

  component identifier
    [ local_generic_clause ]
    [ local_port_clause ]
  end component;
```

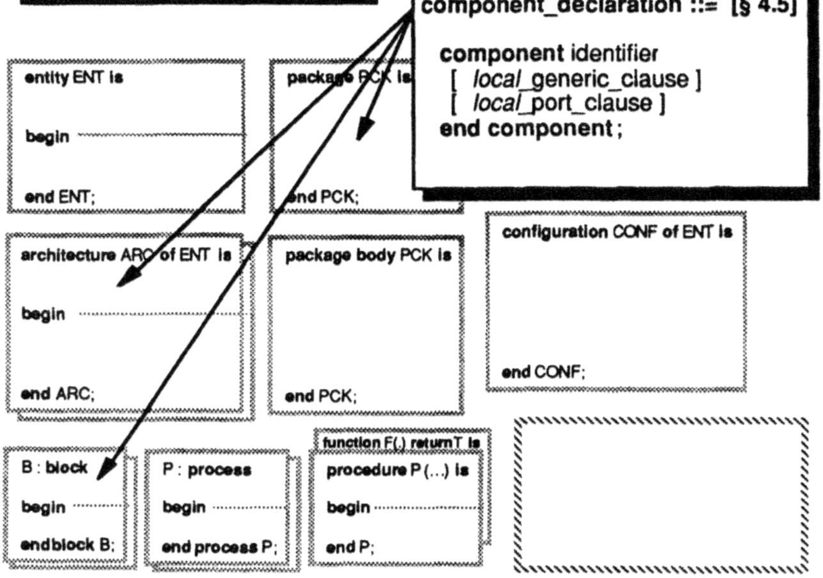


```
component GenClock
  port ( clk : out tri_bit );
end component;
```

```
component FOO
  generic ( Option : A_Type;
            Color  : Colors  );
end component;
```

```
component ADDER_1BIT
  port ( a, b, cin   :  in bit;
         sum, cout :   out bit );
end component;
```

```
component ADDER_8BIT
  port ( a, b  :  in bit_vector(1 to 8);
         cin   :  in bit;
         sum :    out bit_vector(1 to 8);
         cout :  out bit );
end component;
```

```
component ADDER_NBIT
  generic ( N : positive );
  port ( a, b  :  in bit_vector(1 to N);
         cin   :  in bit;
         sum :    out bit_vector(1 to N);
         cout :  out bit );
end component;
```

Declaration

Specifications

Attribute Specification

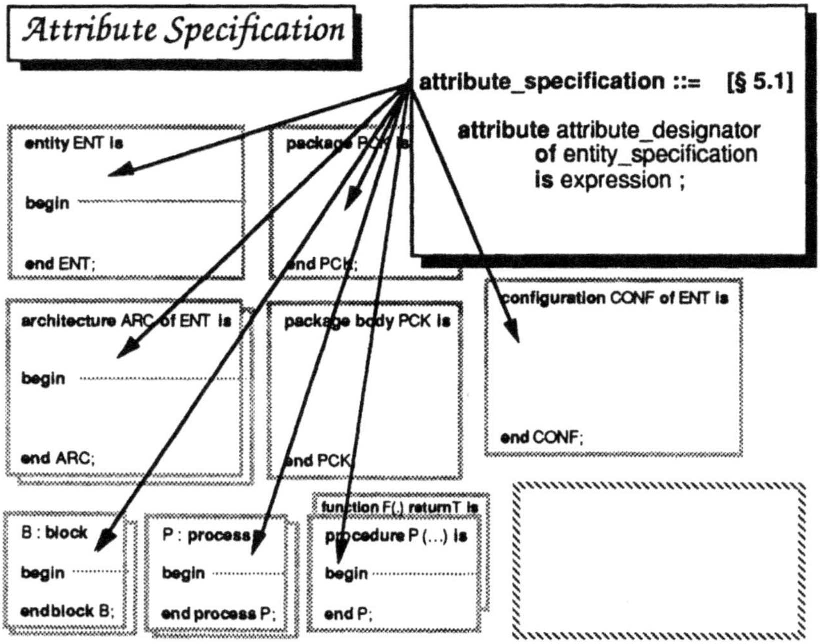


```
attribute PIN_Number of CIN : signal is 10;
attribute PIN_Number of COUT : signal is 5;
```

```
attribute instance_location of ADDER1 : label is (10,15);
attribute instance_location of others : label is (25,77);
```

```
attribute capacitance_value of all : variable is 15pF;
```

```
attribute author of ADD_ENT : entity is "Albert";
attribute is_generic of Cmos_NAND : component is false;
attribute creation_date of ADD_ARC : architecture is (26,Mar,91);
attribute confidentiality of CMOS_PKG : package is restrictive;
attribute safety of arith_conv : procedure is Bug;
```

Specification

Configuration Specification

configuration_specification ::=
for component_specification
use binding_indication ; **[§ 5.2]**

binding_indication ::= **[§ 5.2.1]**
entity_aspect [generic_map_aspect]
[port_map_aspect]

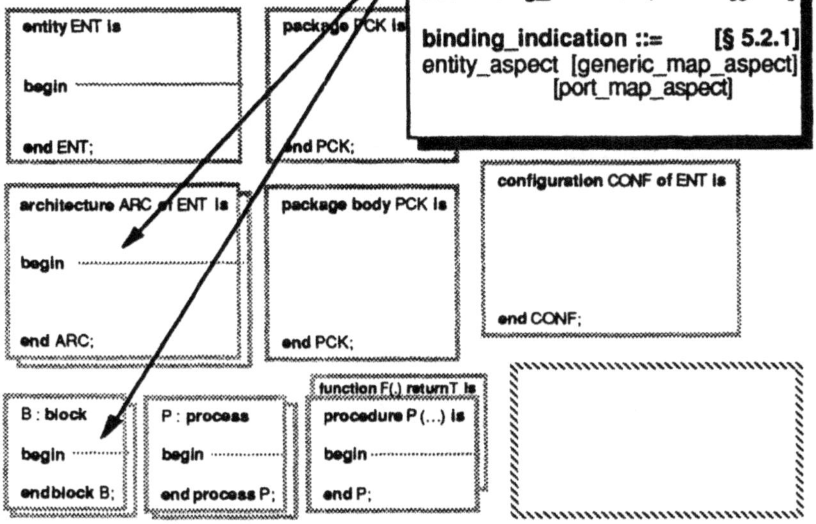


```
for C1, C2, C3 : ADD_COMP  use entity work.ADD1(BEHAV);
```

```
for others : ADD_COMP  use configuration work.ADD_CONF;
```

```
for all : REG_COMP  use open;
```

```
for C(1 to 5) : NAND2_COMP  use entity  NAND(ARC)
                            generic map (N => 2)
                            port map (I(1) => a, I(2) => b, O => s);
```

```
for C(6): NAND2_COMP  use configuration My_Lib.Nand2_Conf
                      port map (a,b,s);
```

Specification

Disconnection Specification

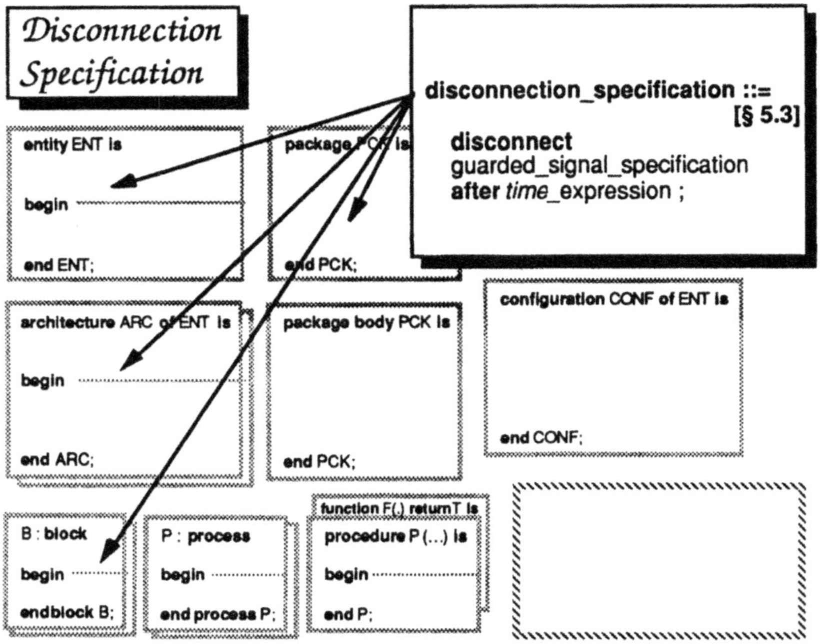


```
signal sig1, sig2, sig3 : resolved_bit register;
disconnect sig1 : resolved_bit after 5 ns;
disconnect others : resolved_bit after 7 ns;
```

```
constant delay : time := 5 ns;
signal busA, busB : resol_zbit_vector32 bus;
disconnect all : resol_zbit_vector32 after delay + 1 ns;
```

```
signal a, b, c : res_type register;
disconnect a, b, c : res_type after 0 ns;
```

Specification

Use & Library Clauses

Use Clause

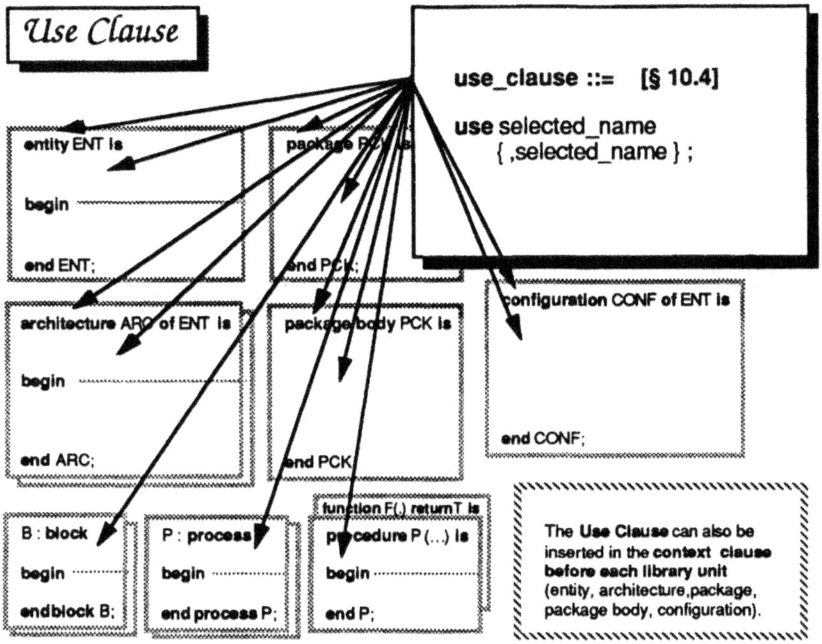


```
use work.all;

use My_library.all;

use work.My_Package.all;

use My_Library.My_Package.all;

use work.My_Package.My_Function;

use My_Library.My_Package.My_Function;

use STD.STANDARD.all; -- implicit !!

use Lib1.Pac1.all, Lib1.Pac2.Proc1, Lib2.Pac3.all;
```

Use Clause

Library Clause

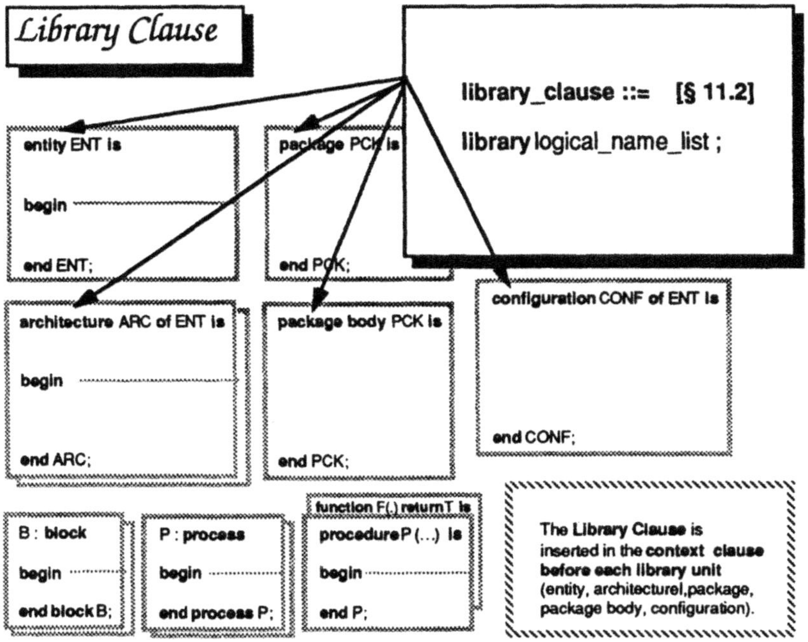

The **Library Clause** is inserted in the **context clause before each library unit** (entity, architecturel,package, package body, configuration).

```
library My_Library;
```

```
library My_Library1, My_Library2, My_Library3 ;
```

```
library STD, WORK ; -- implicit !!
```

Library Clause

Sequential Statements

Wait Statement

```
wait_statement ::=    [§ 8.1]

  wait
    [ sensitivity_clause ]
    [ condition_clause ]
    [ timeout_clause ] ;
```

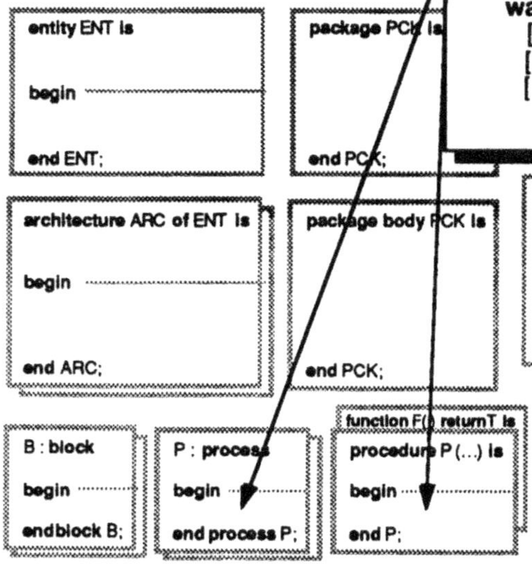

Warning
No wait statement allowed in a **function** body !
No wait statement may be called from a process having a sensivity list.

```
wait;

wait on Sig1, Sig2;

wait until clock = '1';

wait for 5 ns;

wait on data, clock until clock = '1';

wait on Sig1, Sig2 for 10 ns;

wait until data = good  for 25 ns;

wait on Sig1, Sig2, Sig3  until clock ='1' for 12 ns;
```

Sequential Statement

Assertion Statement

```
assertion_statement ::= [§ 8.2]

  assert condition
    [ report expression ]
    [ severity expression ] ;
```

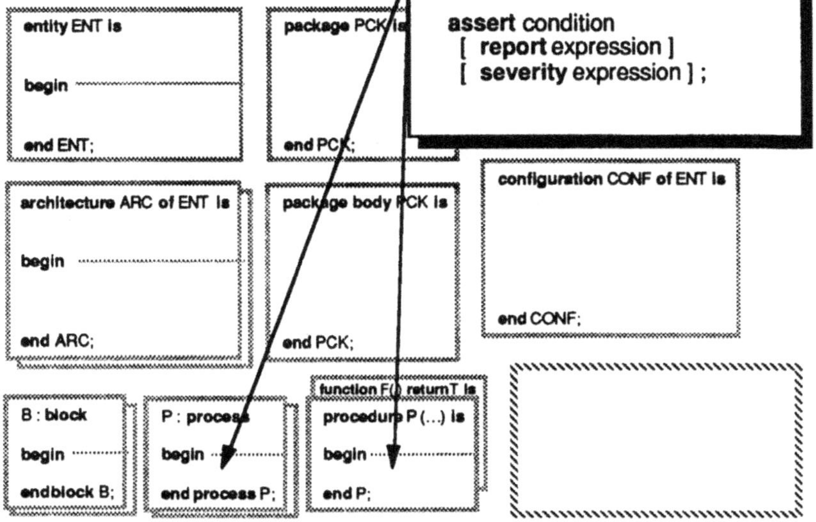


```
assert Signal_reset = '1';
```

```
assert Variable_reset = '1'
report "Reset is active !" ;
```

```
assert Reset = '1'
report "Reset is active !"
severity warning ;
```

```
assert Bool_Sig
report "Data is not valid..."
severity error ;
```

```
assert Reset = '1' or Set = '1'
report "Reset and Set simultaneously active !"
severity failure ;
```

```
assert data = 0
report "Data is equal to" & integer_to_string(data) & " !"
severity note ;
```

Sequential Statement

Signal Assignment Statement

signal_assignment_statement ::= **[§ 8.3]**

target <= [**transport**] waveform ;

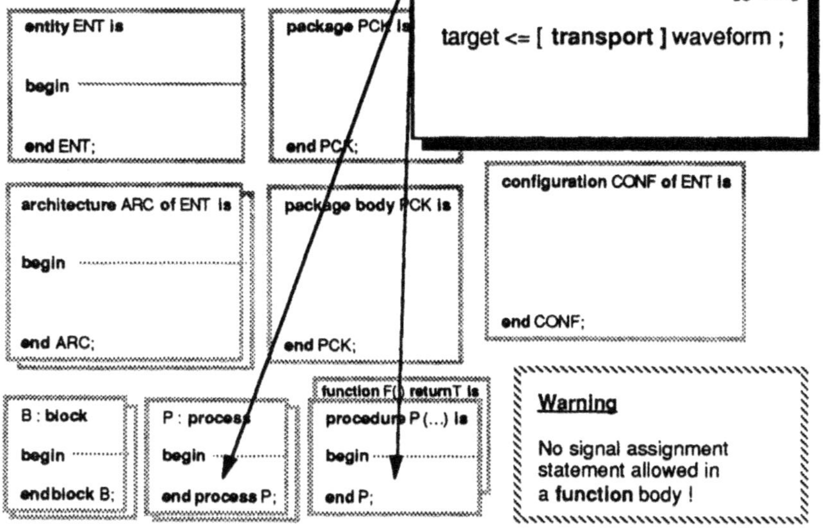


```
A <= B after 5 ns;

A <= transport B after 5 ns;

data<= 2 after 1 ns, 4 after 3 ns, 10 after 8 ns;

(BitA,BitB) <= two_bit'("01") after 5 ns;

A <= My_Funct(data, 4) after 5 ns;

A <= My_Funct(data, 4) after delay ;

A <= My_Funct(data, 4) after delay , 0 after delay + 2 ns;
```

Sequential Statement

Variable Assignment Statement

variable_assignment_statement ::= **[§ 8.4]**

target := expression ;

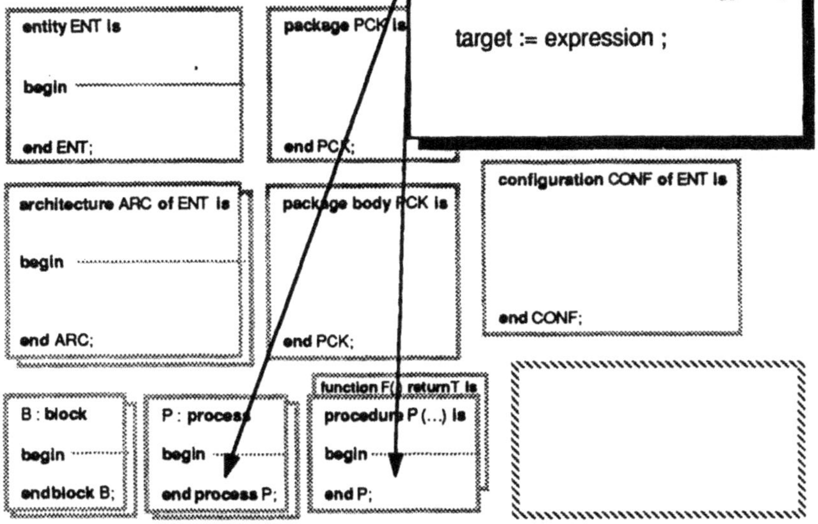


```
VAR := 0;
```

```
A := B;
```

```
StrA := "My Name is " & Get_name & " , and yours ?"
```

```
A := My_Funct(data, 4);
```

```
VarInt := My_Funct(data, 4) + 243 * A_Funct(VarInt) - 2 ** data;
```

Sequential Statement

Procedure Call Statement

procedure_call_statement ::= **[§ 8.5]**

*procedure*_name
[(actual_parameter_part)] ;

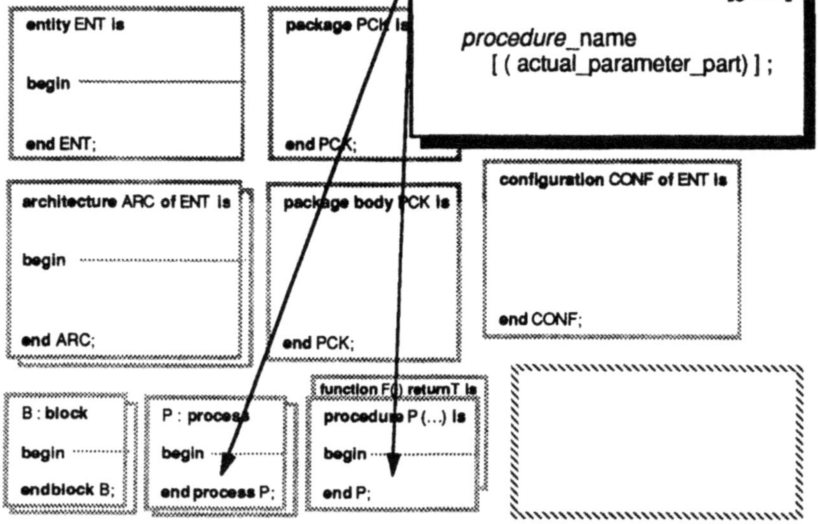


```
A_Proc ;

My_Proc( Sig1, Sig2, Var3) ;

Another_Proc( Var1, Var2, Q => Const3) ;

Setup_Check( clock, data) ;

Register_Procedure( ck => clock, d => RegIn, q => RegOut) ;
```

Sequential Statement

If Statement

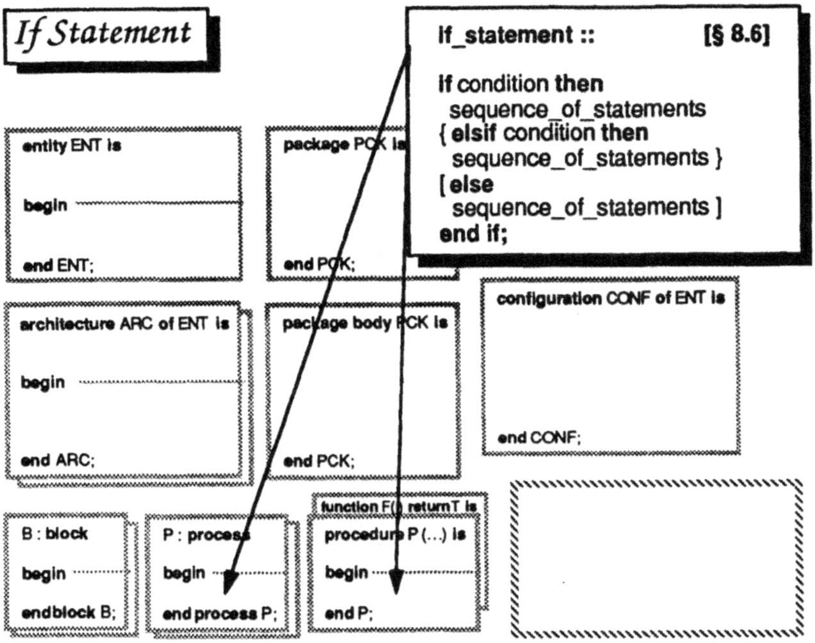


```
if A > 0  then
  B := A;
else
  B := abs(A + 1);
end if;
```

```
if A = '1'  then
  B := '0';
end if;
```

```
if clock'event and clock = '1'  then
  Q <= D  after 5 ns;
end if;
```

```
if reset = '0'  then
  Q <= '0'  after 1 ns;
elsif clock'event and clock = '1'  then
  Q <= D  after 5 ns;
end if;
```

```
if val >= 5  and val < 10  then
  int := 7;
elsif val < 5  then
  int := val + func(val+2);
elsif val < 15  then
  int := func(val);
else
  int := 0;
end if;
```

Sequential Statement

Case Statement

```
case_statement ::=                          [§ 8.7]
case expression is
  case_statement_alternative
  { case_statement_alternative }
end case ;

case_statement_alternative ::=
  when choices =>
    sequence_of_statements
```

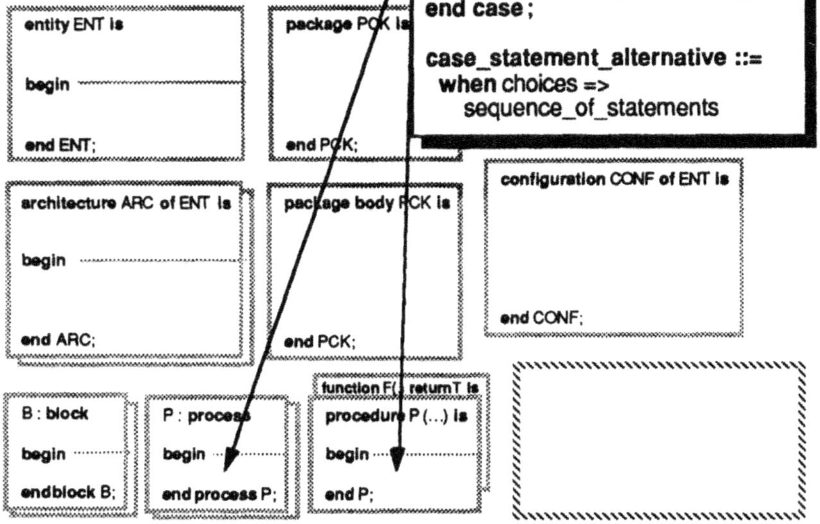


```
case A is
  when '0' => S <= "000"   after 2 ns;
  when '1' => S <= "111"   after 2 ns;
end case;
```

```
case two_bit'(A_enable & B_enable) is
  when "00" => S <= zero   after 1 ns;
  when "10" => S <= A   after 1 ns;
  when "01" => S <= B   after 2 ns;
  when "11" => S <= one   after 1 ns;
end case;
```

```
case int_value is
  when 0          => int := 5;
  when 1 | 2 | 8  => int := int_value;
  when 3 to 7     => int := int_value + 5 ;
  when 9          =>   null;
  when others     => int := 0;
end case;
```

Sequential Statement

Loop Statement

```
loop_statement ::=                          [§ 8.8]
[ loop_label ]
  [ iteration_scheme ] loop
    sequence_of_statements
end loop [ loop_label ] ;

iteration_scheme ::=
  while condition
| for loop_parameter_specification
```

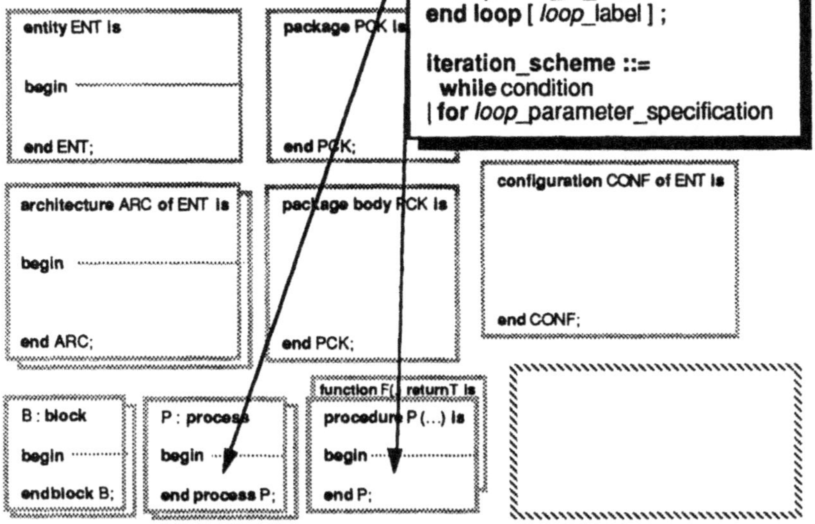


```
loop
  wait until clock = '1';
  q <= d after 1 ns;
end loop;
```

```
lbl : for i in 0 to 5 loop
  var := var + 3;
end loop;
```

```
while value > 0 loop
  next when value = 3;
  tab(value) := value rem 2;
  value := value / 2 ;
end loop;
```

```
for i in 10 downto 2 loop
  for j in 0 to i loop
    table(i,j) := i + j + 3;
  end loop;
end loop;
```

```
L1 : for i in 10 downto 2 loop
  L2 : for j in 0 to i loop
    exit L2 when i = j;
    table(i,j) := i + j + 3;
  end loop L2;
end loop L1;
```

Sequential Statement

Next Statement

next_statement ::= **[§ 8.9]**

next [*loop*_label]
[**when** condition] ;

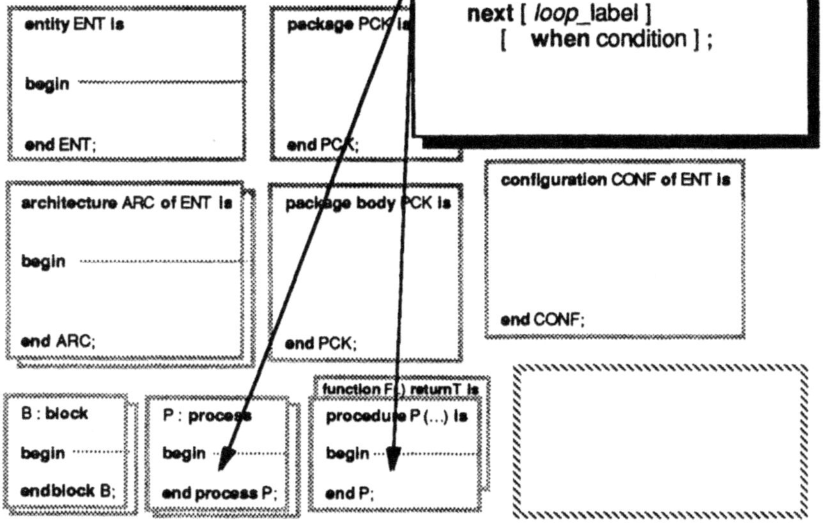


```
while value > 0 loop
  next when value = 3;
  tab(value) := value rem 2;
  value := value / 2 ;
end loop;
```

```
while value > 0 loop
  if value = 3 then
    next;
  end if;
  tab(value) := value rem 2;
  value := value / 2 ;
end loop;
```

```
L1 : for i in 10 downto 2 loop
  L2 : for j in 0 to i loop
    next L2 when i = j;
    table(i,j) := i + j + 3;
  end loop L2;
end loop L1;
```

```
L1 : for i in 10 downto 2 loop
  L2 : for j in 0 to i loop
    next L1 when i = j;
    table(i,j) := i + j + 3;
  end loop L2;
end loop L1;
```

Sequential Statement

Exit Statement

exit_statement ::= **[§ 8.10]**

exit [*loop*_label]
[**when** condition] ;

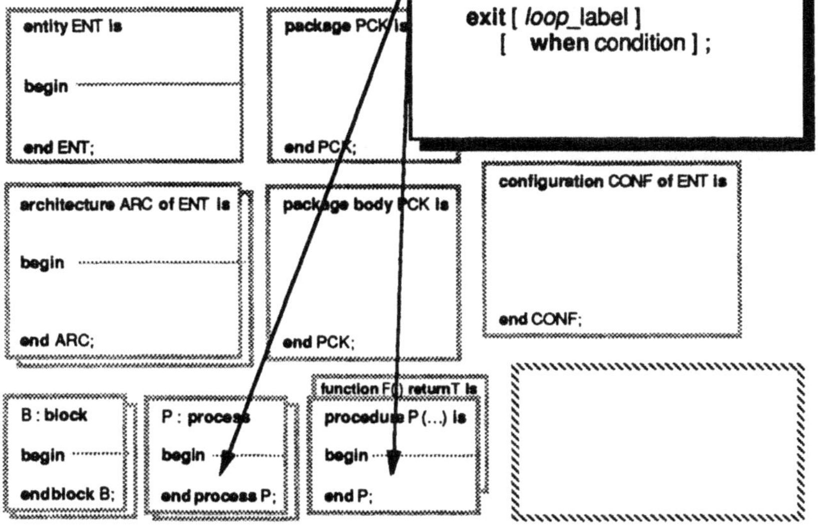


```
loop
  exit when value = 0;
  tab(value) := value rem 2;
  value := value / 2 ;
end loop;
```

```
loop
  if value = 0 then
    exit;
  end if;
  tab(value) := value rem 2;
  value := value / 2 ;
end loop;
```

```
L1 : for i in 10 downto 2 loop
  L2 : for j in 0 to i loop
    exit L2 when i = j;
    table(i,j) := i + j + 3;
  end loop L2;
end loop L1;
```

```
L1 : for i in 10 downto 2 loop
  L2 : for j in 0 to i loop
    exit L1 when i = j;
    table(i,j) := i + j + 3;
  end loop L2;
end loop L1;
```

Sequential Statement

Return Statement

return_statement ::= **[§ 8.11]**

return [expression] ;

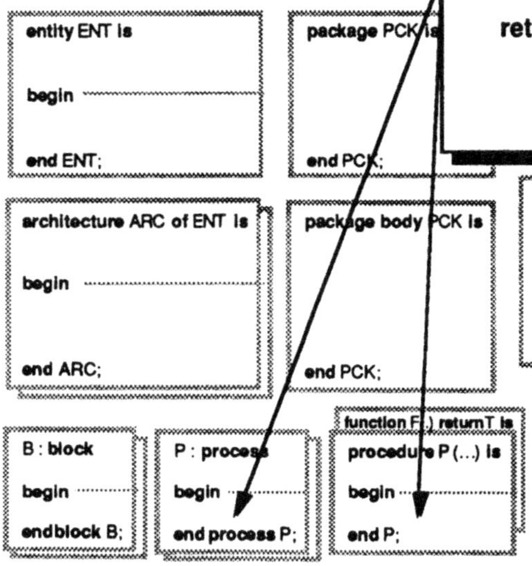

A function cannot go to its "end", and must pass on a return statement.

```
return;
```

```
return Value;
```

```
return A + B + 5 ns;
```

```
return My_Function(data , 5 pF);
```

```
return "Author Name : " & Name;
```

Sequential Statement

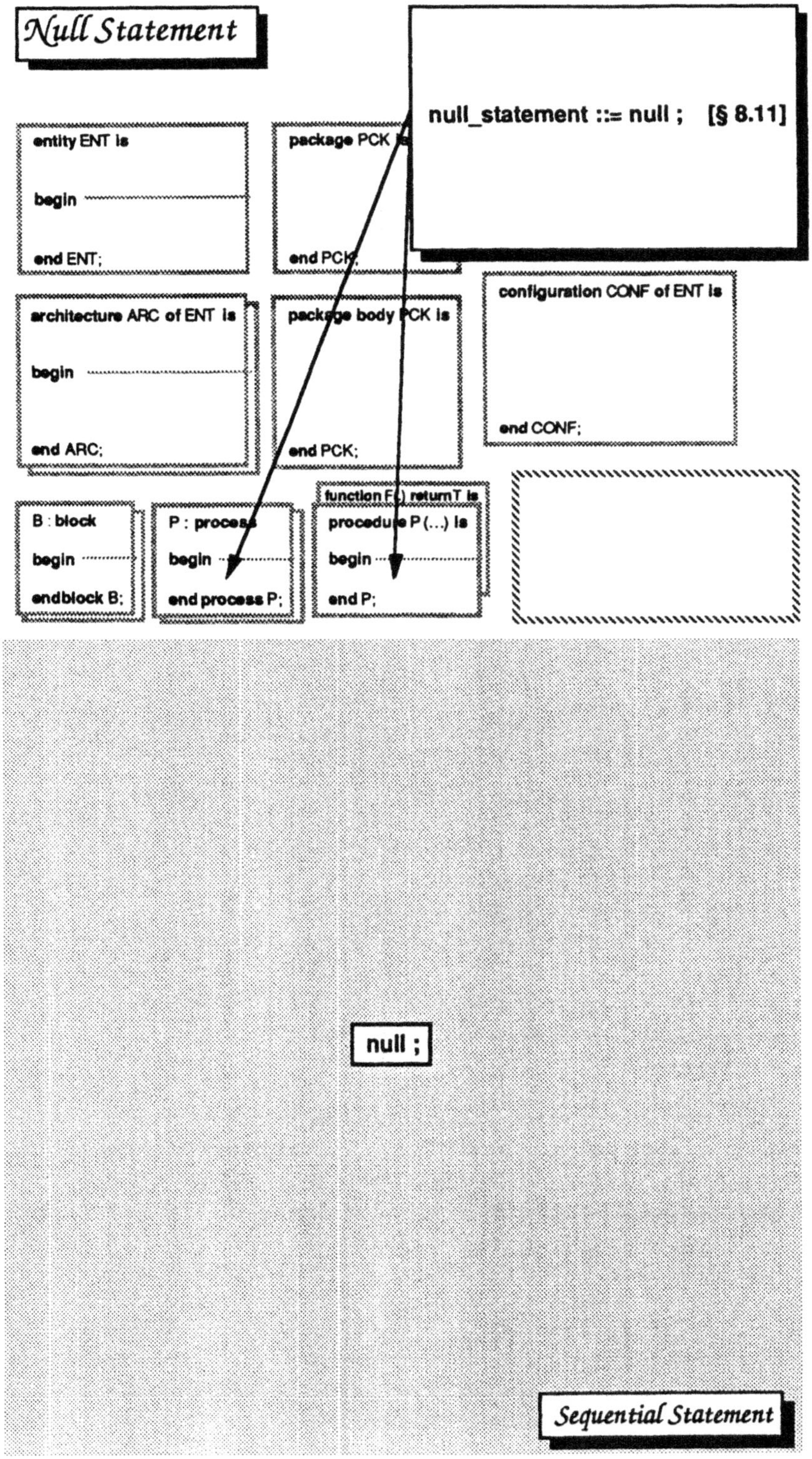
Null Statement
null_statement ::= null ;   [§ 8.11]
entity ENT is
begin
end ENT;
package PCK is
end PCK;
architecture ARC of ENT is
begin
end ARC;
package body PCK is
end PCK;
configuration CONF of ENT is
end CONF;
B : block
begin
endblock B;
P : process
begin
end process P;
function F() return T is
procedure P (...) is
begin
end P;
null ;
Sequential Statement

Concurrent Statements

Block Statement

block_statement ::= **[§ 9.1]**

*block*_label :
block [(*guard*_expression)]
  block_header
  block_declarative_part
begin
  block_statement_part
end block [*block*_label] ;

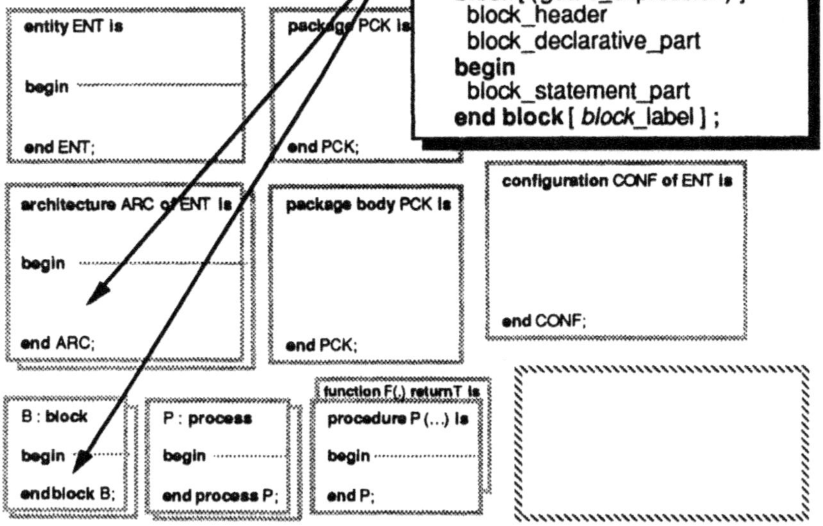


```
B: block
begin
  s <= '1'  after 2 ns;
end block;
```

```
LBL : block
  port (A,B : inout bit);
  port map (A=>S1, B=>S2);
begin
  B <= A  after 1 ns;
  A <= B  after 1 ns;
end block LBL ;
```

```
C: block
  signal int : bit_vector(1 to 3) := "010";
begin
  s <= int  after 5 ns;
end block;
```

```
Latch : block (clock ='1')
begin
  Latch_Output <= guarded Latch_Input after 1 ns;
end block;
```

```
Reg : block (not clock'stable and clock ='1')
  signal Mem : bit;
begin
  Mem <= guarded Reg_Input after 2 ns;
  Reg_Output  <= Mem ;
  Reg_OutputB <= not Mem after 1 ns;
end block Reg ;
```

Concurrent Statement

Process Statement

process_statement ::= **[§ 9.2]**

[*process*_label :]
process [(sensitivity_list)]
  process_declarative_part
begin
  process_statement_part
end process [*process*_label] ;

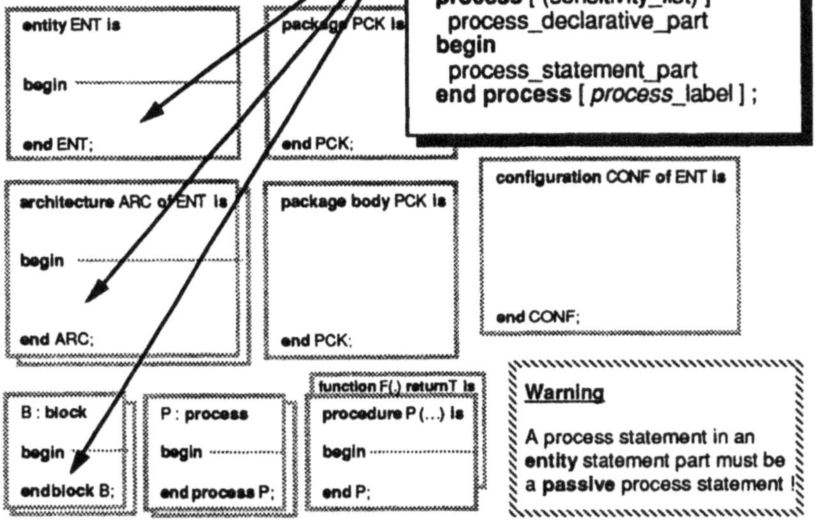


```
process
begin
  if (clock = '1') then
    Q <= D  after 2 ns;
  end if;
  wait on clock;
end process;
```

```
Reg : process (clock)
begin
  if clock = '1' then
    Q <= D   after 1 ns;
  end if;
end process;
```

```
Passive : process
begin
  wait on clock ;
  if clock = '1' then
    assert Reset = '1'
    report "Reset Active !"
    severity warning ;
  end if;
end process Passive ;
```

```
process
  variable V1, V2 : integer := 1;
begin
  Sig1 <= V1  after  1 ns;
  wait until Sig2 = 0;
  V2 := V2 * 2 ;
  V1 := V2 + 4 ;
  Sig1 <= V1  after 1 ns;
  wait on Sig1, Sig2;
  for i in 1 to 3 loop
    V1 := V1 + V2;
  end loop;
end process;
```

Concurrent Statement

Concurrent Procedure Call

concurrent_procedure_call ::= **[§ 9.3]**

[label :] procedure_call_statement

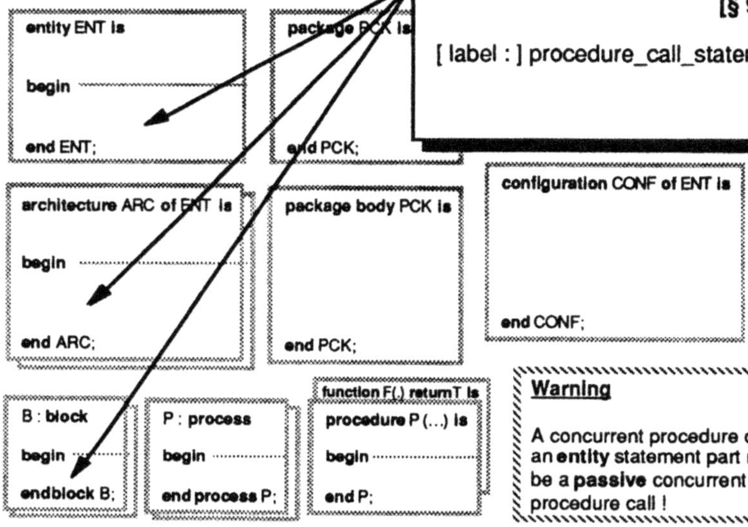

Warning

A concurrent procedure call in an **entity** statement part must be a **passive** concurrent procedure call !

```
A_Proc ;

Lab : My_Proc( Sig1, Sig2, Sig3) ;

Another_Proc( Sig1, Sig2, Q => Sig3) ;

Pass_Proc : Setup_Check( clock, data) ;

Register_Procedure( ck => clock, d => RegIn, q => RegOut) ;
```

Concurrent Statement

Concurrent Assertion Statement

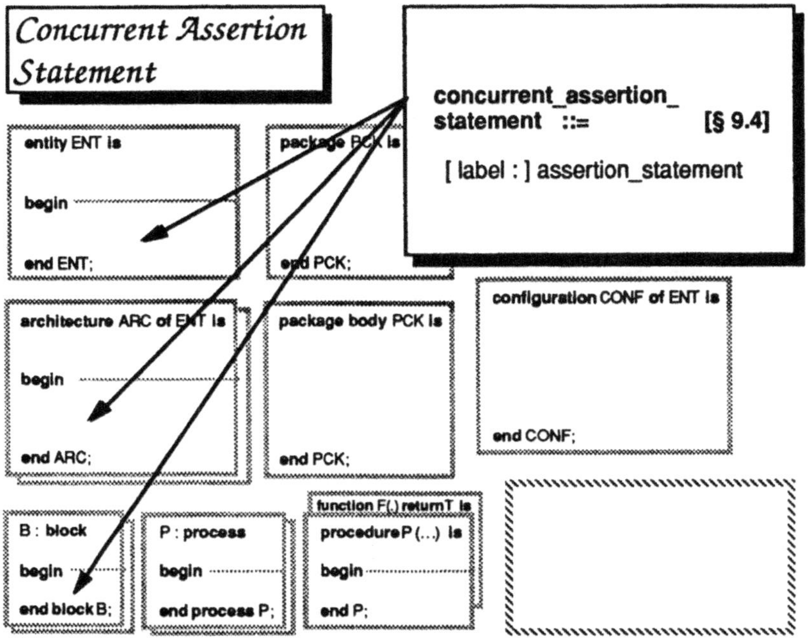


```
assert Reset = '1';
```

```
assert Reset = '1'
report "Reset is active !" ;
```

```
LBL : assert Reset = '1'
report "Reset is active !"
severity warning ;
```

```
assert Bool_Sig
report "Data is not valid..."
severity error ;
```

```
Latch_Check : assert Reset = '1'  or Set = '1'
report "Reset and Set simultaneously active !"
severity failure ;
```

```
LBL : assert data = 0
report "Data is equal to" & integer_to_string(data) & " !"
severity note ;
```

Concurrent Statement

Concurrent Signal Assignment Statement

**concurrent_signal_
assignment_statement ::= [§ 9.5]**

[label:]conditional_signal_assignment
| [label:] selected_signal_assignment

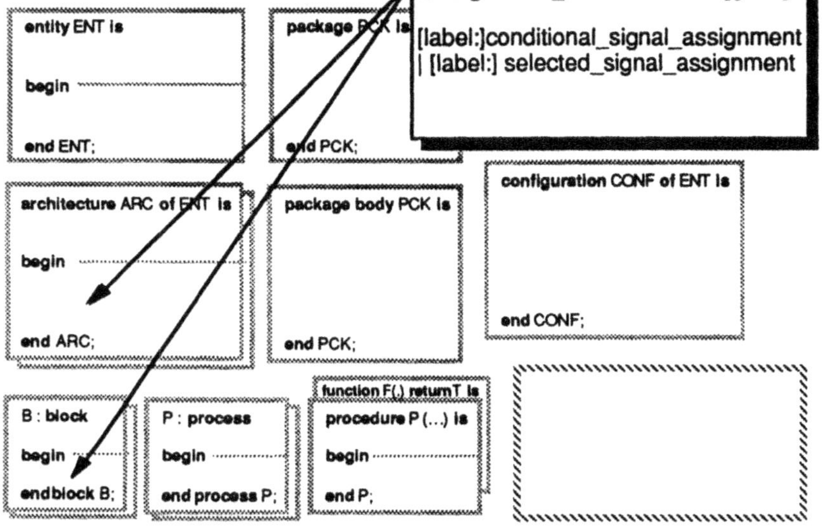


```
A <= B after 5 ns;

data<= 4 after 1 ns, 2 after 3 ns, 10 after 8 ns;

lbl :A <= transport B after 5 ns;

C : DataBus <= guarded RegOut after 2 ns;

(BitA,BitB) <= two_bit'("01") after 5 ns;

A <= '1' after 5 ns when sel = 0 else
     '0' after 3 ns, '1' after 5 ns when sel = 1 else
      'X' after 2 ns;

Lmux : with MuxVal select
 Sig1 <= transport "001" after 2 ns when 0,
                   "110"  after 5 ns when 1,
                   "000"  after 5 ns when 2,
                   "111"  after 5 ns when 3,
                   "XXX"  when others;
```

Concurrent Statement

Component Instantiation Statement

component_instantiation_statement ::= **[§ 9.6]**

*instantiation*_label :
  *component*_name
    [generic_map_aspect]
    [port_map_aspect] ;

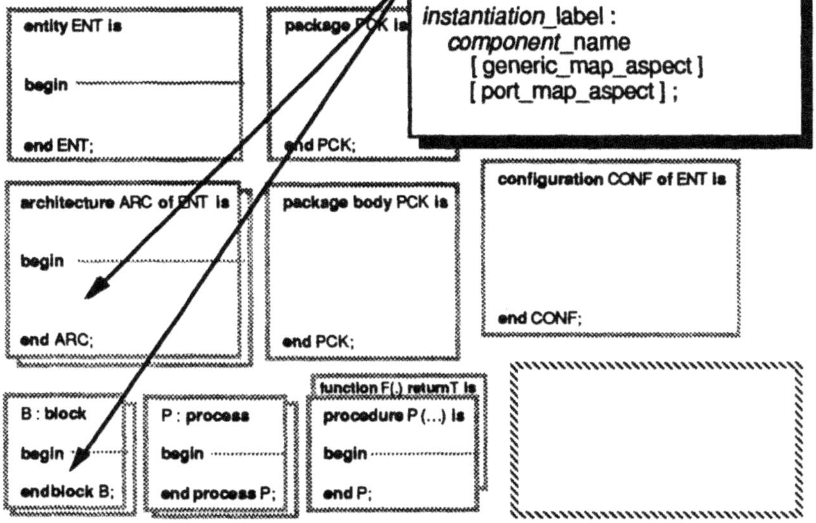

C1 : My_Component ;

C1 : GENCLOCK **port map** (clk => clock);

CADD : ADDER **port map** (a, b, cin, sum, cout);

C : ADDN **generic map** (8)
  port map (vss, A8, B8, sum => S8, cout => Sig);

The_Ram : Ram_Comp
  generic map (nb_data => 8, nb_addr => 5)
  port map (cs => cs, rw => rw, d => data, a => address, ready => **open**);

Concurrent Statement

Generate Statement

generate_statement ::= **[§ 9.7]**

```
generate_label :
   generation_scheme generate
      { concurrent_statement }
   end generate [ generate_label ] ;
```

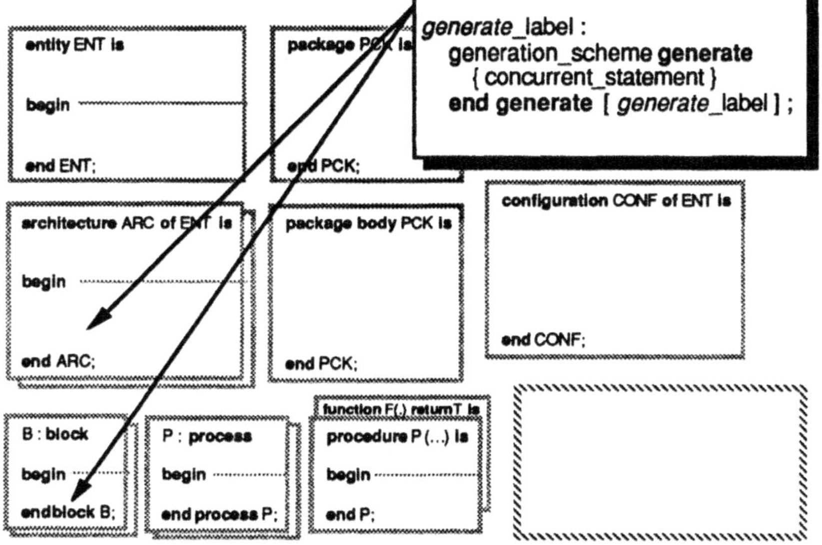


```
Lbl : for i in 1 to 5 generate
 C1 : COMP1 port map (ck => clock, d => a(i), q => a(i+1));
 C2 : COMP2 port map (ck => clock, d => a(i));
end generate;
```

```
Lbl : if N > 0 generate
 Sig1 <= '1' after 5 ns;
 Sig2 <= Sig1 after 2 ns;
 C : My_Comp port map (a,b,c);
end generate;
```

```
cadd : for I in 1 to N generate
  cb : if I =1 generate
    AddB : Add_Begin port map (...) ;
  endgenerate;
  cm : if I > 1 and I < N generate
    AddM : Add_Middle port map (...) ;
  endgenerate;
  ce : if I = N generate
    AddE : Add_End port map (...) ;
  end generate;
end generate cadd ;
```

Concurrent Statement

Miscellaneous

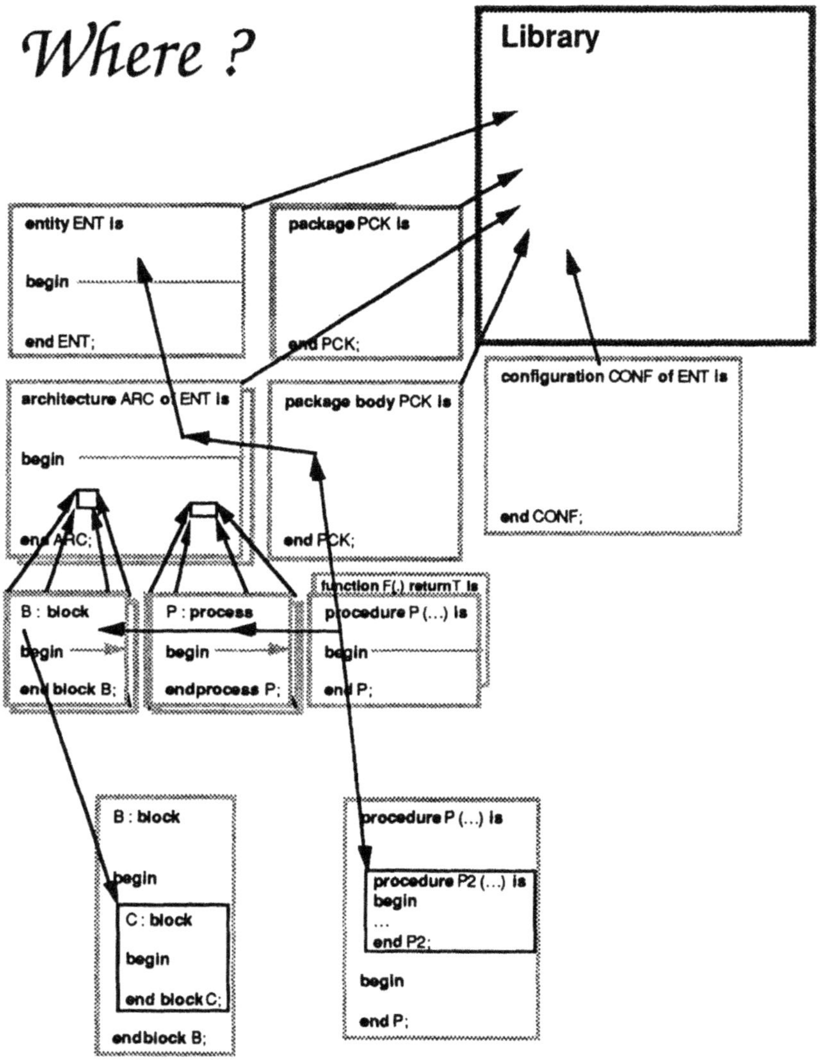

A Block can be found inside another block (*Statement* Part)
A Subprogram can be found inside another subprogram (*Declarative* Part)

Lexical Elements

Keywords

abs	mod
access	nand
after	new
alias	next
all	nor
and	not
architecture	null
array	of
assert	on
attribute	open
begin	or
block	others
body	out
buffer	package
bus	port
case	procedure
component	process
configuration	range
constant	record
disconnect	register
downto	rem
else	report
elsif	return
end	select
entity	severity
exit	signal
file	subtype
for	then
function	to
generate	transport
generic	type
guarded	units
if	until
in	use
inout	variable
is	wait
label	when
library	while
linkage	with
loop	xor
map	

Identifiers

FOO
FOO_2
FOO_ONE
fOo -- (identical to FOO)

Not Identifiers

_FOO
FOO_
FOO__ONE
2FOO
FO$O

Comment

blabla -- Comment till end of line
not in comment -- now in comment

Character Literals

'A', ' ', CR, LF, '''

Not Character Literal

'' -- (two ticks)

String Literals

"this is a string"
"with a "" quote"
"" -- empty string

Reals

0.1
1.4E45
1.2E-2
1.000_001

Not Reals

.3
3.
2

Bit String Literals

"0110_0000_1101"
B"0110_0000_1101"
O"12753"
X"2FAC"
X"2fac"

Not Bit String Literals

"0112_0000_1101"
Z"0110_0000_1101"
O"12F53"

Integer Literals

2
3E4
1000_000

Not Integer Literals

2.0
3.0E4
1000__000
_1
23_

Based Literals

8#7327720#
16#AC3E#
2#1001_1001_0000#
16#2A.3C#E4

Not Based Literals

17#ABCG#
16#2A.3C#E2#10010#

Physical Literals

10 ns
25 pf
4.70 kOhm

Not Physical Literals

10ns -- (the space !)
3Ω

```
package STANDARD is

-- predefined enumeration types
type BOOLEAN is (FALSE, TRUE) ;
type BIT is ('0', '1') ;

type CHARACTER is (
      NUL,SOH,STX,ETX,EOT,ENQ,ACK,BEL, BS, HT, LF, VT, FF, CR, SO, SI,
      DLE,DC1,DC2,DC3,DC4,NAK,SYN,ETB,CAN, EM,SUB,ESC,FSP,GSP,RSP,USP,
      ' ','!','"','#','$','%','&',''','(',')','*','+',',','-','.','/',
      '0','1','2','3','4','5','6','7','8','9',':',';','<','=','>','?',
      '@','A','B','C','D','E','F','G','H','I','J','K','L','M','N','O',
      'P','Q','R','S','T','U','V','W','X','Y','Z','[','\',']','^','_',
      '`','a','b','c','d','e','f','g','h','i','j','k','l','m','n','o',
      'p','q','r','s','t','u','v','w','x','y','z','{','|','}','~',DEL );

type SEVERITY_LEVEL is (NOTE, WARNING, ERROR, FAILURE) ;

-- predefined numeric types:
type INTEGER is range implementation_defined ;
type REAL is range implementation_defined ;

-- predefined physical type TIME:
type TIME is range implementation_defined
      units
         fs;                  -- femtosecond
         ps  =  1000 fs;      -- picosecond
         ns  =  1000 ps;      -- nanosecond
         us  =  1000 ns;      -- microsecond
         ms  =  1000 us;      -- millisecond
         sec =  1000 ms;      -- second
         min =  60 sec;       -- minute
         hr  =  60 min;       -- hour
      end units ;

-- function that returns the current simulation time:
function NOW return TIME ;

-- predefined numeric subtypes:
subtype NATURAL is INTEGER range 0 to INTEGER'HIGH ;
subtype POSITIVE is INTEGER range 1 to INTEGER'HIGH ;

-- predefined array types:
type STRING is array (POSITIVE range <>) of CHARACTER ;
type BIT_VECTOR is array (NATURAL range <>) of BIT;

end STANDARD ;
```

```
package TEXTIO is

-- Type Definitions for Text I/O

type LINE is access STRING ;            -- a LINE is a pointer to a STRING value
type TEXT is file of STRING ;           -- a file of variable-length ASCII records
type SIDE is (RIGHT, LEFT) ;            -- for justifying output data within fields
subtype WIDTH is NATURAL ;              -- for specifying widths of output fields

-- Standard Text Files
file INPUT: TEXT is in  "STD_INPUT" ;
file OUTPUT: TEXT is out "STD_OUTPUT" ;

-- Input Routines for Standard Types
procedure READLINE ( F : in TEXT; L : out LINE ) ;

procedure READ ( L : inout LINE ; VALUE : out THE_TYPE ; GOOD :  out BOOLEAN ) ;
procedure READ ( L : inout LINE ; VALUE : out THE_TYPE ) ;

-- for THE_TYPE  in :   BIT, BIT_VECTOR, BOOLEAN, CHARACTER,
--                      INTEGER, REAL, STRING, TIME

-- Output Routines for Standard Types
procedure WRITELINE ( F : out TEXT ;  L :  in LINE ) ;

procedure WRITE ( L : inout LINE ;  VALUE : in THE_TYPE ;
                  JUSTIFIED : in SIDE := RIGHT ; FIELD : in WIDTH := 0 ) ;

-- for THE_TYPE  in :   BIT, BIT_VECTOR, BOOLEAN, CHARACTER,
--                      INTEGER, STRING

procedure WRITE ( L : inout LINE ;  VALUE : in REAL ;
                  JUSTIFIED : in SIDE := RIGHT ; FIELD : in WIDTH := 0 ;
                  DIGITS : in NATURAL := 0 ) ;

procedure WRITE ( L : inout LINE ;  VALUE : in TIME
                  JUSTIFIED : in SIDE := RIGHT ; FIELD : in WIDTH := 0 ;
                  UNIT : in TIME := ns ) ;

-- File Position Predicates
function ENDLINE ( L : in LINE ) return BOOLEAN ;
 -- function ENDFILE ( F : in TEXT ) return BOOLEAN ;

end TEXTIO ;
```

12. INDEX

A

B

C

D

E

F

G

H

I

K

L

M

T

U

V

W

BIBLIOGRAPHY

[AIR90] *(in French)* "VHDL, du langage à la modélisation"
R. Airiau, J. M. Bergé, V. Olive et J. Rouillard
Presses Polytechniques et Universitaires Romandes, 1990

[ARM88] "Chip Level Modelling in VHDL"
J. Armstrong,
Prentice Hall, 1988

[AUG91] "Hardware Design and Simulation in VAL/VHDL"
L.M. Augustin, D.C. Luckham, B.A. Genhart, Y. Huh, A.G. Stanculescu
Kluwer Academic Publishers, 1991

[BIL91] "STD_LOGIC_1164: Draft Standard PAR 1164" (Version 4.2)
W.D. Billowitch, Chairman
IEEE VHDL Model Standards Group, 20-Jan-1991

[BUL92] "VFORMAL Tutorial"
Commercial Document, BULL SA, 1992

[CAL90] *(in French)* "Spécification et Conception des Systèmes: une Méthodologie"
J.P. Calvez
Editions Masson, 1990

[COE89] "The VHDL Handbook"
D.R. Coelho
Kluwer Academic Publishers, 1989

[CRA91] "HDLs and Logic Synthesis"
M. Crastes de Paulet, A. Fonkoua
ECIPII Esprit Project Report, INPG/91/ECIP2/WP2/SYN.D
Institut Méditerranéen de Technologie, Marseille, France, 1991

[FON89] *(in French)* "Une Application de l'Intelligence Artificielle
à la Synthèse Architecturale des Circuits Intégrés VLSI"
A. Fonkoua
Thèse de Doctorat, INPG/ENSIMAG, 1989

[GOT86] "Design Methodologies"
S. Goto
Elsevier Science B.V., North-Holland, 1986

[HAR91] "Applications of VHDL to Circuit Design"
Edited by R.E. Harr, A.G. Stanculescu,
Kluwer Academic Publishers, 1991

[HUE91] "VHDL Experiments on Performance"
M. Hueber, S. Lasserre, A. de Monteville, Thomson-CSF RGS, France
Proceedings EUROVHDL'91, 8-11 Sept 1991, Stockholm, Sweden

[IEE87] "IEEE Standard VHDL Language Reference Manual"
IEEE Std 1076-1987, IEEE Inc., New York, USA, 1987

[KLI81] "Interpretation Techniques"
P. Klint - Mathematical Center, Netherlands
Software Practice and Experience, Vol 11, 1981

[LEU89] "ASIC System Design with VHDL: A Paradigm"
S. S. Leung, M. A. Shanblatt,
Kluwer Academic Publishers, 1989

[LEV91] "Writing High Performance VHDL Models"
O. Levia - ViewLogic Systems, Inc. USA
Proceedings EUROVHDL'91, 8-11 Sept 1991, Stockholm, Sweden

[LIP89] "VHDL: Hardware Description and Design"
R. Lipsett, C. Schaefer, C. Ussery
Intermetrics, Inc.
Kluwer Academic Publishers, 1989

[MAG90] "VHDL Modeling Guidelines: From Commercial Languages to VHDL"
S. Maginot
ECIPII Esprit Project Report, ECIP2/WP2/IMT/90
Institut Méditerranéen de Technologie, Marseille, France, 1990

[MAG91] *(in French)* "Verilog et VHDL", "UDL/I et VHDL"
S. Maginot
DRET-ELISE Contract Reports
Institut Méditerranéen de Technologie, Marseille, France, 1991

[MEA80] "Introduction to VLSI Systems"
C. Mead, L. Conway
Addison-Wesley, 1980

[MAR90] "Implementing VHDL Simulation Semantics in C"
E. Marschner
VHDL Forum for CAD in Europe, 1990

[MEN89] "M Language Users Guide & Reference"
Version 4.0, May 1989
Mentor Graphics (Silicon Compiler Systems)

[RAM91] "Approaching System Level Design"
F.J. Rammig, Paderborn University, FRG
Proceedings EUROVHDL'91, 8-11 Sept 1991, Stockholm, Sweden

[SAL91] "Formal Semantics of VHDL Timing Constructs"
A. Salem, D. Borrione - Artemis, IMAG, France
Proceedings EUROVHDL'91, 8-11 Sept 1991, Stockholm, Sweden

[SIG90] "Minutes of the Synthesis Interest Group"
VASG Meeting, 1990

[SIN91] "Code Generation for VHDL: Interpreted vs. Compiled Code"
S. Singhani - ViewLogic Systems, Inc. USA
Proceedings EUROVHDL'91, 8-11 Sept 1991, Stockholm, Sweden

[STE90] "Digital Design with Verilog HDL"
E. Sternheim, R. Singh, Y.Trivedi
Automata Publishing Company, 1990

[THO91] "The Verilog Hardware Description Language"
D.E. Thomas, P. Moorby
Kluwer Academic Publishers, 1991

[UDL91] "UDL/I Language Reference"
Draft Version 1.0h4
May 14, 1991
Japan Electronic Industry Development Association

[VER91] "Verilog Hardware Description Language Reference Manual"
Version 1.0, October 1991
Open Verilog International

GPSR Compliance

The European Union's (EU) General Product Safety Regulation (GPSR) is a set of rules that requires consumer products to be safe and our obligations to ensure this.

If you have any concerns about our products, you can contact us on ProductSafety@springernature.com

In case Publisher is established outside the EU, the EU authorized representative is:

Springer Nature Customer Service Center GmbH
Europaplatz 3
69115 Heidelberg, Germany

Batch number: 10761697

Printed by Printforce, the Netherlands